PLASTIC FOAMS
(IN TWO PARTS)

PART II

MONOGRAPHS ON PLASTICS

Volume 1: Plastic Foams (in two parts)
Edited by Kurt C. Frisch and James H. Saunders

IN PREPARATION:

Volume 2: The Breakdown of Plastics
Edited by S. Goldfein

Volume 3: Chemistry and Technology of Poly(Vinyl Chloride) and Related Compositions
Edited by Leonard I. Nass

PLASTIC FOAMS

IN TWO PARTS

Edited by

Kurt C. Frisch and James H. Saunders

Polymer Institute
University of Detroit
Detroit, Michigan

Monsanto Company
Pensacola, Florida

PART II

1973

MARCEL DEKKER, INC. New York

MARCEL DEKKER, INC.
95 Madison Avenue, New York, New York 10016

LIBRARY OF CONGRESS CATALOG CARD NUMBER: 71-157837

ISBN: 0-8247-1219-6

Printed in the United States of America

FOREWORD

The importance of foamed polymers has grown with unusual speed in the last ten years. Many of the more familiar types have increased in usage, and newer varieties have also been developed. In presenting an up-to-date study we have attempted to emphasize an integrated picture of the fundamental principles, technology, and applications of foams, while at the same time giving thorough treatments of specific types. The extent of this effort has led to a division of the information into two parts.

The first of the two parts offers an overview of the entire field in the introductory chapter; an extensive treatment of the fundamental principles of foam formation of nearly all types; then specific chapters on foam varieties which are predominantly flexible. A chapter on testing of all types is included.

Our second part will give coverage of the foam types which are primarily rigid in nature, and a series of chapters on applications of foams. These last chapters are presented from the viewpoint of one who wishes to use a foam, and is concerned with selecting the best properties to match his need, regardless of the chemical structure.

We hope that this dual approach will prove useful to those involved in research and development on foams from specific polymer systems, and equally so to those who wish to use foams for practical applications.

The editors wish to thank many people for assistance in preparing this series on foams. The authors, of course, and the organizations for which they work come first in such consideration. We especially wish to thank the University of Detroit and the Monsanto Company for their encouragement of this effort. Many friends in the foam industry, in addition to the authors, have helped in developing the appreciation for foams and the insight into the needs and contributions of the foam industry which have made the editing of this volume a pleasure.

K. C. Frisch
J. H. Saunders

CONTRIBUTORS TO PART II

J. K. BACKUS, Mobay Chemical Company, Pittsburgh, Pennsylvania

PAUL E. BURGESS, JR., Union Carbide Corporation, South Charleston, West Virginia

J. FOGEL, Research Department, Koppers Company, Inc., Monroeville, Pennsylvania

K. C. FRISCH, Polymer Institute, University of Detroit, Detroit, Michigan

P. G. GEMEINHARDT, Mobay Chemical Company, Pittsburgh, Pennsylvania

R. H. HANSEN, Bell Laboratories, Murray Hill, New Jersey

R. H. HARDING, Union Carbide Corporation, South Charleston, West Virginia

EDGAR E. HARDY, Monsanto Research Corporation, Dayton, Ohio

CARLOS J. HILADO, Union Carbide Corporation, South Charleston, West Virginia

A. R. INGRAM, Research Department, Koppers Company, Inc., Monroeville, Pennsylvania

M. KAPLAN, Allied Chemical Corporation, Buffalo, New York

HENRY LEE, Lee Pharmaceuticals, South El Monte, California

KRIS NEVILLE, Lee Pharmaceuticals, South El Monte, California

ROBERT J. F. PALCHAK, Consultant, Santa Ana, California

ANTHONY J. PAPA, Union Carbide Corporation, South Charleston, West Virginia

STEPHEN C. A. PARASKEVOPOULOS, University of Michigan, Ann Arbor, Michigan

WILLIAM R. PROOPS, Union Carbide Corporation, South Charleston, West Virginia

J. H. SAUNDERS, Monsanto Company, Pensacola, Florida

MARCO WISMER, PPG Industries, Inc., Springdale, Pennsylvania

L. M. ZWOLINSKI, Allied Chemical Corporation, Buffalo, New York

CONTENTS

CONTENTS OF PART I

PLASTIC FOAMS

(IN TWO PARTS)

PART II

Chapter 9

RIGID URETHANE FOAMS

J. K. Backus
and
P. G. Gemeinhardt

Mobay Chemical Company
Pittsburgh, Pennsylvania

I. INTRODUCTION

A. History

Discovered after 1937 by Bayer and co-workers [1], rigid urethane foams, prepared by the addition polymerization of polyols and diisocyanates, were developed slowly during the next two decades. They were put to limited use in Germany during World War II as flotation materials for temporary bridging and as strong, lightweight stabilizing materials in aircraft.

As the German chemical industry rebounded after the end of hostilities in Europe, investigations of the polyurethanes continued with renewed impetus in the laboratories of Bayer AG. At the same time surveys of wartime German developments in plastics, conducted by the Allied Forces, led to the introduction of polyurethanes and rigid foam to the United States [2-4].

News of rigid urethane foam spread rapidly and attracted broad interest in the United States. As a result such companies as Goodyear Aircraft [5], Lockheed Aircraft [6], DuPont, and Monsanto initiated their own research on polyurethanes, particularly rigid foam. Investigations also began in Great Britain [7] and other European countries.

Early German rigid-foam systems were based on polyesters with a high acid number (about 30) and a high viscosity (about 100,000 cP at 25° C). The main mechanism for foaming was the release of carbon dioxide from the reaction between the acid groups on the polyester and isocyanate. Relatively high temperatures (about 125° C) were needed for the reaction. These undesirable conditions for foaming and the handling difficulties caused by the high viscosities of the polyester resins were the main reasons for the slow commercial development of rigid urethane foams. Moreover, the industry saw greater opportunity for flexible foams and concentrated its efforts on them.

It was not until the early 1960s, when flexible foam had made spectacular inroads in markets across the world, that serious attention was again focused on the rigid variety. Since then the growth of rigid urethane foam has been rapid. According to one estimate [8] the U. S. consumption of rigid urethane foam reached 165 million lb in 1968. Furthermore this amount is expected to increase at a rate of about 30% per year through 1975, raising the annual production to at least 500 million lb in the foreseeable future. Several good reviews of rigid-urethane-foam technology have been published [9-14].

B. Chemistry of Foam Preparation

Unlike most systems designed to give a rigid cellular plastic, rigid urethane foams are formed in a single, rapidly occurring step. That is to say, the foaming and formation of the urethane polymer are effected simultaneously by a rapid and controllable release of gas in a system undergoing fast polymerization at the same time. Urethane polymer is formed by the highly exothermic reaction of compounds containing isocyanate (—NCO) and active hydrogen (—XH) functional groups.

The "building blocks" in the preparation of the urethane polymer are polyisocyanates and compounds with two or more groups, such as hydroxyl, amino, and carboxyl, that contain reactive hydrogen atoms. Reactants with high degrees of branching in their chemical structures impart rigidity to the blown urethane polymer by causing crosslinking in the final polymer structure. Compounds terminating in hydroxyl groups are the most common reactants used in conjunction with isocyanate for preparing rigid foam.

The reaction of isocyanate with an alcohol yields a carbamate, a compound known more generally as a urethane (1):

$$\text{RNCO} + \text{R'OH} \longrightarrow \text{RNH}\overset{\text{O}}{\overset{\|}{\text{C}}}\text{OR'} \qquad (1)$$

Polyfunctional components lead to a crosslinked polymer (2) containing many urethane groups, thus explaining the prominent use of the term "polyurethane" in discussions of such polymers.

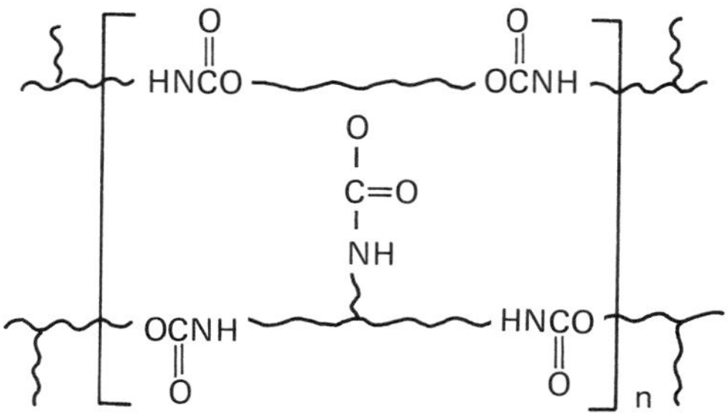

(2)

Compounds terminating in amino or carboxylic acid groups can react with polyisocyanates, but only occasionally form the basis for commercially produced polymers. These reactions give ureas (3) and amides (4).

$$RNCO + R'NH_2 \longrightarrow RHNC(=O)NHR' \quad (3) \qquad (2)$$

$$RNCO + R'COOH \longrightarrow RHNC(=O)R' \quad (4) + CO_2 \qquad (3)$$

Polyfunctional amines and carboxylic acids produce crosslinked polyureas (5) and polyamides (6).

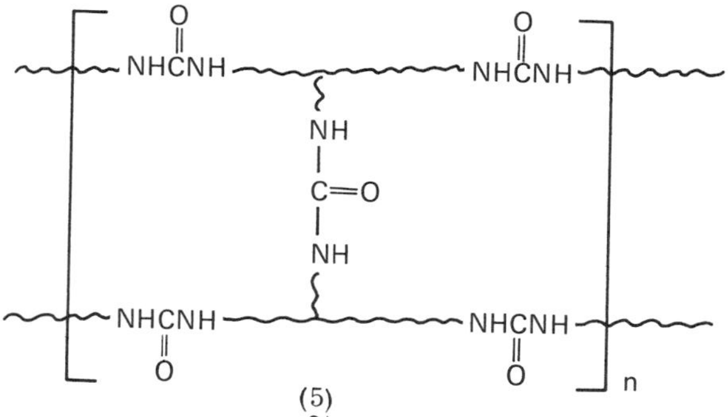

(5)

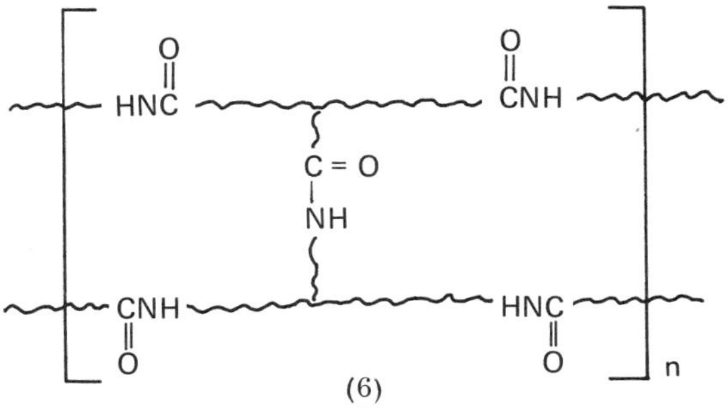

$$\left[\begin{array}{c} \sim\sim \overset{O}{\overset{\|}{HNC}} \sim\sim\sim\sim\sim \overset{O}{\overset{\|}{CNH}} \sim\sim \\ \wr \\ C=O \\ | \\ NH \\ \wr \\ \sim\sim \underset{O}{\underset{\|}{CNH}} \sim\sim\sim\sim\sim \underset{O}{\underset{\|}{HNC}} \sim\sim \end{array} \right]_n$$

(6)

Urethanes will undergo further reaction with isocyanate to give allophanates (7).

$$RHNCOOR' + R''NCO \longrightarrow \underset{\underset{(7)}{CONHR''}}{RNCOOR'} \tag{4}$$

The extent to which this form of crosslinking takes place in a foam is governed by the stoichiometry of the system and the conditions of polymerization, such as the temperature and the type of catalyst employed. Normally this secondary reaction does not occur readily without a catalyst at temperatures below 120° C.

If urea and amide groups are present in the polymerizing system, they can lead to biurets (8) and acyl ureas (9).

$$RHNCONHR' + R''NCO \longrightarrow \underset{\underset{(8)}{CONHR''}}{RNCONHR'} \tag{5}$$

$$RHNCOR' + R''NCO \longrightarrow \underset{\underset{(9)}{CONHR''}}{RNCOR'} \tag{6}$$

In the foaming process the very exothermic nature of these reactions involving isocyanate can be used to advantage. Nonreactive liquid blowing agents are vaporized by the heat of reaction to produce gas, causing the system to foam. Water is less commonly used as a blowing agent in rigid foam systems by reacting with isocyanate to produce carbon dioxide:

$$RNCO + HOH \longrightarrow RNH_2 + CO_2 \qquad (7)$$

The resulting amine is consumed by subsequent chain-extending reactions with the difunctional or polyfunctional isocyanates used.

There are several good reviews concerning the general chemistry of isocyanates as well as that pertaining specifically to the formation of rigid urethane foam [11, 13, 15], and the fundamental aspects of foam formation have been described in Chapter 2.

II. RAW MATERIALS

Rigid urethane foams are made by the well-controlled reaction of a difunctional or polyfunctional isocyanate and a polyfunctional hydroxyl compound usually referred to as a polyol. A blowing agent is required to generate foaming. The process is usually accelerated by a suitable catalyst, and the cell size is usually regulated by the use of a foaming stabilizer. The latter two ingredients are not required in every case. In addition to these raw materials, special properties and foaming conditions can be obtained through the use of numerous miscellaneous additives, such as fillers, plasticizers, flame-proofing agents, and pigments.

In recent years information has been released about highly specialized isocyanate-based foams exhibiting outstanding flame resistance for service at temperatures higher than that recommended for the conventional urethanes [16-18]. More sophisticated isocyanate chemistry is used in their preparation, the details of which are given in Chapter 14.

A. Isocyanates

Tolulene diisocyanate (TDI) is the best known of the isocyanates used in the preparation of rigid urethane foams, but not the only one in commercial use. A family of polyisocyanates derived from aniline-formaldehyde condensation products is gaining rapidly in popularity.

The early German rigid foams were prepared from 2,4-tolulene diisocyanate. Less expensive mixtures of 2,4- and 2,6-isomers have since been developed. In the United States today an 80:20 ratio of the 2,4- and

2,6-isomers is the mixture used almost exclusively, whereas the 65:35 isomer ratio is still used extensively in Europe. An undistilled grade of mixed TDI isomers has grown in commercial interest since its development for rigid-foam preparation in the late 1950s [19]. The typical properties of such a "crude" TDI are summarized in Table 1.

TABLE 1

Physical Properties of a Crude Tolulene Diisocyanate[a]

Color	Brown
Tolulene diisocyanate content, %	83-85
Amine equivalent weight	102-108
Acidity, %	0.20-0.40
Viscosity at 25° C, cP	50-150

[a]Data from Ref. [20].

This product, as seen in Table 1, is approximately 85% TDI and has a high acidity. Since it is undistilled, it contains various by-products from the manufacturing process which are not sufficiently harmful in the foaming process to preclude its use. It is useful particularly in the preparation of one-shot polyether-based rigid foams and offers the chief advantages of lower cost and lower activity over distilled TDI. Several synthetic crude TDI's have also become available.

The development of aniline-formaldehyde condensation products made possible a family of polymeric polyisocyanates designed especially for rigid foams. These polyisocyanates helped greatly to raise interest in urethanes in this country and abroad during the last decade. The polyamines from this condensation are designed to give a mixture of isomers and molecular weights, which, when phosgenated, yields isocyanates of very low freezing point and vapor pressure. The resulting polyisocyanates include such compounds as (10) and (11).

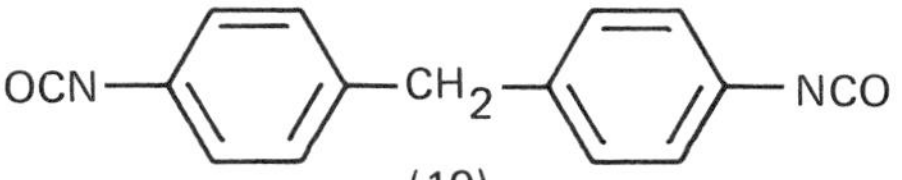

OCN — CH_2 — NCO

(10)

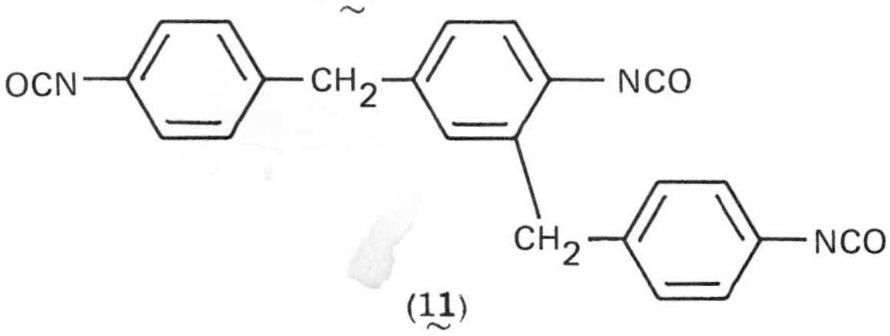

OCN — CH_2 — NCO

CH_2 — NCO

(11)

Assays are in the range 90 to 93% calculated as diphenylmethane diisocyanate (MDI). American suppliers of these isocyanates are the Mobay Chemical Co. and the Upjohn Co., selling them under the trademarks Mondur MR and PAPI, respectively. European manufacturers are Bayer, AG, and Imperial Chemical Industries Ltd., who use the trademarks Desmodur 44V and Suprasec DN, respectively. The typical properties of a polyisocyanate derived from aniline and formaldehyde are listed in Table 2.

TABLE 2

Typical Properties of Polyisocyanate Derived From Aniline and Formaldehyde[a]

Odor	Very slightly aromatic
Color	Dark brown
Isocyanate content, %	31.8
Viscosity at 25° C, cP	200
Hydrolyzable chloride content, %	0.1
Total chloride content, %	0.5
Vapor pressure at 25° C, mm hg	$< 1 \times 10^{-3}$
Pour point, °C	<-20

[a]Data from Ref. [21].

The low vapor pressure and freezing temperature, and consequent easy handling of these polyisocyanates make them very useful and attractive in one-shot systems for spray and pour-in-place applications of rigid urethane foam [21.] In addition, these isocyanates usually provide improved temperature [22, 23] and flame resistance.

These polyisocyanates are more expensive than TDI when costs are expressed as dollars per equivalent of isocyanate, but their unique structure provides important improvements in properties, such as dimensional stability and flame resistance. This offsets the lower cost of TDI in formulating rigid foam. Furthermore these improvements are usually obtainable at the same density or below those possible with TDI and from resins of lower hydroxyl number as well, all of these factors having a direct bearing on the overall cost of making foam [24-28].

B. Polyols

Polyethers and polyesters terminating in hydroxyl groups are well known in the preparation of rigid urethane foam. Both types are branched, low-molecular-weight resins with relatively high viscosities.

Castor oil and its derivatives, a third class of polyols, are less useful in rigid urethane foam. Because of their high equivalent weights and more flexible aliphatic structures, the castor-oil resins do not impart sufficient rigidity to a urethane polymer to make a stable foam when used alone. Usually they find greater application in the preparation of semirigid urethane foam.

The very popular polyether polyols are by far the most widely used class of resin because of their low cost. The best known types are produced by reacting alkylene oxides with polyfunctional alcohols like glycerin, trimethylolpropane, pentaerythritol, sorbitol, α-methylglucoside, and sucrose. The choice of polyfunctional alcohol will govern entirely the average functionality or degree of branching of the resin and has a profound influence on its final room-temperature viscosity.

In most cases the addition of the alkylene oxide serves primarily to change the usual solid state of the alcohol to the more desirable liquid form. Although ethylene oxide is the cheapest alkylene oxide available for this condensation, propylene oxide is preferable because it imparts greater resistance to water absorption. Propylene oxide addition gives a polyol terminated in less reactive secondary hydroxyl groups and thus contributes directly to the degree of chemical activity for reaction with the isocyanate. This has a bearing on the choice of catalyst for the foaming process by necessitating the selection of relatively strong catalysts.

Ethylene oxide is seldom the sole alkylene oxide used in the preparation of polyols for rigid foam. Combinations of ethylene oxide and propylene oxide, however, are becoming more prevalent for obtaining precisely controlled chemical activity within the resin.

Numerous alcohol-based polyether-polyol resins, which are usually classified by the alcohol used in the condensation, are commercially available at several different molecular weights covering a moderately wide, useful range. Table 3 lists the better known manufacturers of these resins, and literature describing their products is readily available.

The products of reaction between polyamines and alkylene oxides constitute a second type of polyether polyols. The more common amines used in their manufacture are ethylenediamine (Wyandotte Chemical), diethylenetriamine (Union Carbide), and piperazine (Jefferson Chemical).

Because of the inherent high chemical activity caused by the presence of tertiary nitrogen, these polyether polyols are rarely used alone in the preparation of foam. Instead, the amine-based polyols are combined in

TABLE 3

Suppliers and Trade Names of Commercial Polyether Polyols

Supplier	Trade name	Base polyhydric alcohol
Atlas Chemical Div., ICI-America, Inc.	Atpol G-	Sorbitol
Dow Chemical Co.	Voranol RS-	Sucrose
	Voranol CP-	Glycerin
Jefferson Chemical Co.	Thanol RS-	Sorbitol
Mobay Chemical Co.	Multranol	
Olin Corp.	Poly-G	Methyl glucoside
PPG Industries Inc.	Selectrofoam	
Union Carbide Corp.	Niax Triol LHT	Hexane triol
	Niax Hexol LS	Sorbitol
	Niax T-	Aromatic
BASF-Wyandotte Corp.	Pluracol TP-	Trimethylolpropane
	Pluracol PeP-	Pentaerythritol
	Pluracol SP-	Sorbitol

formulations with other polyether polyols. They are particularly useful in systems designed for spraying, when a high order of activity is needed for good application.

A third type of polyether polyol that has gained rapid commercial interest is prepared from oxyphosphorous acids and propylene or ethylene oxide. These resins are designed primarily to impart flame resistance to rigid foam. Although foams can be made using this type of resin alone, the phosphorus-containing polyols are most often used as blends with polyols. A list of manufacturers offering this kind of polyol is given in Table 4.

A good review [29] has been published on the general subject of polyether polyols, covering the chemistry, manufacturing, and use of these compounds.

Polyesters are more costly and less widely used than polyethers, but they are excellent resins for rigid urethane foam. They are generally

TABLE 4

Suppliers and Trade Names of Commercial Phosphorus-Bearing Polyols

Supplier	Trade name
Stauffer Chemical Co.	Fyrol 6
Mobil Chemical Co.	Vircol 82
	Vircol 88
Swift Chemical Co.	
Weston Chemical Co.	Tris DPG phosphite
	Tris DPG phosphonate

the condensation products of simple dicarboxylic acids, such as adipic or phthalic acid, and saturated polyfunctional alcohols, such as 1,2,6-hexanetriol, trimethylolpropane, diethylene glycol, and ethylene glycol. The acid-to-alcohol ratio is such that nearly all of the acid groups react, and the hydroxyl groups are in sufficient excess to give a polyester terminating in hydroxyls.

A particularly useful polyester is made from chlorendic acid, better know as Hetacid (Hooker Chemical). Foam prepared from these polyesters exhibits excellent flame resistance [30]. Polyesters based on tetrachlorophthalic and tetrabromophthalic acids also contribute flame resistance.

Of late, the Stepan Chemical Co. has announced a unique polyester-based urethane foam for recovering oil spills, particularly at sea [31]. Stephanol B contains a fatty-acid-derived polyester that is reported to give foam prepared from it distinct advantages over conventional polyesters. These advantages are low water absorbancy and very high oil absorbancy.

C. Blowing Agents

Most rigid-urethane-foam formulations use inert liquids as blowing agents since they can be easily and controllably volatilized during the course of polymer formation. Their common characteristic is the ease by which they are vaporized at or about room temperature; for example, trichlorofluoromethane and 2-chloropropane can be added to either the polyisocyanate or the polyol component without need for high pressure to prevent loss by volatilization.

More volatile liquid blowing agents, such as dichlorodifluoromethane, find applications in "frothing processes" [32]. These compounds need considerable externally applied pressure to keep them in the foam ingredients prior to frothing.

Carbon dioxide, produced in situ by the reaction of isocyanate with water or carboxyl groups, was the first blowing agent used for foaming rigid urethanes. It has been replaced largely by the inert halogenated hydrocarbons, however, and today is used to a much lesser extent.

A poor choice of blowing agent can lead to disastrous foaming results. The main properties to be considered in selecting a blowing agent are the following:

1. How easily the blowing agent in the form of vapor permeates through uncured urethane polymer.

2. How good a solvent it is for the polymer.

It is nearly impossible to produce dimensionally stable foam when the blowing agent permeates the polymer readily. Conversely, the proper choice can impart outstanding properties to foam. For example, the fluorocarbons' popularity in rigid-foam manufacture is due to the low thermal conductivity they impart to the foam.

D. Catalysts

The choice of catalyst for the preparation of rigid urethane foams is governed by many factors, the most important of which are (a) the method of foam application, that is, spraying, foaming in place (molding), or slabbing[21, 33, 34]; (b) the specific polyol or polyisocyanate used [35-37]; and (c) the activity of the catalyst itself in promoting urethane or gas formation [37-42]. Two general classes are very well known. The first is composed of tertiary amines, of which 1,4-diazabicyclo 2.2.2 octane (DABCO), tertiary alkylamines, substituted morpholines, piperazines, guanidines, and substituted hydroxy amines are representative types. A wide range of activities is represented in this class, but stronger catalysts are sometimes needed to promote the reaction of secondary hydroxyl groups with isocyanates. Stronger catalysis is supplied by catalysts of the second class: organotin compounds like dibutyltin dilaurate and stannous octoate. The reaction of secondary hydroxyl groups can be effectively promoted by these potent catalysts. Representative catalysts and their suppliers are listed in Table 5.

In rigid-foam systems a catalyst is not always necessary since these systems have a moderate degree of inherent chemical activity. Some systems perform even better when a deactivator is used in combination with catalysts, particularly in such applications of high-rise foaming as filling

TABLE 5

Manufacturers and Trade Names of Common Catalysts for Rigid Urethane Foams

Manufacturer	Trade name or abbreviation	Chemical name
	Amines	
Abbott Laboratories	--	N, N-Dimethylcyclohexylamine
American Cyanamid	TMG	1, 1, 3, 3-Tetramethylguanidine
Amer. Laboratories	TMEDA (T-23)	Tetramethylethylenediamine
Air Products and Chemicals Inc.	DABCO (solid) 33 LV (liquid)	Triethylenediamine 33% DABCO in Tripropylene glycol
Jefferson Chemical Co.	--	N, N'-Dimethylpiperazine
Rohm and Haas Co.	TMEDA	Tetramethylethylenediamine
Union Carbide Corp.	TMBDA	N, N, N', N'-Tetramethyl-1, 3-butanediamine
BASF-Wyandotte Corp.	-- DHP-MP	Trimethylpiperazine 1, 4-Bis(2-hydroxylpropyl)-2-methylpiperazine
Barlow Chemical Industries	TMEDA	Tetramethylethylenediamine
	Organometallics	
Carlisle Chemical Co.	Carstan-18 Carstan-8	Stabilized stannous oleate Stannous octoate

TABLE 5 (Continued)

Manufacturer	Trade name or abbreviation	Chemical name
	Carstan DBTDL	Dibutyltin dilaurate
	--	Tetrabutyl titanate
	--	Dibutyltin bis(butylmalonate) (very low activity)
M and T Chemical Co.	T-6	Stannous salt of fatty acids with more than 18 carbon atoms
	T-9	Stannous octoate
	T-10	Stannous octoate and plasticizer, 50:50
	T-12	Dibutyltin dilaurate
	T-18	Stannous oleate
Union Carbide Corp.	D-22	Dibutyltin dilaurate
Nuodex Products Co.	--	Chelated metal catalysts; regular tin catalysts
Witco Chemical Co.	C-2	Stabilized stannous octoate
	C-4	Blend of C-2 and dioctyl phthalate, 50:50
Aceto	--	Stannous octoate
LONZA (Baird Chemical Industries, Inc.)	--	Stannous octoate

the walls of tractor trailer bodies, refrigerated railroad boxcars, and curtain-wall panels [43] . In these cases acidic compounds, such as hydrogen chloride, benzoyl chloride and alkyl acid phosphates, are effective deactivators.

E. Foaming Stabilizers

As their name implies, foaming stabilizers are added to impart stability to the foaming process. They are intended to control the cell structure by regulating the size and, to a large extent, the uniformity of the cells. The choice of surfactant, as it is commonly called, is governed primarily by the type of polyol used and by the method of foam preparation.

Not all rigid foam systems require a foaming stabilizer. Higher viscosities at any time during the foaming tend to diminish the need for a stabilizer at that time. Since rigid foam systems tend to be higher in viscosity at the time of foaming and build viscosity quickly during foaming, it is possible to prepare foam without stabilizer, depending, of course, on the system being used.

The most common surfactants for polyether-based foams are silicones, particularly the polyalkylsiloxane-polyoxyalkylene copolymers. These usually give extremely fine-cell foam of excellent uniformity and high-closed-cell content. Because copolymers containing Si-O-C bonds tend to hydrolyze slowly under normal conditions of storage and use, the more hydrolytically stable types containing direct Si-C bonds have gained in popularity and use.

Silicone oils, polydimethylsiloxanes of relatively low viscosity (10-100 cP at 25° C) have been used in polyether systems prepared and foamed as prepolymers.

Polyester-based systems can often be used without added foaming stabilizers because of their usually high initial viscosities. In these systems the silicones can have an effect opposite to stabilizing the foaming mass and in fact can cause bubble instability. Stabilizers employed for polyesters are usually of the ionic or nonionic type and include sulfonated castor oils, amine esters of fatty acids, and polyoxyalkylene derivatives of acids or alcohols. A list of common foaming stabilizers for rigid urethane foam is given in Table 6. Silicone stabilizers are also supplied by Bayer AG, Imperial Chemical Industries Ltd., Th. Goldschmidt AG, Midland Silicones Ltd., and Shin-Etsu Chemical Industry Co. Ltd.

There are times when it is desired that the cells of rigid foam be more open than closed. An effective additive for opening the cells of rigid foam is a special Alcoa aluminum leafing powder (No. 427), which is added to a foam formulation in concentrations of about 0.5% by weight [20].

TABLE 6

Common Silicone-Polyol Blockcopolymer Stabilizers for Rigid Urethane Foams

Manufacturer and silicone block copolymer	Comments
Dow Corning:	
DC-193	Liquid; hydrolytically stable; approximately 2% hydroxyl content; soluble in polyols and TDI, limited solubility in polymeric isocyanates
DC-195	Liquid; hydrolytically stable; approximately 2% hydroxyl content; soluble in isocyanates and polyols
General Electric:	
SF-1066	Liquid; hydrolytically stable ester; polyol block is ethylene oxide-propylene oxide copolymer; soluble in polyols and TDI; normally used in flexible urethane foams but useful in some rigid-urethane-foam systems
SF-1079	Higher-molecular-weight version of SF-1066; more effective stabilizer; higher viscosity
SF-1109	Conventional rigid-foam stabilizer containing ethylene oxide-based polyol block; stable except in contact with organotin catalysts
SF-1158	Stable in all systems; contains ethylene oxide-based polyol block
Union Carbide:	
L-530	Liquid; hydrolytically stable; useful in systems based on sorbitol-, aromatic-, and amine-initiated polyether polyols and on halogenated polyester polyols

TABLE 6 (Continued)

Common Silicone-Polyol Block-Copolymer Stabilizers for Rigid Urethane Foams

Manufacturer and silicone block copolymer	Comments
L-531	Liquid; hydrolytically stable; for use in systems based on sorbitol-initiated polyether polyols
L-5310	Soft wax (mp 32-35° C); for use in systems based on polyether polyols initiated by sorbitol, sucrose, aromatic compounds, and amines; isocyanate soluble
L-5320	Liquid solution of L-5310; hydroxyl number 160
L-5340	Preferred product for use in isocyanate side of all fluorocarbon-blown and partially water-blown systems; liquid; stable in isocyanate solution
L-5410	For use in polyol side of all rigid-foam systems; stable in polyol solution

F. Miscellaneous Additives

Flame retardants are the most widely used additives in the preparation of rigid foam. They are usually unreactive organic phosphorous and/or halogen compounds or inorganic materials. Typical examples of organic flameproofing agents are tris(chloroethyl) phosphate, tris(2,3-bromopropyl) phosphate, chlorowaxes, and the Phosgards (Monsanto). Such inorganic materials as antimony oxide, ammonium phosphate or carbonate, and red phosphorus [44] are effective alone or in combination with the organic additives.

Although effective for flameproofing foam, these compounds will usually adversely affect other foam properties, especially thermal conductivity, water-vapor permeability, and dimensional stability. For this reason polyols containing phosphorus or halogen are preferable for imparting flame resistance to rigid foam.

Generally, no real advantage is gained by adding unreactive fillers to rigid foams. Processing is more difficult, design densities are harder to achieve, and overall properties are poorer. Some exceptions have been reported, however [45-47]. Excellent use has been made of inexpensive extenders obtained as waste products in the conversion of wood to paper. Two of these are a special crude tall oil (Emtall 907, Emery Industries) and Vinsol modified with ethylene oxide (polyol S-1076, Hercules Powder). Each contains active hydrogen atoms, which, when reacted with isocyanate, become an integral part of the polymer structure. These liquid or low-softening-temperature reactive extenders have been used in amounts of 40 to 60% by weight of the total amount of resin used. Properties of the resulting foams were good at a saving of about 20% of normal raw material costs.

III. FOAMING PROCESSES

A. Method of Foaming

Two basic methods of rigid-foam preparation are in general use: the quasiprepolymer, or semiprepolymer, method and the one-shot method. The quasiprepolymer method gained widespread use during the 1950s, but with the advent of crude TDI and the polymeric isocyanates, the one-shot method began growing rapidly.

In the quasiprepolymer method a large molar excess of the isocyanate is reacted with the polyol to give a product with a high percentage of free isocyanate groups (20-30%). The process is preferably carried out under well-controlled conditions of temperature and time (e.g., 45-66° C for 2 h in a completely dry environment). Because the preparation of quasiprepolymers employs large excesses of isocyanate, they can also be made by merely cold-blending the isocyanate and polyol and allowing sufficient time to elapse for complete reaction to occur. A period of at least 24 h is required for this longer process.

On completion of the preparation of the prepolymer, it is combined with more polyol, blowing agent, catalyst, and surfactant for foaming. In foaming the same polyol as in the prepolymer preparation or different polyols can be used.

The one-shot procedure consists of mixing together all the ingredients for making foam at a single time. Since most one-shot systems are mixed at room temperature, it is necessary that all components for foaming be liquid at this temperature. It is also highly desirable that the ingredients have a moderate degree of compatibility with each other, thereby facilitating the mixing step.

B. Equipment for Foaming

Successful preparation of rigid urethane foam is most often the result of using the proper equipment [48]. In principle a foam machine consists of two parts: a pumping unit capable of accurately metering two or more components and a continuous mixer capable of blending and efficiently mixing the various components. The equipment often is very similar to that used for flexible foam (see Chapter 3).

A single pumping or metering unit, of the two or more necessary for the preparation of foam, consists essentially of a storage tank, a pump, a pump-drive system, miscellaneous valves, instruments, piping, and heat exchangers in special cases. Numerous types and makes of pumps are available. Gear pumps and piston-type pumps with an accuracy of ±0.5% are obtainable and adequate. In multicomponent machines where recycle of a component is not required, high-pressure fuel-injection pumps are very acceptable for low-viscosity components. Accurate proportioning of the one or more components employed for foaming is possible through variable-speed drives on gear pumps or adjustable stroke lengths on piston pumps.

The mixing is just as important as the metering of the foam ingredients since poor mixing nullifies proportioning by the most elaborate pumping and metering devices. There are two general mixers for pouring rigid foam. The continuous-pour mixer usually provides low-shear agitation and is capable of operating for long periods. An example is shown in Fig. 1. It consists of an agitator with pinlike projections on a central shaft. The intermittent-pour mixer is necessary in processes involving foaming in place. It provides high-shear mixing, is self-cleaning, and operates on an on-off cycle. Fig. 2 illustrates an example of this type.

Intermittant mixers without any moving parts are available and are being put to good use. An example is shown in Fig. 3. Their use is somewhat limited, however, since efficient mixing is a direct function of component viscosity, and these intermittent mixers do not operate well when the viscosities of foam ingredients exceed 1000 cP at 25° C.

All of the mixers described here are used for the preparation of rigid foam. In selecting the type to use one must seriously consider what is expected of the mixer and have good familiarity with the foaming ingredients to be mixed. There are many manufacturers of foaming equipment, and detailed information is available from them. Table 7 lists some of the better known manufacturers.

C. Methods of Manufacture

Rigid urethane foams can be manufactured by a batch process, by continuous slabbing, by foaming in place, and by spraying. Of these, the foaming in place method is the most common because it makes the best use of the main advantages of a urethane system: the ease of handling liquid foam components under ambient conditions and the spontaneity of

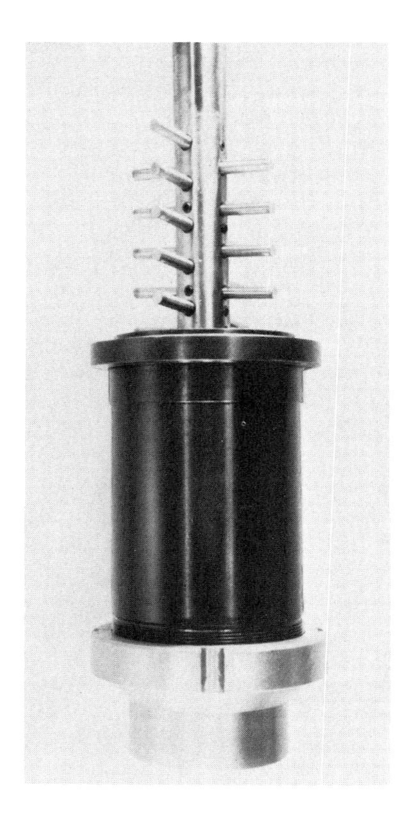

FIG. 1. Low-shear mixing head and pin type agitator used in urethane-foam-slab machines.

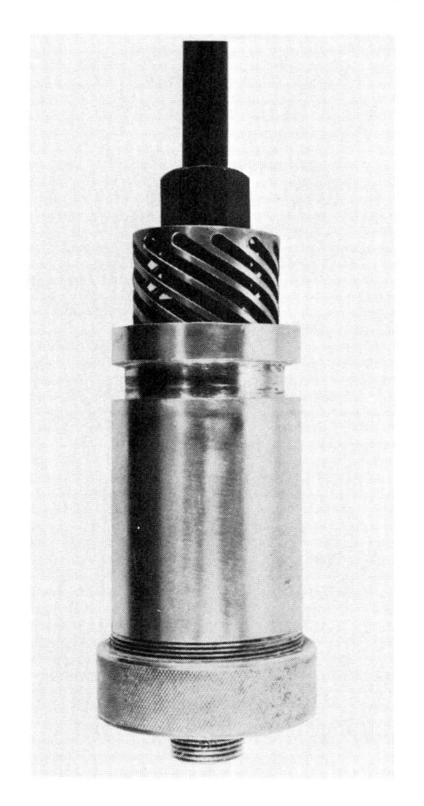

FIG. 2. High-shear mixing head and agitator used in intermittent-pour machines for urethane-foam molding.

foaming under these same conditions. Spraying is a very economical method of applying rigid foam, but one must be expert at spraying to achieve the economics. It is rapidly gaining popularity.

1. Batch Process

Rigid foam can be crudely but simply processed by batch means, using buckets, pails, or beakers as mixing containers and paddles or mechanized laboratory-type stirrers as mixing devices. Good results can be achieved with this method by carefully and consistently employing

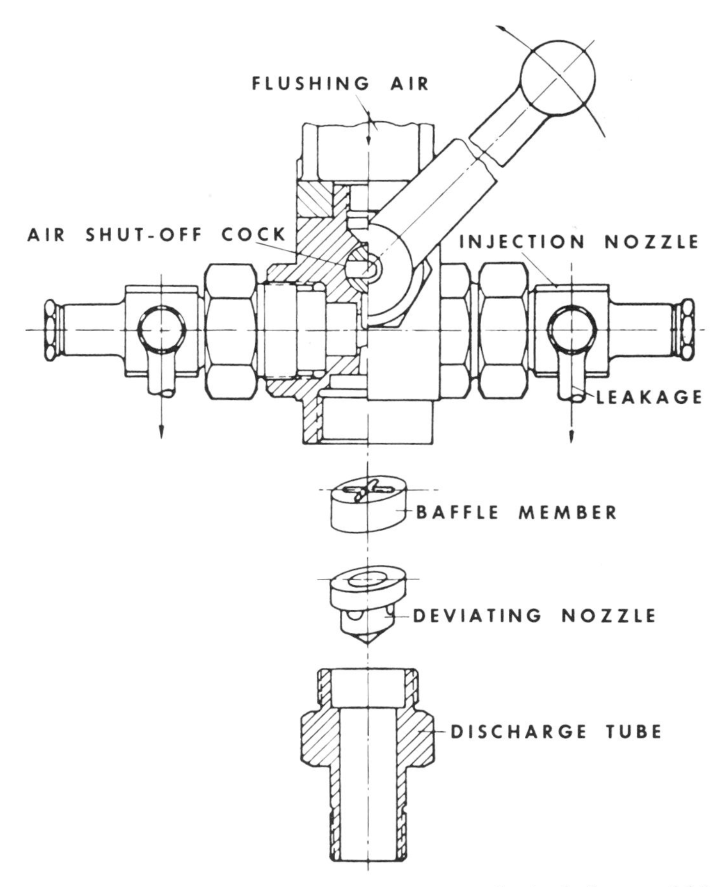

FIG. 3. Two-component mixing head using the turbulence of high-pressure injection to disperse raw materials. Courtesy of Maschinenfabrik Hennecke GmbH.

well-developed conditions (e.g., temperature) and procedures (e.g., mixing times). The method has been used successfully in pours of 50 lb or more. Lack of uniformity in the foam and high scrap losses due to foam adhering to the sides of the mixing container are the method's main drawbacks [49, 50].

2. Continuous-Slab Production

Continuous production of rigid-foam board stock involves the accurate metering of all foam ingredients to a common mixer operating

TABLE 7

Manufacturers of Equipment for Processing Rigid Urethane Foam

Supplier	Location	Foaming application
United States:		
Martin Sweets Co.	Louisville, Kentucky	Molding
Admiral Machine Co.	Akron, Ohio	Molding
Lake Erie Machine Co.	Toledo, Ohio	Molding
Binks Manufacturing Co.	Chicago, Illinois	Spraying
Guzmer Co.	Union, New Jersey	Spraying
Kornylak Corp.	Hamilton, Ohio	Laminating (board stock)
Leon Corp., Div. of Mobay Chemical Co.		
Europe:		
Maschenfabrik Hennecke GmbH	Birlinghoven, West Germany	Slab, molding, laminating, spraying
Viking Engineering Ltd.	Manchester, England	Slab, molding, laminating, spraying
Pla-ma	Norway	Molding

at a high rotational speed (2000-5000 rpm). The intimately mixed chemicals are poured from the mixer onto a paper-lined conveyor moving in a direction away from the mixer. At the same time the mixer of the foam machine is swung in a reciprocating motion from one side of the conveyor to the other. In this fashion the liquid prefoam mixture is deposited as a thin layer on the conveyor and subsequently foams to form a continuous block of foam. Such a foam-producing machine is shown in Fig. 4. At the end of the 60 to 100 ft long conveyor the foam is cut into sections by a vertical cutter that moves at the same speed as the conveyor. After the sectionalized blocks have been cured (this takes place away from the production machine) the foam can be fabricated to specific thicknesses and shapes by cutting, heat forming, and other mechanical means of shaping.

FIG. 4. Production of urethane foam in continuous-slab form.

A modification of this process restricts the thickness of the foam produced with a top conveyor or belt moving at the same speed as the bottom conveyor. In this manner foam sheets of specific thickness (1-2 in. usually) are produced continuously. On the board-stock manufacturing machine shown in Fig. 5 coils of paper, asphalt-treated paper, or metal foil can be fed between the double conveyors so that the foam produced will be permanently attached to the facing materials. If no facing is desired, release paper is used instead. In another version of this type of production rigid sheets of various materials, such as steel, aluminum, or plastic, are fed between the double conveyor and serve as facings permanently bonded to the foam.

3. Foaming in Place

Foaming in place involves pouring the liquid or froth foam mixture after mixing directly into a cavity or mold prepared to contain the foam [51, 52]. Whether the cavity is formed by the walls of a refrigerator,

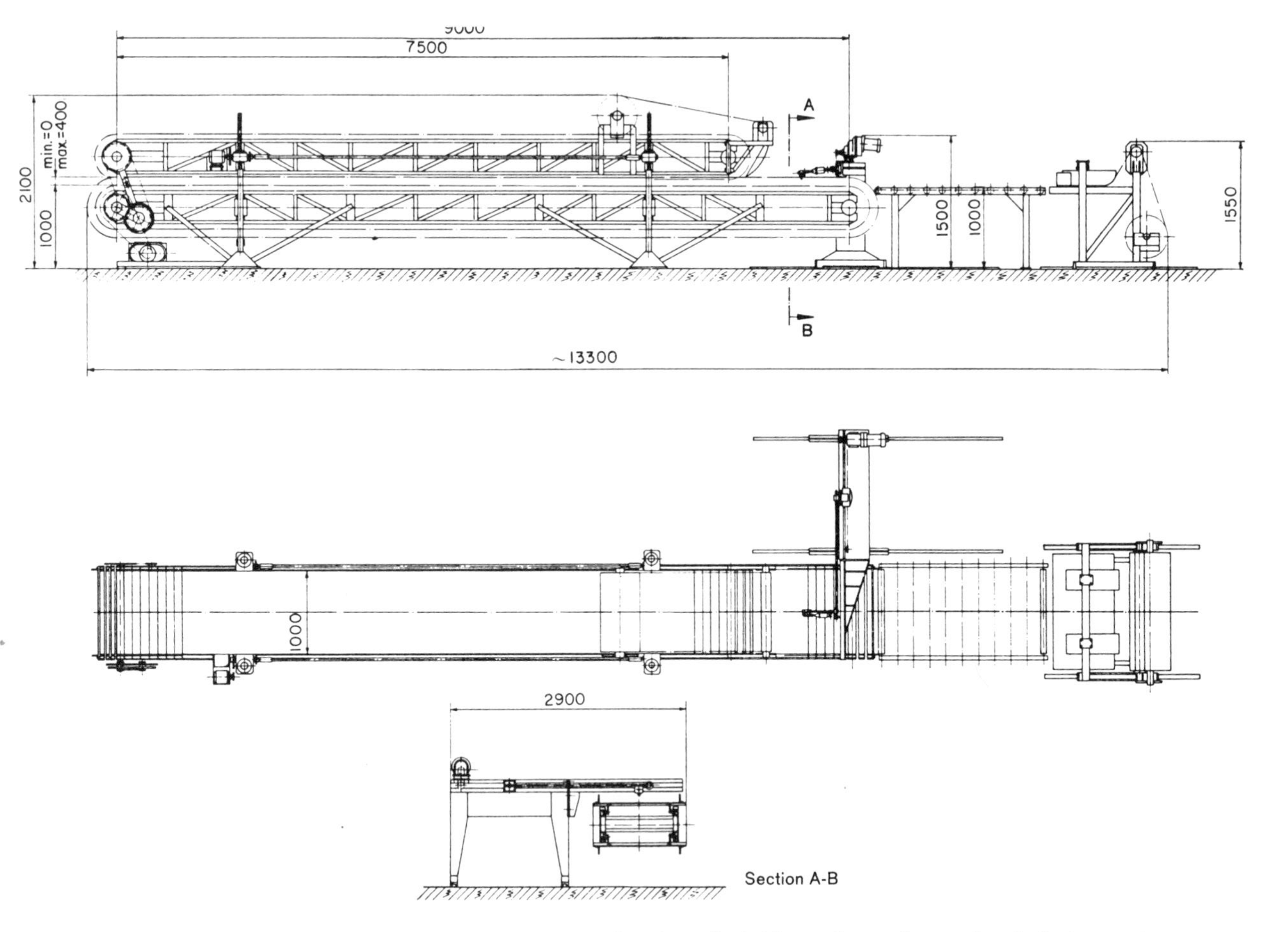

FIG. 5. Laminating machine for the continuous production of rigid-urethane-foam sheets between two conveyors. Courtesy of Maschinenfabrik Hennecke GmbH. Dimensions in mm.

truck body, or building, the resulting foam completely fills the cavity and naturally develops good adhesion to its sides. To improve the adhesion it is sometimes desirable to use a primer [51], which can be applied by brushing or spraying. Under any circumstances, best adhesion is usually obtained on a clean, oil-free, dry, and roughened surface.

The frothing [32] of rigid-foam prefoam mixtures is becoming increasingly more popular in applying foam by the foaming-in-place method. The process consists of incorporating a volatile liquid, such as dichlorodifluoromethane (Fluorocarbon 12), as blowing agent into the foam reactants under a pressure adequate to maintain the mixture in a liquid state. When dispensed from a mixer designed to operate at pressures much higher than atmospheric, frothing occurs as the volatile liquid, contained in the foam mixture, vaporizes on the reduction in pressure. Usually the pressure within the mixer at the time of pouring the foam mixture into a cavity is on the order of 125 psi. This approximate pressure must be maintained throughout the feed system of the foaming component containing Fluorocarbon 12 in order to achieve a successful frothing operation. In the absence of this condition, the efficiency of blowing is seriously hampered, making it nearly impossible to achieve design densities in foams. Also, poor dispersion or lack of complete solubility of the Fluorocarbon 12 in the foam component and mixture will lead to a highly irregular foam cell structure. This condition naturally calls for a much greater sophistication in the design of equipment.

The advantages of a frothing operation are (a) lower mold pressures, (b) simplified filling of voids or cavities, and (c) greater consistency in foam density throughout any given part and more reproducible uniformity of all size and structure. The method is applicable to either prepolymer or one-shot systems.

A two-stage expansion can be achieved by using a combination of Fluorocarbon 12 and Fluorocarbon-11 (trichlorofluoromethane). In this process the lower boiling Fluorocarbon 12 produces immediate frothing on pouring, giving the first stage of foaming, and the higher boiling Fluorocarbon 11 completes the foaming process in the cavity or a second stage. The equipment needs for two-stage foaming are the same as for total frothing.

Fig. 6 illustrates the foaming-in-place process. When the cavity used is a mold to give the foam a specific shape, it is necessary to use a release agent like wax or silicone on the inside mold surfaces for separating the foam from the mold.

4. Spraying

Spraying involves the same important first steps of metering and mixing as required by all rigid-foam processes. It differs from the other procedures, however, by dispensing the prefoam mixture as a spray rather

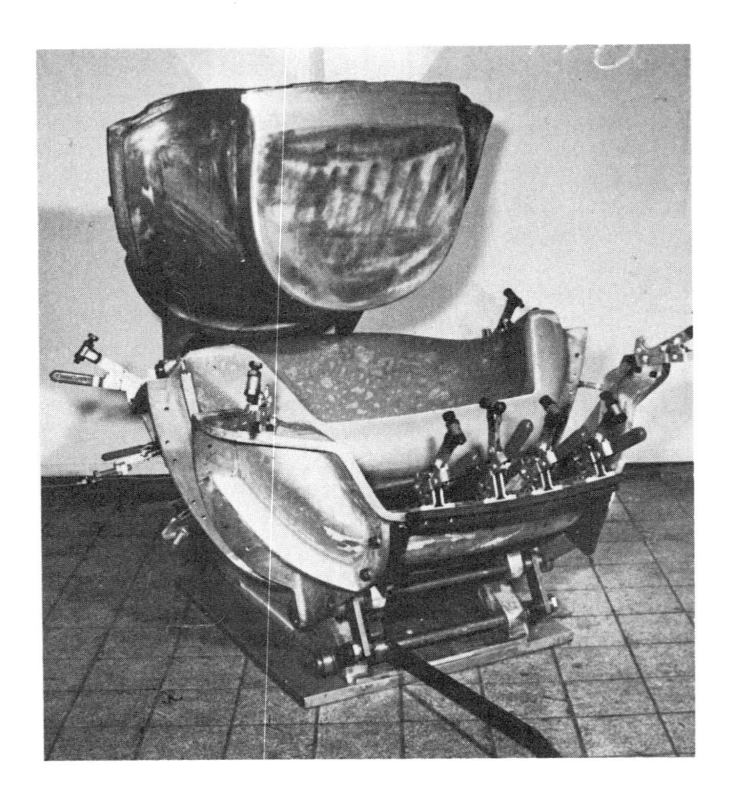

FIG. 6. Molding of rigid-urethane-foam shells for chairs of the type shown in Fig. 27. Courtesy of Bayer AG.

than as a stream (see Fig. 7). By this process the spray-foam mixture is directed to a surface, such as a concrete block wall or the outside of a metal storage tank, and foaming occurs without restriction.

Systems designed for spraying generally consist of two components, each exhibiting relatively low viscosity to facilitate mixing and atomization. The components are usually formulated so that they are used at a nearly equal weight or volume ratio.

Catalysis of the system is a most important consideration since the success of spraying is largely governed by the rate at which the liquid-foam mixture expands and sets. In general, spray systems are highly catalyzed to effect a very rapid foaming and gelation on the surface being sprayed.

The extent of catalysis is instrumental in governing the character of the produced foam. Too much catalyst causes foaming and possibly

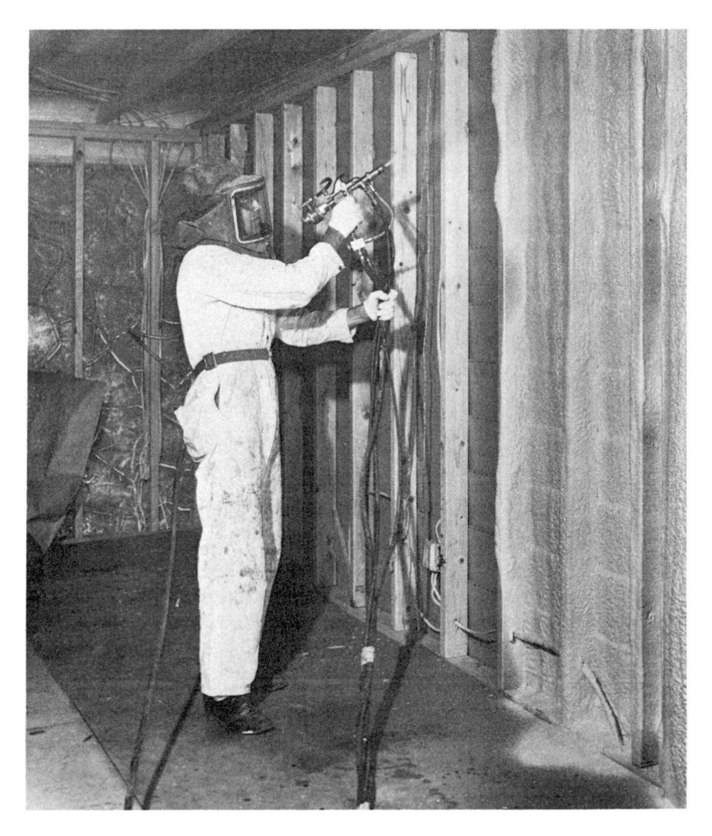

FIG. 7. Spray application of rigid urethane foam on inner wall of frame residence. Courtesy of Callery Chemical Co., Division of Mine Safety Appliances.

gelation of the particles after emission from the spray gun, but before reaching the surface being sprayed. This degree of catalysis produces a very irregular and unsightly surface and causes problems of poor adhesion on the substrate and between successive foam layers. The optimal degree of catalysis is that allowing the particles of the sprayed system to be deposited on the surface as a liquid followed by very rapid foaming and setting. This enables the prefoam particles to flow together on the surface being sprayed, forming a continuous, uniform film of foam mixture and giving excellent control of layer thickness and uniformity.

Best spraying results are obtained on surfaces that are clean, dry, and heated slightly above room temperature (25-30°C); however, these conditions are not always possible in practice. Good results can often be obtained on wet or cold surfaces with an initial application of a flash coat of the system being sprayed. This, in effect, primes the surfaces and provides an insulating barrier against a cold substrate so that subsequent application can proceed normally.

A new process producing rigid urethane foam with a thick, high-density skin and a cellular core, called Duromer urethane, has been reported by Piechota [53]. The all-urethane sandwich foam moldings are produced from special polyols and isocyanates in a single-step operation on conventional equipment. In some respects, the process is similar to the injection molding of thermoplastics in that the reactive foam ingredients are injected into heavily jigged molds wherein the combination of the special ingredients and the foaming pressure produces foam structures with massive skins. It differs from typical injection molding by operating at appreciably lower injection and mold pressures. Mold residence time is short in comparison with conventional methods used for molding rigid foam. The resulting sandwich-like foam structure exhibits high strength and stiffness, and should find wide use in applications requiring these properties, such as construction and furniture manufacture.

IV. PROPERTIES

The properties of foams, unlike those of solid materials, depend on both the properties of the solid and the structural form of the foam. The latter depends in turn on the foam density, the method by which the foam is produced, and the viscoelastic changes that occur as the foam is formed. Thus the properties of rigid urethane foam produced by a process of simultaneous polymerization and expansion differ from those of foamed polystyrene produced by a process of thermal softening and expansion, not only because of the difference in polymers but also because of structural differences associated with the foam-forming methods. Similarly the properties of rigid urethane foam applied by spraying are different from those of similar molded foams because of differences in cell shape and orientation.

The most common types of rigid urethane foams are those based on tolylene diisocyanate or polymeric isocyanates and hydroxyl-terminated polyethers (see Section II). The inert blowing agent trichlorofluoromethane (fluorocarbon 11) is most often used because of the low thermal conductivity of its vapor trapped in a closed-cell structure. Emphasis on flame resistance has made the addition of phosphorus- and halogen-containing compounds to foam systems common. Formulations and general physical properties of typical flame-resistant one-shot rigid urethane foams are listed in Tables 8 and 9. The foams were prepared in block form and have properties typical of the low-density rigid urethane foams used as thermal insulation and in construction.

TABLE 8

Formulation, Foaming Characteristics, and Physical Properties of Rigid Urethane Foam Based on a Polymeric Isocyanate and a Sucrose-Type Polyether Resin[a]

	Foam type	
	A	B
Formulation, parts by weight:		
Sucrose-initiated polyether polyol (OH No. 410)	70	70
Flame retardant, O, O-diethyl-N, N-bis(2-hydroxyethyl) amino methyl phosphonate	30	30
Trichlorofluoromethane	30	40
Silicone stabilizer	1.0	1.0
Tertiary amine catalyst	1.5	1.5
Catalyst diluent	1.5	1.5
Polymeric polyisocyanate	108	108
Foaming characteristics:		
Cream time, sec	10	8-9
Rise time, sec	130-140	150-160
Tackfree time, sec	140-150	160-170
Maximum exotherm, 10- to 14-in. block, °C	160	140
Physical properties:		
Density, lb/ft^3	2.0	1.4
Compressive strength, psi at yield:		
Parallel to rise	31-33	15-19
Perpendicular to rise	16-20	9-10
Tensile strength, psi	33	29
Closed cells, %:		
Uncorrected for cut surface	89	86
Corrected approximately	>94	>91
"k" Factor, Btu/(h)(ft^2)(°F/in.), original value at 23°C, 50% R.H.	0.11	0.11
Cell size, cells/per inch	60-65	50-55
Flammability (ASTM D1692):		
Burning length, in.	0	0
Rating	S.E.[b]	S.E.[b]
Water Absorption, lb/ft^3	0.033	0.044
4 x 4 x 1-in. sample, $g/1000\ cm^3$	18.9	26.4
Water-vapor permeability, perm.-in.		
Wet cup	6.3	14.3
Dry cup	1.5	2.8

TABLE 8 (Continued)

Formulation, Foaming Characteristics, and Physical Properties of Rigid Urethane Foam Based on a Polymeric Isocyanate and a Sucrose-Type Polyether Resin[a]

	Foam type	
	A	B
Dimensional stability, ΔV, %:		
At 70° C, 100% R. H.:		
1 week	7	4
2 weeks	7	8
At 100° C, ambient R. H.:		
1 week	5	8
2 weeks	5	10
At -40° C, ambient R. H.:		
1 week	No change	Shrinkage

[a]Date from Ref. [21].

[b]Self-extinguishing.

Although the use of polyether polyols is widespread in the manufacture of rigid urethane foam, the use of polyester polyols is known. In particular, special chlorinated polyesters impart outstanding flame resistance to polyurethanes. An example of a one-shot foam employing such a polyester is shown in Table 10.

The quasiprepolymer, or partial-prepolymer, method is used in producing some rigid urethane foams in order to reduce the maximum temperature created by the exothermic reaction and provide somewhat greater control of foam formation. Quasiprepolymer systems based on TDI of the type shown in Table 11 are particularly common.

A. Static Mechanical Properties

The specific properties of rigid urethane foams can be varied within limits by the choice of raw materials and methods of applications. Density, however, is the main variable controlling most strength properties and has an effect much larger than that of possible changes in polymer structure.

TABLE 9

Formulation, Foaming Characteristics, and Physical Properties of Foam with an Aromatic-Based Polyether Resin[a]

	Foam type	
	A	B
Formulation, parts by weight:		
Aromatic-initiated polyether poloyl (OH No. 380)	64	64
Amine-initiated polyether poloyl (OH No. 700)	16	16
O, O-Diethyl-N, N-bis(2-hydroxyethyl) Amino methyl phosphonate	20	20
Trichlorofluoromethane	30	40
Silicone stabilizer	1.0	1.0
Tertiary amine catalyst	0.5	0.5
Catalyst diluent	0.5	0.5
Polymeric polyisocyanate	114	114
Foaming, characteristics:		
Cream time, sec	11-12	10-11
Rise time, sec	110-120	140-150
Tackfree time, sec	120-130	150-160
Maximum exotherm, 10- to 14-in. block, °C	162	140
Physical properties:		
Density, lb/ft^3	1.9	1.6
Compressive strength, psi at yield:		
Parallel to rise	30-35	25-28
Perpendicular to rise	19-20	12-14
Tensile strength, psi	33	37
Closed cells uncorrected for cut surface %	89	88
"k" factor, Btu/(h)(ft^2)(°F/in.)	0.13	0.11
Cell size, cells per inch	65-70	55-60
Flammability (ASTM D1692):		
Burning length, in.	0.4	0
Rating	S.E.[b]	S.E.[b]
Water absorption, lb/ft^2	0.027	0.033
Water-vapor permeability, dry cup, perm.-in.	1.5	1.9
Dimensional stability, ΔV, %:		
2 weeks at 70°C, 100% R.H.	5	10
2 weeks at 100°C, ambient R.H.	5	10
1 week at -40°C, ambient R.H.	No change	No change

[a] Data from Ref. [21]. [b] Self-extinguishing.

TABLE 10

Formulation, Foaming Characteristics, and Physical Properties of Foam Prepared from a Halogenated Polyester[a]

Formulation, parts by weight:	
Halogenated polyester (OH No. 360)	100
Trichlorofluoromethane	35
Surfactant	2.0
Polymeric polyisocyanate	92
Foaming characteristics:	
Cream time, sec	10-12
Rise time, sec	150-160
Tackfree time, sec	160-170
Physical properties:	
Density, lb/ft^3	1.6
Compressive strength, psi:	
Parallel to rise	24.7
Perpendicular to rise	8.7
Closed cells (uncorrected for cut surface), %	88
Flammability (ASTM D1692):	
Burning length, in.	0
Rating	S.E.[b]
Water-vapor permeability, wet cup, perm.-in.	3.7
Dimensional stability, ΔV, in 2 weeks, %:	
At 70° C, 100% R.H.	12.2
At 100° C, ambient R.H.	9.5
At -40° C, ambient R.H.	Shrinkage

[a] Data from Ref. [21].

[b] Self-extinguishing.

Several publications [54-56] have noted that strength properties of rigid urethane foams are related to density by a function of the form

$$\log (\text{strength property}) = \log A + B \log (\text{density}). \tag{8}$$

A survey of a large number of rigid foams prepared by using differing formulations and application techniques showed the linear log-log plots of such a relationship. Those relating compressive, flexural, and shear strengths to density [54] are shown in Figs. 8, 9, and 10. In each plot the

TABLE 11

Formulation, Foaming Characteristics, and Physical Properties of Rigid Polyether Urethane Foam Prepared by the Quasiprepolymer Method[a]

Formulation, parts by weight:	
Component A:	
Prepolymer (polyether[b] + TDI)	40.9
Trichlorofluoromethane	12.7
Component B:	
Polyether (OH No. 656)[b]	25.8
Silicone stabilizer	0.4
Tris(dichloropropyl) phosphate	19.6
Tetramethylguanidine	0.6
A/B Ratio	1:1.132
NCO/OH Ratio of the quasiprepolymer	4.5:1.0
NCO/OH Ratio of the foam	1.05:1.00
Foaming characteristics:	
Rise time, min	3.33
Tackfree time, min	4.30
Hardness time, min	8.30
Physical properties:	
Core density, lb/ft^3	2.3
Open cells, %	5.9
Compressive strength at 10%, psi	38.0
Shear strength, psi	25.0
Tensile strength, psi	44.4
Water absorption, 48 h, vol %	4.91
"k" Factor, (Btu/h)(ft^2)(°F/in.)	0.11
Dimensional stability at 70° C, 100% R.H., 1 week ΔV, %	+3.0
Flammability rating (ASTM D1692)	S.E.[c]

[a]Data from Ref. [15].

[b]Polyoxypropylene adduct of sorbitol (OH No. 490).

[c]Self-extinguishing.

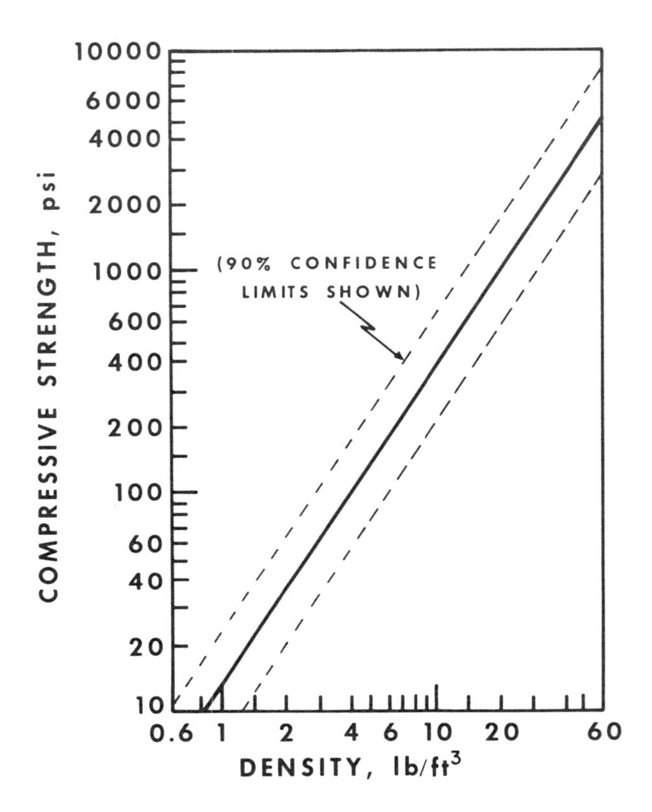

FIG. 8. Effect of density on the compressive strength of rigid urethane foams. Data from Ref. [54].

solid line is the best fit of the data as determined by least-squares analysis, and the broken lines enclose with 90% confidence all data for the types of foam considered.

Reported values for the constants of Eq. (8) are shown in Table 12. The values of the constant B determined from the slope of the log-log curves lie in the range 1 to 2 for compressive, flexural, tensile, and shear properties expressed as either ultimate strength or modulus. For the same polymer system and property the value of B is essentially constant and can be used in Eq. (8) to determine the effect of a density change. The constant A, determined from the intercept of the log-log plot at a density of 1 lb/ft^3, is more widely dependent on the specific foam type, and the values of Table 12 are general averages.

B. Effects of the Environment

The chemical resistance of the highly crosslinked polyurethanes prepared from aromatic isocyanates and hydroxyl-terminated polyethers

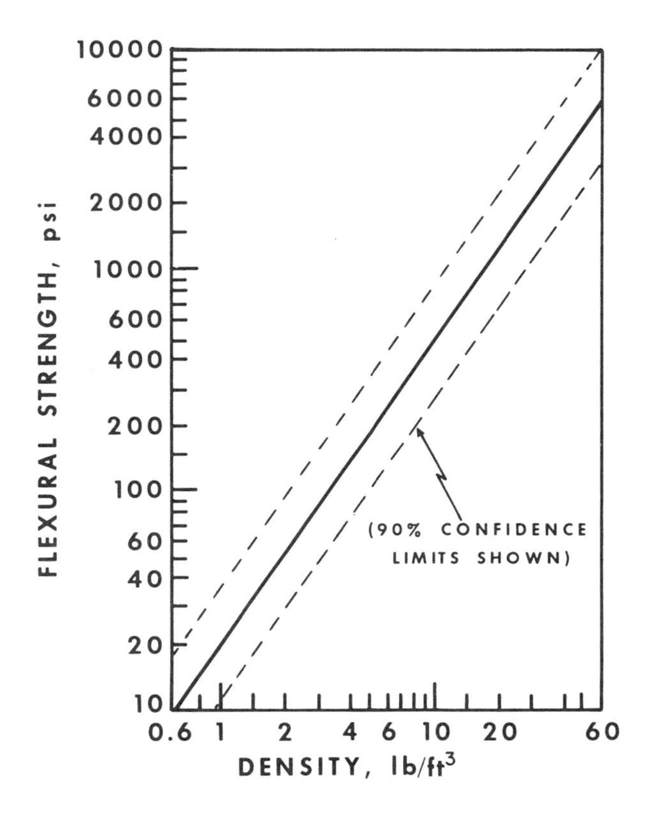

FIG. 9. Effect of density on the flexural strength of rigid urethane foams. Data from Ref. [54].

combine with the normal closed-cell structures of rigid urethane foams to limit interactions with surroundings to the surface layer and produce a chemically and physically stable material. Effects produced by chemical agents depend both on the chemical and on the permeability of cell membranes. Solubility of the chemical in the foam polymer affects both permeability and swelling.

1. Chemical Resistance

Rigid urethane foams are resistant to the action of most organic solvents and are seriously degraded only by strong acids, oxidizing agents, and similarly corrosive chemicals. Only the stronger polar solvents, which significantly swell the polymer, lead to shrinkage of the foam structure. Evaporation of the solvent normally returns the polymer to its original state. The resistance of a rigid polyether-based urethane foam to a number of chemicals is shown in Table 13.

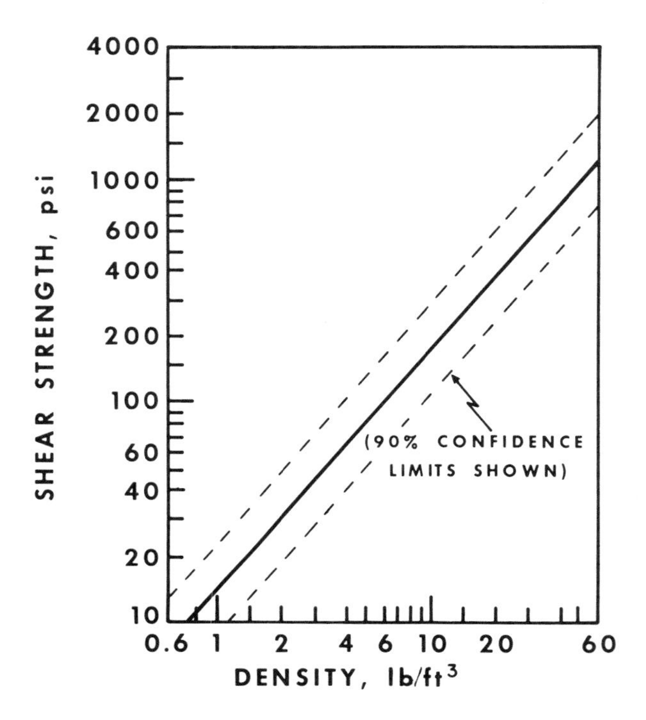

FIG. 10. Effect of density on the shear strength of rigid urethane foams. Data from Ref. [54].

2. Water-Vapor Permeability

The driving force for transmission of water vapor across rigid urethane foam used as thermal insulation is large because of the low water content of cold air and the relatively high water content of warm air. Absorption of water raises the thermal conductivity and can plasticize the more hydrophilic foam polymers, justifying protection of rigid-urethane-foam insulation by a vapor barrier.

The values of water-vapor permeability given in Tables 8 and 9 are typical of rigid polyether-based urethane foams. The differences between the two common test methods, wet cup and dry cup, and the great influence of any open-cell structure confuse the meaning of quoted values. The effect of the more important characteristics of foam and polymer structure have been related [58] to water-vapor permeability by multiple-regression analysis of data for a large number of foams prepared from polyether polyols and tolylene diisocyanate. The results, summarized in Table 14, show that moisture-vapor permeability can be lowered most effectively by raising the density and aromatic structure of the foam. Similar results

TABLE 12

Experimental Values of the Constants in Eq. (8)

Strength property, psi = A (density, lb/ft^3)^B

Strength property	Constant			
	A[a]	B[a]	B[b]	B[c]
Compressive strength	12.8	1.42	1.6	1.6
Flexural strength	19.0	1.38	1.4	1.3
Tensile strength	23.0	1.11	1.3	1.2
Shear strength	14.9	1.08	1.1	1.3
Compressive modulus	293.8	1.45	1.6	1.8
Flexural modulus	186.3	1.76	1.8	1.7
Tensile modulus	573.5	1.15		
Shear modulus	169.9	1.08		1.7

[a] Data from Ref. [54].

[b] Data from Ref. [55].

[c] Data from Ref. [56].

were reported by another author [59]. The adverse influence of absorbed moisture on thermal conductivity has also been studied [60].

3. Permeability to Gases Other than Water

The permeability of rigid urethane foams to air and common fluorocarbon blowing agents has been of concern because of its effect on the composition and internal pressure of the gas within normally closed cells. Organic polymer foams are discontinuous dispersions of a large volume of gas in a small volume of polymer and depend mainly on the gas composition for thermal insulating properties [61]. In addition, internal pressures differing greatly from atmospheric create stresses that can produce gradual swelling or shrinkage of the foam.

Two authors [62, 63] chose different experimental and theoretical techniques to study the changes in gas composition within the cells of rigid urethane foams blown by trichlorofluoromethane. Their conclusions

TABLE 13

Chemical Resistance of Rigid Polyether-Based Urethane Foams[a]

Active material	Resistance Rating[b]	
	At 24° C	At 52° C
Diesel oil	E	E
Motor oil	E	E
Regular gasoline	G	-
Turpentine	E	-
Kerosene	G	G
Linseed oil	G	G
Benzene	E	-
Toluene	E	-
Methylene chloride	F	-
Ethyl alcohol	F	F
Methyl alcohol	F	F
Carbon tetrachloride	E	E
Methyl ethyl ketone	P	-
Orthodichlorobenzene	E	E
Acetone	F	-
Perchloroethylene	E	E
Water	G	G
Brine, saturated	G	G
Brine, 10%	E	G
Sulfuric acid, concentrated	S	S
Sulfuric acid, 10%	G	G
Nitric acid, concentrated	S	S
Hydrochloric acid, concentrated	S	S
Hydrochloric acid, 10%	G	G

TABLE 13 (Continued)

Chemical Resistance of Rigid Polyether-Based Urethane Foams[a]

Active material	Resistance rating[b]	
	At 24° C	At 52° C
Ammonium hydroxide, concentrated	G	-
Ammonium hydroxide, 10%	G	G
Sodium hydroxide, concentrated	E	E
Sodium hydroxide, 10%	E	G

[a]Data from Ref. [57].

[b]Key to ratings: E, excellent resistance; G, good resistance; F, fair resistance; P, poor resistance; S, severe solvent action or chemical attack--not recommended for use.

were similar. The diffusion constant of trichlorofluoromethane in urethane foam is smaller than those of components of air (Table 15), causing nitrogen and oxygen to diffuse into the cell more rapidly than trichlorofluoromehtane diffuses out. The internal cellular pressure rises as shown in Fig. 11, theoretically reaching 1.7 atm after approximately 2.5 years in the case of a 12 x 12 x 2-in. slab of 2-lb/ft^3 foam with all faces exposed 63 . The effect on thermal conductivity is shown in Fig. 12. Gradual swelling also normally occurs. The changing properties are not observed in carbon dioxide-blown rigid foams of similar polymer structure.

The rate of gas diffusion in and out of urethane-foam cells is retarded by increasing density and particularly by enclosing the foam within impermeable facings as in refrigeration cabinets and metal-faces panels.

The diffusion of gas from carbon dixoide-blown rigid urethane foams under vacuum has also been studied [64] and found to occur in agreement with conventional diffusion theory. Diffusion coefficients were found to be on the order of 10^{-6} to 10^{-5} cm^2/sec between 22 and 81° C. Loss of gas from closed-cell foams must be seriously considered in aerospace and some electrical encapsulating applications.

TABLE 14

Water-Vapor Permeability of Rigid Urethane Foam[a] at 23° C[b]

Direction	Water-vapor permeability[c] (perm. - in.)	Correlation coefficient	Standard error of estimates (perm. - in.)
Parallel to foam-rise direction	$2.03 - 0.0080h + 2.69/f + 5.39/(r + 1) - 0.859d + 0.390V^{1/3} + 9.28F + 0.642R$	0.94	±0.43
Perpendicular to foam-rise direction	$1.98 - 0.0033h + 2.16/(r + 1) + 0.601d + 0.201V^{1/3} + 5.44F + 4.07R$	0.90	±0.24

Nomenclature

Symbol	Definition	Determined by
d	Foam-specimen density, lb/ft^3	ASTM D1622-59T[a]
f	Average functionality of polyol (blend) foamed: the number of reactive groups per average molecule	Functionality of starting materials
F	Volume fraction of interconnected (open) cells within a foam (dimensionless)	Air displacement
h	Average hydroxyl number of a polyol, mg KOH/g polyol	ASTM D1638-60T
r	Average number of aromatic rings per molecule of polyol (blend) foamed	Starting composition

R	Ratio of 25° C compressive strengths within a foam (parallel/perpendicular) (dimensionless)	ASTM D1621-59T
V	Gas volume of the average closed cell within a foam, 10^{-4} cm^3	Air displacement

[a] Polyether reacted with tolulene diisocyante and blown with trichlorofluoromethane.

[b] Reprinted from Ref. [58] by courtesy of Interscience.

[c] Grains of water vapor transferred per hour per square foot (moisture-vapor-pressure differential per inch of thickness); the driving force for mass transfer (vapor-pressure differential) is expressed in inches of mercury. Measured by ASTM C355-59T, wet-cup method.

TABLE 15

Diffusion Constants in 2.2-lb/ft^3 Urethane Foam[a]

Gas	Diffusion constant (cm^2/sec)
Trichlorofluoromethane	2.25×10^{-9}
Nitrogen	6.27×10^{-8}
Oxygen	1.12×10^{-7}

[a]Data from Ref. [62].

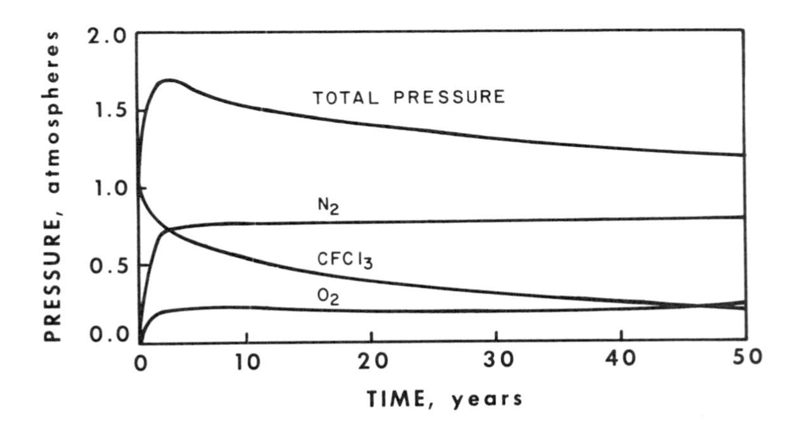

FIG. 11. Theoretical change in cell gas pressure with time at 25°C. Sample 12 x 12 x 2 in., all faces exposed; foam density 2.2 lb/ft^3. Data from Ref. [63].

C. Thermal Effects

The mechanical properties of rigid urethane foams change with temperature in the same manner as those of many crosslinked polymers. The abrupt softening shown in the Vicat softening and Clash-Berg torsional stiffness curves of Fig. 13 [22] falls over a range of 85 to 200° C, depending on the structure and crosslink density of the polymer. The low-temperature transition reported at -20 to -60° C for flexible urethane foams and urethane elastomers does not appear in curves for the tightly crosslinked polymers, and only a gradual stiffening of the rigid foam occurs with falling temperature.

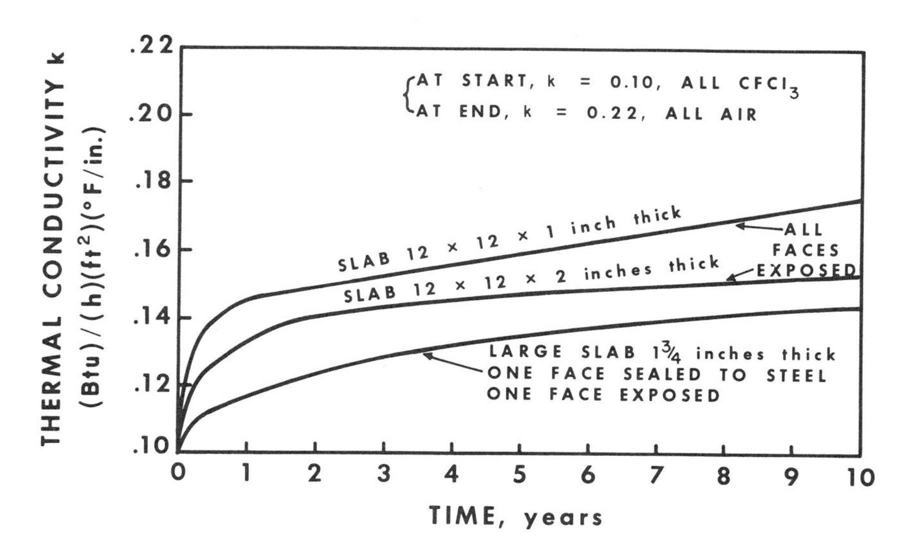

FIG. 12. Theoretical increase in thermal conductivity with time at 25° C. Foam density 2.2 lb/ft^3. Data from Ref. [63].

High softening temperatures have been found in rigid urethane foams based on highly branched aromatic or alicyclic raw materials [22, 65]. Low-temperature properties have not been of great concern, and rigid foams that are dimensionally stable at low temperatures generally have adequate mechanical properties for cryogenic insulating applications.

Changes in the weight, volume, and compressive strength of polyether-based urethane foam described in Table 8, when used at moderate to high temperatures, were studied by exposing foam samples to a given temperature until constant weight and dimensions were obtained [20]. The results are shown in Fig. 14. Some volume increase occurred with minor loss in weight and compressive strength up to temperatures of about 120° C. At higher temperatures the properties were seriously degraded. The changes are consistent with gradual softening of the polymer and simultaneously increasing gas pressure as the temperature increases.

Loss of weight from a rigid urethane foam prepared from an aliphatic polyether triol and a polymeric isocyanate is shown by the thermogravimetric analysis curves of Fig. 15. The curves were obtained by measuring the sample weight during a rapid linear temperature rise, and the major weight-loss temperatures are higher than those that would be observed under isothermal conditions.

Serious thermal degradation of rigid urethane foams does not normally occur at temperatures below 150° C [66-68]. The initial degradation

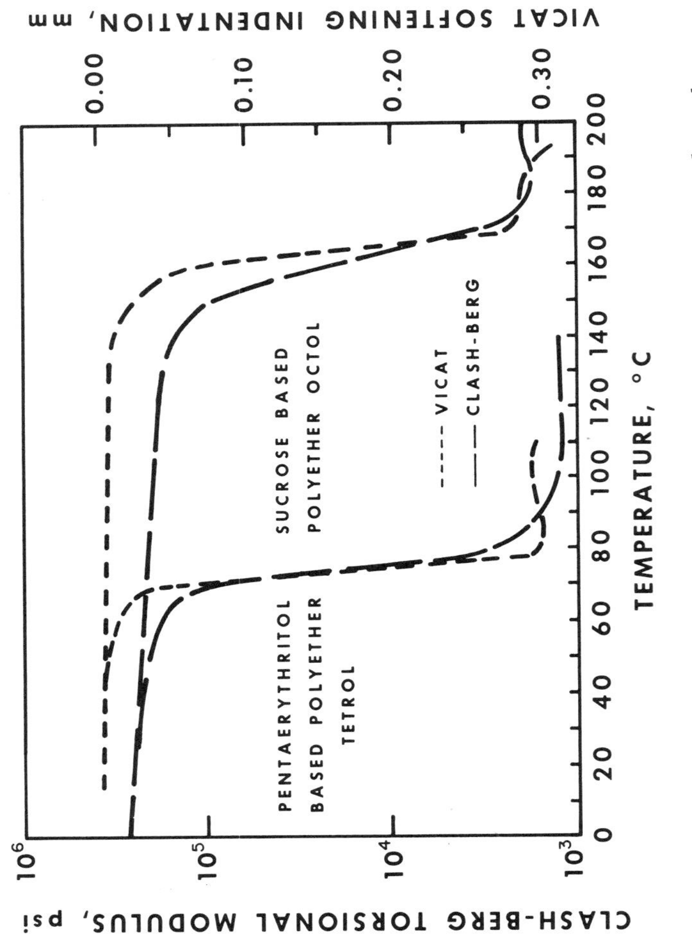

FIG. 13. Vicat softening and Clash-Berg torsional stiffness curves for solid polyurethanes prepared from polyether polyols and TDI. Data from Ref. [22].

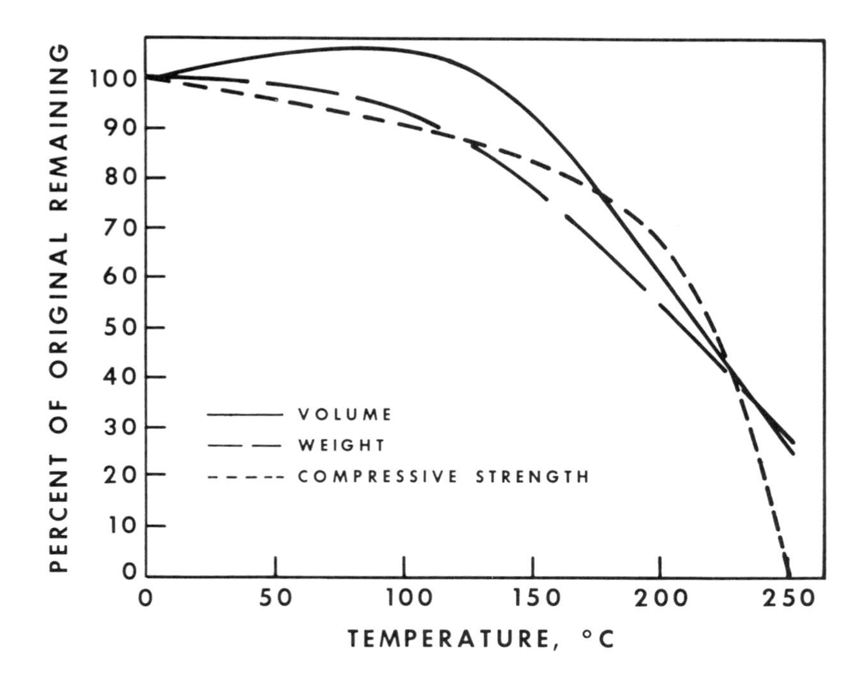

FIG. 14. Effect of temperature on the equilibrium volume, weight, and compressive strength of 2-lb/ft^3 rigid urethane foam. Data from Ref. [20].

process of Fig. 15 is similar in air and helium, and involves breaking of urethane bonds and loss of volatile fragments derived from aliphatic structures. A more thermally stable carbonaceous char is formed from the degrading polymer at higher temperatures. Vaporization of the char at still higher temperatures occurs by a combination of oxidation, further thermal breakdown to small fragments, and physical attrition. The greater stability of the char in helium, (Fig. 15) indicates the importance of oxidation in char weight loss.

The thermal stability of rigid urethane foams was improved by the use of branched aromatic polyisocyanates [68]. Polyether-based urethane foams have been reported to be less thermally resistant than polyester-based urethane foams because of the oxidative cleavage of the polyether

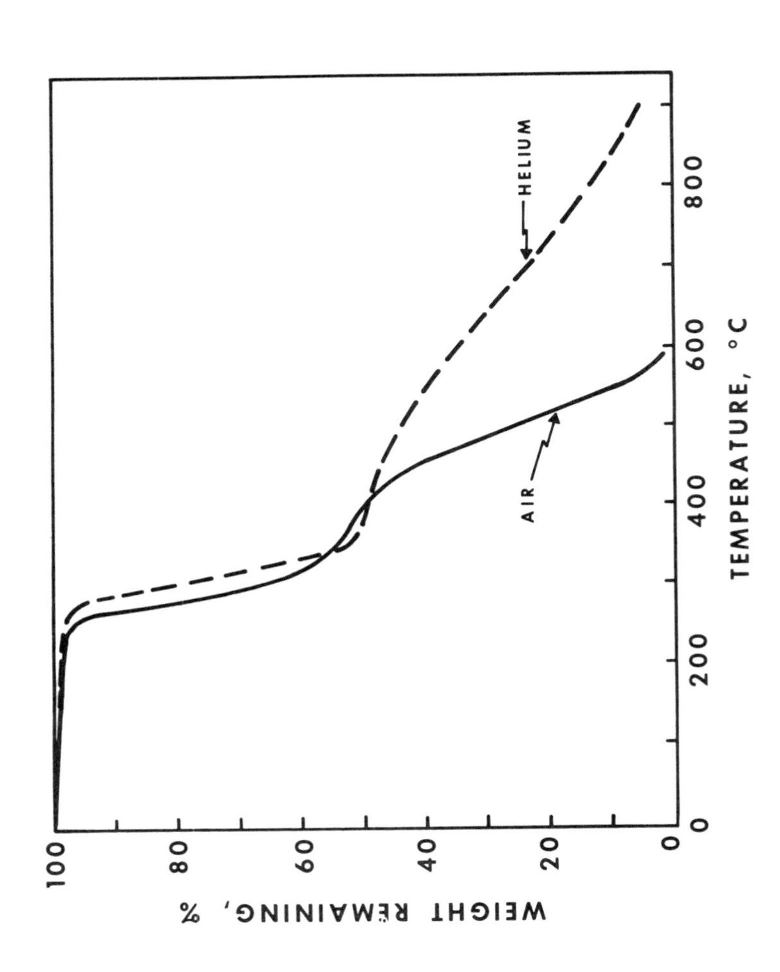

FIG. 15. Theromgravimetric analysis curves for polyurethane prepared from an aliphatic polyether triol and a polymeric isocyanate. Heating rate 10° C/min; dynamic gas flow 0.35 standard ft^3/h. Data from Ref. [68].

structure at high temperatures [69]. Thermal stability generally has been improved by the incorporation of imide, isocyanurate, carbodiimide, and similarly stable bond structures [68, 70-74].

D. Thermal Conductivity

The thermal conductivity, or k factor, of rigid urethane foam is among its most important commercial properties and is normally reported in the refrigeration units of Btu/(h)(ft^2)(° F/in.). The k factor has been attributed [61] mainly to the low thermal conductivity of the gas filling the foam cells. Contributions from conduction through the solid polymer and from radiation are small and constant at room temperature for a given foam density. The contribution from convection through the cell gas is also small if cell size is small.

From the standpoint of thermal conductivity, rigid urethane foams must be classed into two groups: those blown with trichlorofluoromethane and those blown with carbon dioxide. The k factor of 2-lb/ft^3 trichlorofluoromethane-blown foams at room temperature is 0.10 to 0.12 in freshly made samples but rises with time in samples with exposed cut surfaces to about 0.16, as shown in Fig. 12. The increase is caused by the diffusion of air into the closed cells, coincident with slower diffusion of trichlorofluoromethane out of the cells, as discussed in Section IV.B.3. Foams enclosed by steel or other impermeable facings retain k factors on the order of 0.12 for long periods and in most cases for the useful life of the insulated item.

The permeation of polyurethanes by carbon dioxide is quite rapid, and the initial gas of carbon dioxide-blown foams is exchanged for air within a relatively short period, producing a 2 lb/ft^3 a room-temperature k factor of about 0.24, which remains at that level.

In addition to the type of blowing agent, age, and fraction of open cells, the thermal conductivity of rigid foams varies with foam density and temperature. Although the k factor should theoretically increase linearly with density [63], experimental curves [54] pass through a minimum at about 2 lb/ft^3 because of the difficulty in maintaining closed-cell structures at lower densities. The increase in the k factor of carbon dioxide-blown foams is linear at higher densities, but the slope of the curve decreases at densities below about 10 lb/ft^3 [54].

The increase in k factor with increasing temperature has been attributed [61] to the sum of a small increase in the conductivity of the solid polymer, a small to large increase in the radiation contribution, and a medium to large increase in conductivity through the gas contained in the cells. The variation for trichlorofluoromethane-blown foams as shown in Fig. 16 passes through a minimum in k factor at approximately the boiling point of the blowing agent. The increase in k factor with temperature is monotonic for carbon dioxide- or air-containing foams.

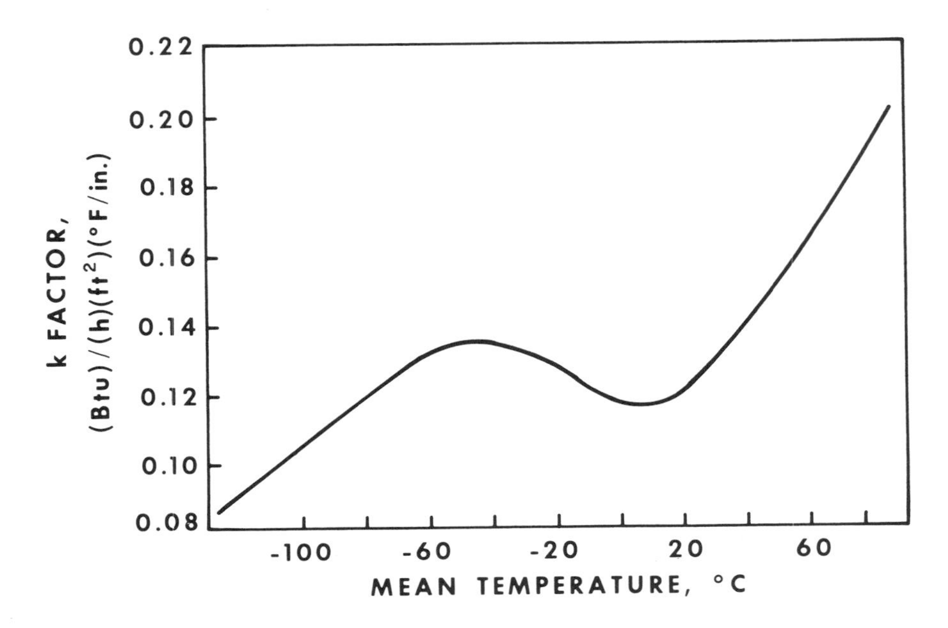

FIG. 16. Effect of temperature on the thermal conductivity of rigid urethane foam blown with trichlorofluoromethane. Data from ref. [54].

E. Dimensional Stability

1. Short-Term Volume Change

The reasons for dimensional changes in rigid urethane foams are complex and related to the molecular structure of the polymer, its modulus-temperature relationship, the type and pressure of the blowing agent remaining in the foam cells, and the foam density. The volume changes noted in Tables 8, 9, and 10 are common for rigid urethane foams exposed to the conditions noted. The thermal softening of the foam polymer and increase in the cell pressure of fluorocarbon-blown foams has been already mentioned. Dimensional expansion at high relative humidities is higher than that under dry conditions because of the absorption of moisture, which plasticizes the foam polymer.

The dimensional stability of fluorocarbon-blown foams at low temperatures is usually better than that at high temperatures, as the reduced pressure caused by cooling and liquefaction of the blowing agent is counteracted by an increase in the modulus of the polymer. Weaker foams may shrink badly at 0 to -20°C, but be dimensionally stable if cooled rapidly to -40°C or below.

The volume of rigid-urethane-foam specimens may pass through a maximum or minimum during the first few days after preparation [75]. The data of Fig. 17 trace the volume change of 2-in. cubes of a typical 2-lb/ft^3 rigid foam exposed to different environmental conditions and show, in each case, a maximum or minimum volume during the first week of exposure. The maximum or minimum volume may have more significance than the equilibrium volume, particularly for foamed-in-place applications, which do not allow preconditioning of the rigid foam before installation. The maximum volume change observed in high-temperature dry exposure of several commercial rigid urethane foams of 2-lb/ft^3 density is shown in Fig. 18. In all cases swelling becomes serious at approximately 93° C.

Foam cells are usually elongated in the direction of foam rise, and dimensional growth, when it occurs, is greater in the direction perpendicular to foam rise-- that is, in the direction of the short cell axis. The deviation from spherical cells is greater in molded foam items than in large blocks, particulary when rising foam must flow long distances through narrow cross sections. The high-temperature dimensional stability of rigid urethane foams improves with increasing density and, in equivalent foams, with the use of carbon dioxide as the blowing agent.

2. Coefficient of Linear Thermal Expansion

Reported values of the coefficient of linear thermal expansion of rigid urethane foams lie between 4.5 and 7.2 x 10^{-5} in./in. - °C for

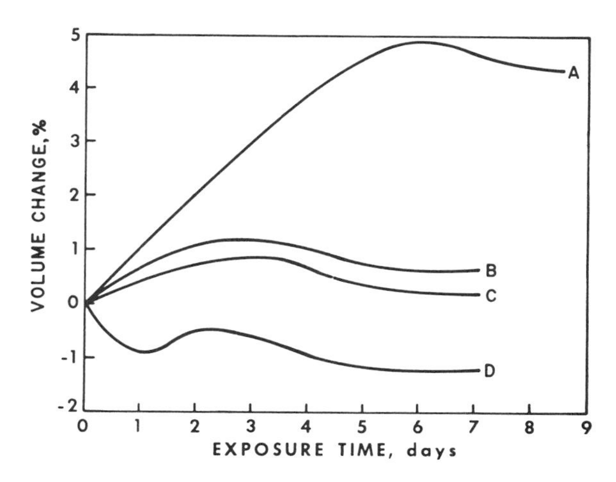

FIG. 17. Volume change of commercial 2-lb/ft^3 rigid-urethane-foam samples as a function of time at various conditions. Curve A, 121° C; curve B, 32° C and 90% relative humidity; curve C, 93° C; curve D, -23° C. Data from Ref. [75].

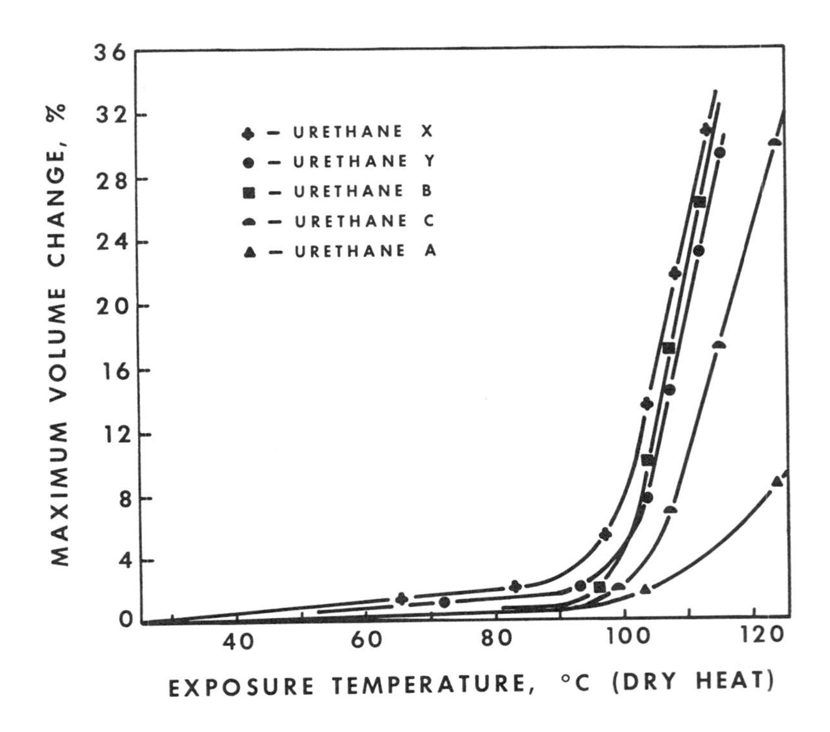

FIG. 18. Maximum volume change in commercial 2-lb/ft^3 rigid-urethane-foam samples as a function of temperature. Data from Ref. [75].

densities below 12 lb/ft^3. The expansion coefficient is almost constant over a wide temperature range and tends to increase slightly with foam density.

The thermal warp of rigid-foam-filled sandwich panels may be appreciable as a result of the large temperature gradients to which such panels are exposed in structural and thermal insulating applications. An analysis of panel warping [76] has pointed out that the panel faces may be held at widely different temperatures and are restrained from expanding relative to each other by the core material. Panels with thin, flat faces are best designed with a thick, low-density foam core capable of reversible shear deformation. Even better is the use of thick, deeply formed faces.

3. Creep

The creep, or continued slow deformation, of components made from low-density rigid polyether-based urethane foam and loaded for long periods of time has been investigated for shear loads. Foam-filled beams with 20-gauge steel facings were exposed to shear stresses by supporting the ends while loading in the middle [77]. An immediate deflection was followed by additional deflection at an exponentially decreasing rate. After 1 to 3 months, depending on experimental conditions, creep became too slow to measure, and deflection was two to four times the immediate deflection after loading. Typical creep curves are shown in Fig. 19. After removal of the load, recovery at an exponentially decreasing rate occurred for up to 1 year. Some permanent set was noted after recovery had become complete. Creep increased with both temperature and relative humidity, and decreased with the crosslink density of the rigid foam. The results were consistent with analysis based on a spring-dashpot model.

F. Flammability

Discussion of the flammability of rigid urethane foams is complicated by the large number of test methods and variety of foam types involved. The more general aspects of cellular plastic flammability, its determination

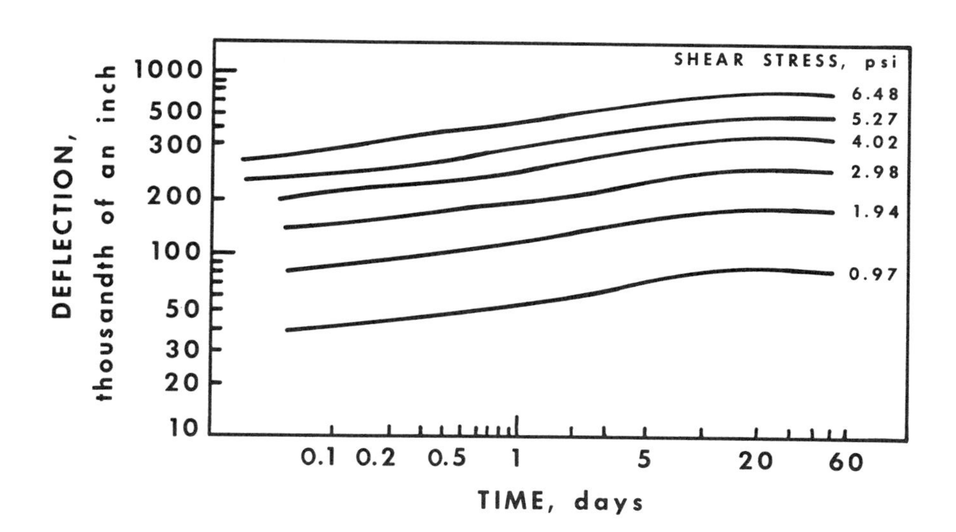

FIG. 19. Creep of 1.8-lb/ft^3 rigid urethane foam loaded in shear and exposed at 23° C and 50% relative humidity. The fluorocarbon-blown foam was based on an α-methyl glucoside-initiated polyether and a polymeric isocyanate. Data from Ref. [77].

and meaning, are considered in Chapter 18. The particular problems associated with rigid-urethane-foam flammability warrant discussion in this section.

From the standpoint of flame resistance, rigid urethane foams can be divided into several classes:

1. General purpose products that contain no or few flame-retarding additives and are used in enclosed applications where burning is unlikely.

2. Flame-resistant products that incorporate sufficient quantities of phosphorus- and halogen-containing raw materials to ensure that combustion started by an outside source will extinguish itself when the source is removed.

3. Extra-flame-resistant foams that are specially modified by large amounts of flame-retarding additivies and chemical structures that themselves contribute to flame resistance.

Groups 1 and 2 include almost all of the rigid urethane foams now used commercially. Active development, particularly of rigid foams for construction applications, falls in group 3 and is much more varied in its approaches. Several rigid urethane foams modified with amide, imide, isocyanurate, and carbodiimide structures have been shown to have inherent flame resistance [68, 70, 71, 78, 79].

A particularly interesting urethane-foam modification [79] contained high concentrations of polyvinyl chloride or polyvinylidene chloride and potassium fluoroborate, KBF_4. The large amounts of hydrogen chloride and boron trifluoride vapors released when the foam was exposed to a flame actually extinguished a fire of JP-4 jet fuel in the cockpit of a test fighter plane.

Exceptionally flame-resistant composites have also been prepared by bonding foam glass beads together with rigid-urethane-foam system [80].

The meaning of the large number of flammability tests in use and their correlation with actual fires continue to be discussed and investigated. The more common tests have been compared in a recent publication [81]. Most measure one or more of four factors: flame spread, smoke generation, fuel contribution, and char durability. Results that can be expected when testing foam of each of the three groups by one of three common tests are summarized in Table 16. The Underwriters' Laboratories tunnel test, ASTM E84, simultaneously measures flame spread, fuel contribution, and smoke density and bases ratings on asbestos cement board as 0 and red-oak flooring as 100; ASTM D1692 measures flame spread along a horizontal strip under mild conditions; the U. S. Bureau of Mines torch penetration test is primarily a measurement of char formation and durability. None of these or other flammability tests is considered to give an overall measurement of rigid-foam flammability, and most organizations use a combination of several tests to rate their products.

TABLE 16

Rigid-Urethane-Foam Flammability

Test method	Rigid urethane foam		
	Group 1	Group 2	Group 3
ASTM E84, tunnel test:			
Flame-spread rating	>1000	50-300	15-40
Fuel contribution	> 500	100-500	15-50
Smoke-density factor	> 500	100-500	15-50
ASTM D1692:			
Rating	Burning	Self-extinguishing	Self-extinguishing
Burning length, in.	2	< 2	0
U.S. Bureau of Mines			
Torch penetration test:			
Burn-through time, min	< 1	2-15	10-90

G. Electrical Properties

A combination of low mass, mechanical strength, and favorable electrical properties has lead to the use of rigid urethane foams in radomes and encapsulation of electronic components, among other electrical applications. The dielectric constant ϵ and loss tangent tan δ are only slightly above those of the gaseous blowing agent for low-density foams. The values increase and approach those of solid polyurethanes with increasing density. Ranges of a number of values reported in the literature are summarized in Table 17.

The dielectric constant of rigid urethane foams changes very little with temperature over the range -50 to +150°C [54], although the loss tangent increases rapidly with increasing temperature in the same range. Increasing frequency above 1 MHz causes a decrease in the dielectric constant and an increase in the loss tangent.

TABLE 17

Dielectric Constant and Loss Tangent of Rigid Urethane Foams and Solid Polyurethanes at Room Temperature and Frequencies Below 1 MHz[a]

Polyurethane	Dielectric constant ϵ	Loss tangent tan δ
Foam, 1-5 lb/ft^3	1.02-1.5	0.0005-0.005
Solid	7-8	0.05

[a]Data from Refs. [12] and [50].

H. Acoustical Properties

Investigations of the acoustical properties of both flexible and rigid urethane foams [82] have shown that open-cell types are effective sound absorbers, whereas the common closed-cell rigid foams are relatively ineffective. Sound-absorption coefficients of both open- and closed-cell rigid foams are listed in Table 18.

Sound absorption has sometimes been confused with sound transmission and sound deadening. Open-cell urethane foams, though effective sound absorbers, are poor in reducing sound transmission, a property that varies inversely with mass. Some designers of rigid-foam-containing building panels have added dense materials to the foam in order to increase mass and decrease sound transmission.

Rigid urethane foams have been frequently used to deaden the sound produced in metal cabinets by stiffening the cabinet walls and reducing the possibility of resonant vibration of the walls and the enclosed cavity.

I. Resistance to Microbial Attack

Published data (see, for example, Refs. [54] and [83]) and experience for almost 20 years indicate that rigid urethane foams are not attractive either for consumption by vermin or as a medium for fungal growth. In one series of tests [54], foam samples inoculated with *Aspergillus niger*, *A. flavus*, *A. terreus*, *Penicillium citrinum*, *Spicaria violacea*, (*Paecilomyces marquandii*), and *Trichoderma T1*, according to Military Specification MIL-F-13927, showed no growth on the specimens after 28 days. Soil burial according to ASTM D684-54 similarly showed no evidence of fungi on foam samples, and tests by Federal Specification CCC-T-191B, Method 5760, showed only minor trowth on foam samples placed for 28 days on cotton wicks inoculated with mineral salt solutions of *Aspergillus*

TABLE 18

Sound-Absorption Coefficients of Rigid Urethane Foams[a]

Cell type	Sound-absorption coefficient measured at indicated frequency (Hz)							NRC[b]
	125	250	350	500	1000	2000	4000	
Open	0.14	0.22	0.31	0.69	0.52	0.83	0.73	0.56
Closed	0.12	0.18	0.20	0.27	0.19	0.62	0.22	0.32

[a]Reprinted from Ref. [82], p. 266, by courtesy of Official Digest.

[b]NRC (noise-reduction coefficient) is an average of the absorption coefficients of 250, 500, 1000, and 2000 Hz. A high value is desirable.

clavatus, A. niger, Chaetomium globosum, Penicillium funo-glaucum, and Trichoderma T1. Other tests showed that starved rats would nibble on rigid urethane foam but would not swallow the pieces.

The resistance of micro organisms is not completely in agreement with recent investigations of urethane elastomers. In one series of experiments [84], linear polyurethanes prepared from one of four diisocyanates and one of a series of simple diols, polyether diols, and adipic acid polyester diols either uncatalyzed or in the presence of 0.025% ferric acetylacetonate, were inoculated with A. niger, A. flavus, A. versicolor, Penicillium finiculosum, Pullularia pullulans, Trichoderma sp., and Ch. globosum. Observations after 1, 2, and 3 weeks at 30° C showed limited growth on the polyether-based urethanes, but heavy growth on the polyester-based urethanes. Degradation of polyester-based urethane elastomers in some specialized applications has also been attributed to microbial growth.

The rigid-foam and elastomer results are not contradictory but do indicate that additional experiments are needed to determine whether the apparently better resistance of the rigid foams is caused by the particular polyols used, the higher crosslink density, or residual fungicidal foam catalysts.

The mass of reported information indicates that common rigid urethane foams are physiologically inert and will neither support biological growth nor inhibit growth in contact with it. In some other applications rigid foams have been treated with fungicides. The practice may have some value if the application requires contact with another material that supports growth of fungi.

J. Toxicity

Although, as noted in the preceding section rigid urethane foams are themselves not toxic, some of the materials from which they are made have caused serious problems of industrial hygiene. This is particularly true of the tolylene diisocyanates (TDI) whose relatively high vapor pressure leads to high concentrations of vapor in poorly ventilated areas. Short-term toxic effects of the asthmatic type occur after exposure but normally leave no permanent damage. Sensitization to TDI occurs in a limited number of persons, who then must be removed from further contact [85]. Similar effects have been noted less frequently for the less volatile polymeric isocyanates. A maximum allowable concentration of 0.02 ppm is accepted for both types and makes good ventilation or adequate respirators mandatory in areas in which these products are used.

Strong tertiary amine catalysts are also used in many rigid urethane formulations and have produced dermatitis and visual distortion in workers not wearing adequate protective equipment.

High concentrations of fluorocarbon vapors have produced serious toxic effects and should be avoided. Smoking in the presence of even low concentrations of these vapors is dangerous because of possible oxidation to phosgene or carbonyl fluoride.

V. STRUCTURE-PROPERTY RELATIONSHIPS

Because of the variety of rigid urethane foams that have been prepared during their 25-year history, the previous discussion of foam properties was based primarily on the fluorocarbon-blown polyether-based urethane foams, which are now most common. The use of other reactants and modification of both molecular and foam structures have produced products with different properties that have been useful in some applications. The contributions of some of the important raw materials to foam properties are summarized in Tables 19, 20, and 21.

A. Resin Components

The polyether polyols were preceded in earlier stages of rigid-urethane-foam development by adipic and phthalic acid polyesters similar to these now used in urethane elastomers and coatings. As shown in Table 19, the polyester-based urethane foams, particularly when blown by the addition of water, have excellent mechanical properties and better chemical resistance than do polyether-based urethane foams. The hydrolytic insta-

TABLE 19

Trends in Rigid-Urethane-Foam Properties with Type of Resin Component[a]

Property	Polyol or nonreactive ingredient			
	Polyethers	Polyesters	Castor oil	Flame retardants
Mechanical strength	+	+	-	--
Environmental stability:				
Oxidative	-	+	-	=
Hydrolytic	+	-	+	=
Solvent swelling	=	+	-	=
Thermal softening	=	+	-	-
Dimensional stability	+	+	-	-
Thermal conductivity	=	=	=	=
Flame resistance	-	=	-	++
Ease of application	+	-	=	=

[a] Rating key: ++, property greatly improved; +, property improved; =, no effect; -, property degraded; --, property greatly degraded.

bility of the polyester has not been a serious problem in the tightly crosslinked, closed-cell structure, and foams of this type are being used for some aerospace and other specialty applications. The most serious adverse characteristics of the polyester polyols have been higher viscosity and cost in comparison with polyether polyols.

Castor oil has been used as a major and minor component of rigid-urethane-foam systems, but it generally detracts from foam properties (Table 19).

Reactive and nonreactive flame retardants are frequent components of urethane-foam resin blends. Both types increase the flame resistance of the foam but normally decrease the crosslink density and lower the softening and decomposition temperatures.

TABLE 20

Trend in Rigid-Urethane-Foam Properties with Type of Isocyanate Component[a]

Property	Polyisocyanate		
	Distilled TDI	Crude TDI	Polymeric MDI
Mechanical strength	=	=	+
Environmental stability:			
Oxidative	=	=	=
Hydrolytic	=	=	=
Solvent swelling	=	=	+
Thermal softening	-	=	+
Dimensional Stability	-	=	+
Thermal conductivity	=	=	=
Flame resistance	-	=	+
Ease of application	-	-	++

[a]Rating key: ++, property greatly improved; +, property improved; =, no effect; -, property degraded.

B. Isocyanates

The foam property contributions of the three types of isocyanate most frequently used in rigid foams are summarized in Table 20. Distilled TDI, most often the commercial 80:20 mixture of 2,4- and 2,6-tolulene diisocyanates, has been used to produce rigid foams with acceptable properties but has been hindered in rigid-foam applications by the toxicity of its vapors and the relatively greater flammability of its foams. Crude TDI contains some branched polyisocyanates and has been used in applications, such as refrigerator insulation, in which the foam is protected against fire.

The trend in rigid urethane foam is toward types based on one of the polymeric polyisocyanates prepared from aniline-formaldehyde condensates. The low vapor pressure of these products makes vapor toxicity less of a problem during application, and low viscosity makes possible spray, molding, and slab application with a variety of machines. The normal 2.5 to 3.0 functionality and thermally stable aromatic molecular structure have led to good mechanical strength and inherent flame resistance.

TABLE 21

Trends in Rigid-Urethane-Foam Properties with Type of Blowing Agent and Cell Structure[a]

Property	Blowing agent		Cell structure	
	$CFCL_3$	[b]CO_2	Open	Closed
Mechanical strength	=	+	=	=
Environmental stability:				
Oxidative	=	=	=	=
Hydrolytic	=	=	=	=
Solvent swelling	=	=	=	=
Thermal softening	=	+	=	=
Dimensional stability	-	+	++	-
Thermal conductivity	++	=	=	+
Flame resistance	=	=	=	=
Ease of application	+	-	-	+

[a]Rating key: ++, property greatly improved; +, property improved; =, no effect; -, property degraded.

[b]Produced by the reaction between isocyanate and water.

C. Blowing Agents and Cell Structure

Although the majority of rigid urethane foams are of the closed-cell type, expanded by trichlorofluoromethane, the properties of open-cell foams and foams blown with carbon dioxide (Table 21) are worth noting. Open-cell foams have all the properties of closed-cell foams except the very low thermal conductivity and low permeability to liquids and gases. These trends may be countered by the better dimensional stability of the foams and by using the foams in filtration applications.

The carbon dioxide-blown foams are prepared by adding water and additional polyisocyanate to the formulation and often have better dimensional stability than trichlorofluoromethane-blown foams, both because of more rapid diffusion of carbon dioxide from the foam cells and strengthening of the polymer structure by the aromatic urea structures formed at the same time as the carbon dioxide. Mechanical and thermal properties are good, and only the thermal conductivity is adversely affected.

D. Elements of Molecular Structure

Table 22 classifies the influence of the important elements of molecular structure on rigid-urethane-foam properties. If the discussion is limited to common starting materials, crosslink density has the greatest effect, determining whether a urethane foam will be flexible or rigid, tough or friable. In the rigid-foam area increased crosslinking produces greater mechanical strength and stiffness, with accompanying greater resistance to chemical and thermal changes and lower permeability. If crosslinking is too great, however, rigid foams become too friable for practical use.

From the standpoint of foam preparation, increased branching of the foam ingredients causes gelation of the foaming system at an earlier stage of reaction and increases the severity of the flow problems that are encountered in rigid-foam molding. A rigorous investigation [21] of polyurethane systems showed that increased crosslinking could lead not only to premature gelation but ultimately to incomplete reaction as a result of steric effects. Increasing softening temperatures with crosslink density were also noted.

Aromatic and alicyclic components, because of the bulk, stiffness, and both chemical and thermal stability they contribute to urethane polymers, are desirable in theory. Practically, they are often high in viscosity and difficult to handle.

Application is limited to aromatic polyisocyanates and a few aromatic- or heterocyclic-based resins used as crosslinking agents. The aliphatic polyethers are widely varied in molecular weight and functionality, produce foams generally with good properties, and are easily processed. Aliphatic hydrocarbon-based materials, such as hexamethylene diisocyanate, are infrequently used in nonyellowing specialty foams but are poor choices for rigid polymers.

Several other authors [65, 86-88] have reported on aspects of the structure-property relationship of polyurethanes.

E. Application Technique and Property Variations

The physical structure and hence the properties of rigid urethane foams vary with the application technique, mainly because of cell size and orientation effects. Some of the more important differences are listed in Table 23.

Rigid foam produced in large slabs or blocks is not highly oriented in the direction of foam rise, cells are almost spherical, and compression strength is normally only a little higher in the rise direction. Sprayed foam, on the other hand, rises rapidly from the substrate, and cells are usually highly oriented in the direction of rise, producing a compressive strength approximately twice that in the perpendicular direction. The cell orientation in molded items varies, the cells being elongated in the direction

TABLE 22

Trends in Rigid-Urethane-Foam Properties with Increasing Concentrations of Structural Elements Introduced in Reactive Components[a]

Property	Crosslink density	Aliphatic structure			Alicyclic or aromatic structure	
		Ether	Ester	Hydrocarbon	Single ring	Multi-ring
Mechanical strength	+	-	-	-	+	++
Friability	-	+	+	+	-	--
Environmental stability:						
Oxidative	+	-	=	-	+	+
Hydrolytic	+	+	-	+	+	+
Solvent swelling	+	-	=	-	+	++
Thermal softening	+	-	-	-	+	++
Dimensional stability	+	-	-	-	+	++
Water-vapor permeability	+	-	-	+	+	+
Thermal conductivity	=	=	=	=	=	=
Flame resistance	=	--	-	-	=	++
Ease of application	±[b]	+	=	+	=	-

[a]Rating key: ++, property greatly improved; +, property improved; =, no effect; -, property degraded; --, property greatly degraded.

[b]Best processing occurs at an optimum crosslink density.

TABLE 23

General Characteristics of Rigid-Urethane-Foam Application Methods

Characteristic	Method of application			
	Slab	Spray	Molding	Froth molding
Foam properties isotropic	Can be	Highly oriented	Oriented	Oriented
Foam internally stressed	Slightly	Highly	Yes	Yes
Maximum temperature of exotherm	High	Low	Moderate	Moderate
Pressure exerted	Expansion not restricted	Expansion not restricted	Relatively high in closed molding	Lower than normal molding
Ease of application:				
As sheets	Good	Poor	Poor	Poor
Over odd surfaces	Poor	Good	Good	Good
In complex shapes	Poor	Poor	Good	Good
In-plant	Good	Moderate	Good	Good
Field operation	Poor	Good	Moderate	Moderate
Machine expense	High	Low	Moderate	Moderate

of flow. These flow patterns often produce large variations in foam properties from place to place in molded products. The froth technique, by partially expanding the foam before introducing it into a mold, eliminates part of the flow and produces a more uniform product in some cases.

The heat developed in rigid urethane foam as it is formed depends on the minimum dimension of the foam mass and is therefore higher in blocks and slabs than in sprayed or most molded items. Stress patterns in foam products follow the flow and orientation patterns as well as temperature distribution and may cause dimensional instability.

VI. APPLICATIONS

Rigid urethane foam has been used in diverse applications, its choice, in most cases, being dependent on four characteristics that distinguish it from similar materials.

1. Urethane foams are easily prepared from liquid raw materials.
2. Properties can be changed by selecting raw materials and foam density.
3. Thermal conductivity is the lowest among common insulating materials.
4. Mechanical strength is high at low density.

The most important single application has been thermal insulation. Many home refrigerators and freezers depend on urethane-foam insulation for their thin-wall design. Urethane-foam-insulated refrigerated trucks, railroad cars, and shipping containers carry food products and other temperature-sensitive materials long distances with little or, in some instances, no refrigeration after loading. Industrial tanks and piping in which heat transfer must be controlled have also been efficiently insulated. Urethane-foam insulation has been applied in the form of precut slabs and by on-site foaming using spraying and pour-in-place techniques. Closed refrigerator cabinets and railroad tank cars have been insulated by pouring the proper amount of foam raw materials mixture into the space between the inner and outer walls. The expanded foam fills the cavity and bonds firmly to the walls, giving both thermal insulation and physical reinforcement.

Cold-storage warehouses and large storage tanks have been insulated by spraying rigid urethane foam onto exposed surfaces. A vapor barrier to minimize moisture transmission completes a durable installation.

Rigid urethane foam has been used extensively in construction, and the potential for future growth in this market is considered great. Applications have increased because of needs to reduce construction costs and to provide more effective thermal insulation for air-conditioned and electrically heated buildings. Curtain-wall panels containing rigid foam have been widely used for facing steel-frame and reinforced-concrete-frame buildings. Most have been foam-core sandwich structures fabricated in specially equipped plants and then shipped to job sites. The urethane foam provides stiffness and thermal insulation to the thin-skin material, but is light in weight for simple handling during installation. The panels are not load bearing, and in some cases the inner skin was omitted and urethane foam was sprayed directly on the inner surface of the outer skin. Facing panels of the Manufacturers and Traders Trust Co. Building in Buffalo, New York, were prepared in this way (Figs. 20 and 21).

FIG. 20. Manufacturer and Traders Trust Co. Building, Buffalo, New York. Architects, Minoru Yamasaki and Associates; general contractor, the John W. Cowper Co., Inc. Courtesy of the John W. Cowper Co., Inc.

FIG. 21. Spray-applied urethane-foam backing of panels for Manufacturers and Traders Trust Co. Building, Buffalo, New York. Courtesy of the John W. Cowper Co., Inc.

Urethane-foam insulation has been extended to all building surfaces, except windows, by spraying onto panel supports and filling joints and gaps by spray or pour-in-place application. Skidmore, Owings, and Merrill, architects of the 35-story Equitable Building in Chicago, successfully placed pipes for the heating-and-cooling system in urethane-foam-insulated exterior mullions rather than in an interior shaft. In a different design Vosbeck-Vosbeck and Associates specified rigid urethane foam sprayed onto the inner surface of the concrete-block wall of the 16-story, 444-unit 4600 Duke Apartments in Alexandria, Virginia. Plaster was applied directly over chicken wire fastened to the foam. Farm buildings, such as that shown in Fig. 22, warehouses, and other sheet-metal industrial buildings have been made environmentally more pleasant by installing urethane foam sprayed on the wall surfaces or fabricated as the core of wall panels. The skin-and-foam combination has been applied to residential construction in aluminum siding, which was stiffened and insulated by urethane foam sprayed on the back.

FIG. 22. Agricultural building assembled from sheet-metal panels spray coated with rigid urethane foam. Courtesy of Behlen Manufacturing Co., Inc., Columbus, Nebraska.

Urethane foam has been a particularly useful roof insulation. Applied as preformed slab stock, often with skins of asphalt-treated paper, urethane foam has been combined with other roofing materials to form roofs that reduce heat loss, condensation, and leakage. Spraying has been useful in insulating the inner surfaces of contoured roofs and has been used for exterior applications in which a further protective barrier has been applied.

Prefabrication of small structures and building components containing rigid urethane foam has increased. Mobile homes and campers have been formed from light metal or plastic facings reinforced and insulated by urethane foam in operations similar to that shown in Fig. 23. In the portable school building shown in Fig. 24 sandwich-panel construction was used to combine strength, durability, and light weight. In temporary construction entire buildings have been fabricated of urethane foams. Erection of barriers in coal mines by spraying urethane foam on a cloth substrate has become a standard technique. The variety of urethane-foam application methods has attracted a number of architects and design engineers to new construction techniques, some of which are discussed in Chapter 20.

Use of rigid urethane foam in marine applications has grown. Blocks of urethane foam and foam-core sandwich construction have commonly been used to create buoyancy in small boats, and dead spaces in large naval ships have been filled with foam to prevent flooding should outer hulls be ruptured. Foam-filled floats and buoys have become well known, and urethane foam has been used in salvage operations in which liquid raw materials were pumped into submerged hulls where they expanded and displaced the water.

Rigid urethane foam absorbs energy when crushed and has been used for protective packaging of electronic components and military hardware.

FIG. 23. Spray application of rigid urethane foam to the wall section of trailer. Courtesy of the Vesely Manufacturing Co., Lapeer, Michigan.

Protective barriers in automobiles have also been studied. Aerospace applications include protection of missile components and rigidization of aircraft components and fuselage structures. Inflatable satellites and sections of orbital space stations based on the production of urethane foam in space have been designed. These applications are discussed in greater detail in Chapter 21.

New application technology has sought areas that make use of the mechanical strength of both low- and high-density rigid urethane foams. The experimental sports car of Fig. 25 convincingly demonstrated uses of urethane foams and other plastics in automobile design. The car was based on a sandwich-type chassis fabricated from rigid urethane foam and fiberglass-reinforced epoxy-resin skins. The hood and trunk lids were molded from special high-density rigid urethane foams by a process that developed thick, dense skins. The process involved 53 polished-metal molds and high-output foam machinery. Termed "reaction injection molding," it produced large structural parts without the cost of large injection-molding

FIG. 24. Relocatable school modules based on sandwich panels with rigid-urethane-foam cores. Courtesy of Vinnel Steel, Oakland, California.

FIG. 25. Experimental plastic automobile designed and fabricated by Bayer AG. Courtesy of Bayer AG.

presses or tedious operations needed to produce fiber-glass-reinforced polyester parts.

The use of foam in decorative applications has also increased. Simulated rustic beams, molded from rigid urethane foam in flexible molds prepared from hand-hewn natural wood beams, cannot be distinguished from the original after staining to an aged, weathered color. Similar reproductions of wood and stone carvings have appeared.

The largest decorative use of rigid urethane foam has developed in furniture. A combination of appreciation of highly carved styles with a shortage of hardwood and skilled carvers has made replacement with molded urethane foam desirable. Pieces such as that shown in Fig. 26 are excellent simulations.

FIG. 26. Intricately styled molded furniture prepared from rigid urethane foam. Courtesy of the Drexel Furniture Company.

Rigid urethane foam has also been used for structural support in upholstered chairs. Low-density foam, molded in contoured shells, is only covered with cushioning material and fabric covering before legs are attached to form a chair of the type shown in Fig. 27.

Foaming has added the dimension of density to material specifications, and rigid urethane systems have made foaming in place a practical technological operation. Both concepts are expected to be more widely used in the future.

FIG. 27. Chairs fabricated around rigid-urethane-foam shells. Courtesy of the Burris Manufacturing Co.

REFERENCES

1. O. Bayer, Angew. Chem., A59, 257 (1947).
2. Publication Board (PB) Reports Nos. 33561, 45246, 67237, 67244, 70175, 70336, 81831.

3. J. M. DeBell, W. C. Goggin, and W. E. Gloor, German Plastics Practice, DeBell and Richardson, Cambridge, Mass., 1946.
4. G. M. Kline, Mod. Plastics, 23, 152A (1945).
5. Publication Board Reports 104463-104470.
6. E. Simon and F. W. Thomas (to Lockheed Aircraft), U.S. Patents 2,577,279, 2,577,280, and 2,577,281 (1951).
7. J. M. Buist and A. Lowe, Trans. J. Plastics Inst., 27, 67 (1959).
8. A Review of the Urethane Industry, Society of Plastics Industry, July 1968.
9. J. M. Buist and H. Gudgeon, Advances in Polyurethane Technology, John Wiley and Sons, New York, 1968.
10. B. A. Dombrow, Polyurethanes, 2nd ed., Reinhold, New York, 1965, Chapter 3.
11. T. H. Ferrigno, Rigid Plastic Foams, 2nd ed., Reinhold, New York, 1967, Chapter 1.
12. J. H. Saunders and K. C. Frisch, Polyurethanes, Technology, Part II, Interscience, New York, 1964.
13. R. A. Stengard, Handbook of Foamed Plastics, Section 9, Lake Publishing, Libertyville, Ill., 1965.
14. R. Vieweg and A. Hoechtler, Kunststoff Handbuch, Vol. VII, Hansen, Munich, 1966.
15. J. H. Saunders and K. C. Frisch, Polyurethanes, Chemistry, Part I, Interscience, New York, 1962.
16. Imperial Chemical Industries Ltd., Tech. Inf. Bull. U81, November 1967.
17. Imperial Chemical Industries Ltd., Tech. Inf. Bull. U82, November 1967.
18. Mobay Chemical Co., Tech. Inf. Bull. 101-F36.
19. National Aniline Division, Allied Chemical Co., Tech. Data Bull. One-Shot Polyether Rigid Polyurethane Foams Based on Nacconate 40-40, May 1960.
20. Mobay Chemical Co., unpublished data.
21. P. G. Gemeinhardt, J. K. Backus, W. C. Darr, and J. F. Szabat, J. Cell. Plastics, 3, 210 (1967).
22. W. C. Darr, P. G. Gemeinhardt, and J. H. Saunders, Ind. Eng. Chem. Prod. Res. Develop., 2, 194 (1963).
23. J. M. Buist, J. Cell. Plastics, 1, 101 (1965).
24. The Upjohn Co., Tech. Bull. 101.
25. F. K. Brockhagen, Kunststoffe, 51, 246 (1961).
26. J. M. Buist, Rubber J. Intern. Plastics, 135, 792 (1958).
27. R. Hurd, Bull Intern. Inst. Refrig., Annex 1957-1, pp. 195-214.

28. Mobay Chemical Co., Sales News Letter FCS-21.
29. A. G. Gaylord, Polyethers, Part I, Polyalkylene Oxides and Other Polyethers, Interscience, New York, 1963.
30. Hooker Chemicals Corp., Bull. No. 43, March 1964.
31. Stepan Chemical Co., Tech. Inf. Bull., Stepanol for Oil Recovery, October 1967.
32. R. E. Knox, Chem. Eng. Progr., 57, 40 (1961).
33. E. A. Dickert, W. A. Himmler, D. E. Hipchen, M. Kaplan, H. A. Silverwood, and R. Zetter, Chem. Eng. Progr., 33 (September 1963).
34. Mobay Chemical Co., Tech. Inf. Bull. No. 76-F30.
35. Dow Chemical Co., Urethane Tech. Bull. V-2, September 1960.
36. Dow Chemical Co., Urethane Tech. Bull. V-3, September 1960.
37. J. W. Britain and P. G. Gemeinhardt, J. Appl. Polymer Sci., 4, 207 (1960).
38. J. W. Britain, Ind. Eng. Chem. Prod. Res. Develop., 1, 261 (1962).
39. G. P. Mack, Chem. Proc., 23 (4), 30 (1960).
40. B. G. Alzner and K. C. Frisch, Ind. Eng. Chem., 51, 715 (1959).
41. A. Farkas, G. A. Mills, W. E. Erner, and J. B. Maerker, Ind. Eng. Chem., 51, 1299 (1958).
42. J. W. Baker and J. B. Holdsworth, J. Chem. Soc., 724 (1945).
43. Mobay Chemical Co., System Suppliers' Newsletter No. 1.
44. H. Piechota, J. Cell. Plastics, 1, 186 (1965).
45. P. G. Gemeinhardt, W. C. Darr, and J. H. Saunders, Ind. Eng. Chem. Prod. Res. Develop., 1, 92 (1962).
46. W. C. Darr and J. K. Backus, Ind. Eng. Chem. Prod. Res. Develop., 6, 167 (1967).
47. F. A. Coglianese, J. Cell. Plastics, 1, 42 (1965).
48. P. Hoppe, E. Weinbrenner, C. Muhlhausen, and K. Breer (to Farbenfabriken Bayer), U.S. Pat. 2,764,565 (1965).
49. M. E. Bailey and R. C. Kuder, SPE Journal, 14, 31 (1958).
50. E. I. du pont de Nemours & Co., Elastomer Chemicals Dept. Bull. Rigid Urethane Foams-Methods of Application, June 1957.
51. R. E. Knox and R. A. Stengard, E. I. du Pont de Nemours & Co. Bull. Molding Rigid Urethane Foam, October 28, 1960.
52. R. E. Jones, Plastic Technol., 7 (10,11), 27, 43 (1961).
53. H. Piechota, paper presented at the 2nd SPI Intern. Cellular Plastic Conf., New York, November 6-8, 1968.
54. E. I. du Pont de Nemours & Co., Properties of Rigid Urethane Foams, September 1966.
55. R. K. Traeger, J. Cell. Plastics, 3, 405 (1967).
56. F. M. Kujawa, J. Cell. Plastics, 1, 400 (1965).

57. The Dow Chemical Co., Urethane Tech. Bull. VR-1, October 1963.
58. C. J. Hilado and R. H. Harding, J. Appl. Polymer Sci., 7, 1775 (1963).
59. C. M. Barringer, SPE Journal, 15, 961 (1959).
60. M. M. Levy, J. Cell. Plastics, 2, 37 (1966).
61. R. E. Skochdopole, Chem. Eng. Progr., 57 (10), 55 (1961).
62. R. H. Harding, J. Cell. Plastics, 1, 224 (1965).
63. F. J. Norton, J. Cell. Plastics, 3, 23 (1967).
64. E. F. Cuddihy and J. Moacanin, J. Cell. Plastics, 3, 73 (1967).
65. R. H. Harding and C. J. Hilado, J. Appl. Polymer Sci., 8, 2445 (1964).
66. J. K. Backus, W. C. Darr, P. G. Gemeinhardt, and J. H. Saunders, J. Cell. Plastics, 1, 178 (1965).
67. J. H. Saunders and J. K. Backus, Rubber Chem. Technol., 39, 461 (1966).
68. J. K. Backus, D. L. Bernard, W. C. Darr, and J. H. Saunders, J. Appl. Polymer Sci., 12, 1053 (1968).
69. A. Singh, L. Weissbein, and J. C. Mollica, Rubber Age, 98 (12), 77 (1966).
70. Belgian Pat. 680,380 (to Imperial Chemical Industries) (1966).
71. E. Windemuth, G. Braun, and P. Hoppe (to Farbenfabriken Bayer), British Pat. 908,337 (1962).
72. H. Nohe, R. Plate, and E. Wegner (to BASF), Belgian Pat. 657,835 (1964).
73. L. Nicholas and G. T. Gmitter, J. Cell. Plastics, 1, 85 (1965).
74. H. E. Frey (to Standard Oil Co. of Indiana), U.S. Pat. 3,300,420 (1967).
75. W. E. Voisinet, Given before the Philadelphia Chapter, the Thermal Insulation Society, January 1965.
76. J. A. Hartsock, J. Cell. Plastics, 1, 291 (1965).
77. J. A. Hartsock, J. Cell. Plastics, 3, 81 (1967).
78. J. K. Finke and G. F. Wilson, paper presented at the 2nd SPI Intern. Cellular Plastics Conf., New York, November 6-8, 1968.
79. J. A. Parker, S. R. Riccitiello, W. J. Gilwee, and R. Fish, paper presented at the 25th Ann. SPI Conf., San Francisco, Calif., May 1968.
80. I. N. Einhorn, J. Cell. Plastics, 1, 25 (1965).
81. C. J. Hilado, Ind. Eng. Chem. Prod. Res. Develop., 6, 154 (1967).
82. G. L. Ball, III, M. Schwartz, and J. S. Long, Off. Digest, 32 (425), 817 (1960).
83. R. Hurd, Insulation Rev., 11 (6), 18 (1958).
84. R. T. Darby and A. M. Kaplan, App. Microbiol., 16, 900 (1968).
85. H. C. Bruckner, S. B. Avery, D. M. Stetson, V. N. Dodson, and J. J. Ronayne, Arch. Environ. Health, 16, 619 (1968).
86. B. L. Williams, Polymer Conf., Wayne State University, 1967.

87. J. H. Engel, Jr., S. L. Reegen, and P. Weiss, J. Appl. Polymer Sci., 7, 1679 (1963).
88. K. C. Frisch, J. Cell. Plastics, 1, 321 (1965).

Chapter 10

POLYSTYRENE AND RELATED THERMOPLASTIC FOAMS

A. R. Ingram
and
J. Fogel

Research Department
Koppers Company, Inc.
Monroeville, Pennsylvania

I. MARKET SITUATION

Most rigid cellular plastics today are made from polystyrene [1]. In 1969 the total market for polystyrene foams in the United States was 320 million lb, consisting of 202 million lb of foam molded from expandable polystyrene beads and 118 million lb of foam produced by extrusion. The next most common material for rigid foams was polyurethane, of which 201.5 million lb were used. The markets for polystyrene-foam materials are broadly outlined in Table 1.

Molded-bead foams will be considered first. The raw material usually consists of particles of polystyrene with pentane trapped among the rigid chains. Expandable polystyrene in bead form was introduced in the United States by the Koppers Co. [2] in 1954. Significant growth of this market did not start until 1958, when the total sales amounted to 10 million lb. In 1969 there were four major suppliers of expandable polystyrene in the United States: BASF Corp., Dow Chemical Co., Foster Grant Co., and Sinclair-Koppers Co. As will be explained later, several types of expandable polystyrenes are offered, varying in particle size and shape, as well as composition, depending on the end use. The most significant property responsible for the commercial acceptance of expandable polystyrene is its ability to be steam-molded into a great variety of useful, lightweight, low-cost objects. The molding of foams involves a two-step expansion process. First, the particles are expanded in a continuous upward flow of steam to yield free-flowing foams of the desired density. After an aging period of several hours during which air diffuses into the expanded particles, they are placed in a slightly vented mold. The particles are then heated by steam injected into the cavity so that they expand to fill the interstices and fuse together. The molding operation is completed by circulating or spraying cold water in the jacket of the mold until the foam no longer exerts pressure.

TABLE 1

Polystyrene-Foam Markets in the United States[a]

Market	Quantity (lb x 10^6)		
	1967	1968	1969
Expandable polystyrene beads:			
Packaging	40	50	69
Cups	40	43	47
Boards and billets	36	43	49
Custom molding, export, etc.	29	33	36
Miscellaneous	2	1	1
Subtotal	147	170	202
Extruded foam sheet from crystal polystyrene	35	55	70
Extruded polystyrene-foam board	43	45	48
Total	225	270	320

[a]Reprinted from Ref. [1] by courtesy of Modern Plastics.

An important market for molded-bead foam is in packaging. Figures 1, 2, and 3 show typical packaging applications. Molded-bead foams, usually in the desity range 1.25 to 1.75 lb/ft^3, are used to package a multitude of industrial, consumer, and military items. These molded-polystyrene-foam packages have contributed outstanding savings through reduced damage, shipping weight, and labor costs. Furthermore the eye appeal, product identification, and reusability of the packages are added dividends that help to sell the packaged product. An outstanding example of the rapid growth of the molded-polystyrene-bead market is the foam cup, particularly for hot drinks. From essentially nothing in 1960 this application has grown to 47 million lb or at least 8 billion cups in 1969. An important early

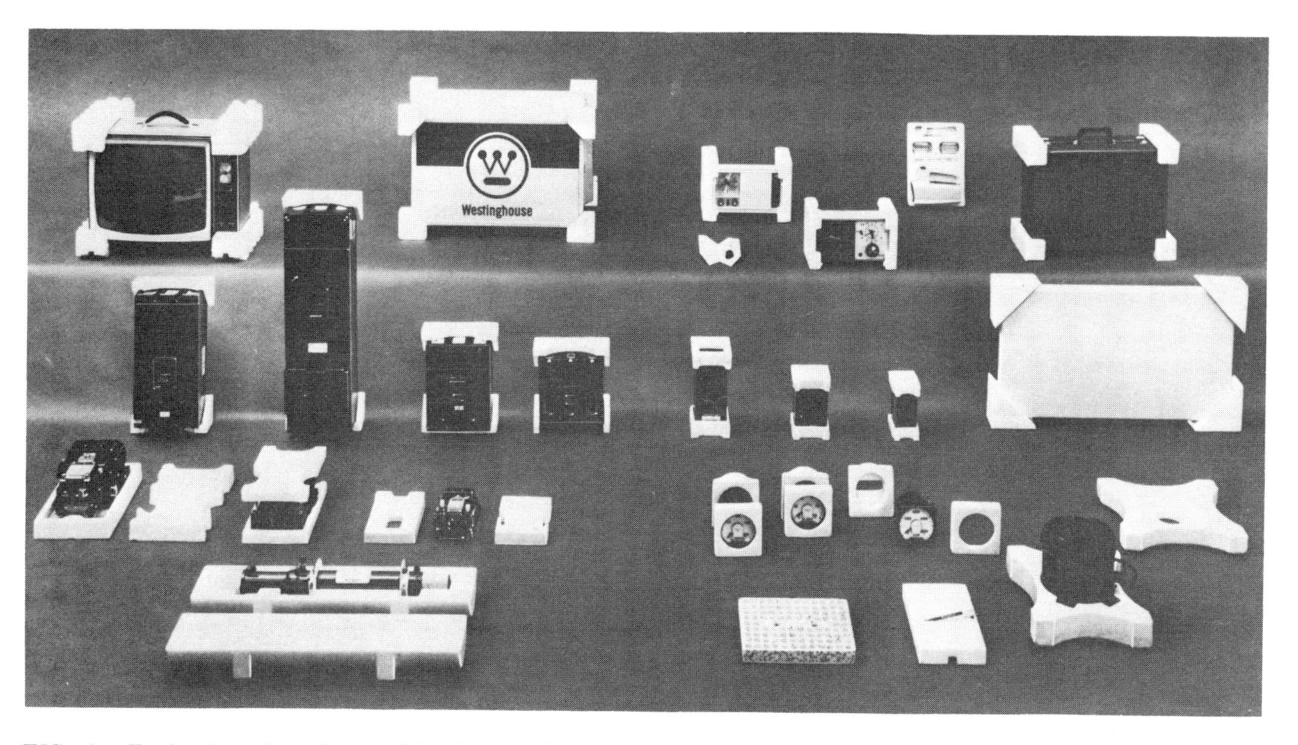

FIG. 1. Packaging of products of the Westinghouse Electric Corp. in molded Dylite polystyrene bead foams. Reprinted by courtesy of the Sinclair-Koppers Co.

application for molded-bead foam was insulation board cut from molded billets. In 1969 this market amounted to 49 million lb, a large portion of which was of the self-extinguishing variety. The main uses for insulation board are in cold-storage refrigeration, housing insulation, and cut ceiling tiles. A substantial quantity, 36 million lb, of polystyrene foam was molded in the category of "custom molding, export, etc." This consisted largely of picnic coolers, ice buckets, components of refrigerators and air conditioners, displays, toys, housewares, and flotation items. Under the miscellaneous category is expandable polystyrene that was extruded to form a sheet. This amounted to only 1 million lb in 1969.

At present the fastest growing market is that of foam sheet extruded from crystal polystyrene, amounting to 55 and 70 million lb in 1968 and 1969, respectively. Polystyrene-foam sheet is made by extruding through an annular die a mixture of polystyrene and nucleating agents, along with a volatile liquid blowing agent that is injected into the extruder. The sheet ranges in density from 5 to 12 lb/ft^3 and is thermoformed into meat trays, produce trays, and egg cartons. Dow, Mobil, Sinlcair-Koppers, Gulf States, Amoco, and W. R. Grace plants manufacture both sheet and thermoformed products.

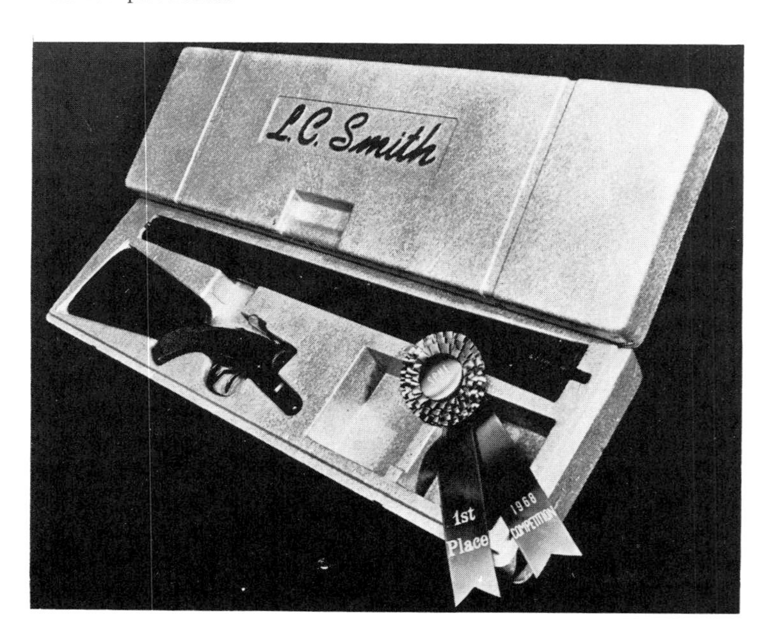

FIG. 2. Award-winning expandable-polystyrene package by Marlin for L. C. Smith shotgun. Molder: J. R. Sexton, Inc., Meriden, Conn. Reprinted by courtesy of the Sinclair-Koppers Co.

Polystyrene foam was first introduced in the United States in the form of extruded board, Styrofoam, by The Dow Chemical Co. in 1944. Styrofoam is produced by extrusion in the form of billets, which may be cut into board stock, or it can be furnished with continuous integral polystyrene skins to a desired thickness. The Styrofoam process has been developed so that foams can be tailor-made to exhibit specific structural properties and thermal conductivities by varying density, cell size, and blowing agent. Uses include comfort and low-temperature insulation in residential and commercial construction as well as in decorative, novelty, buoyancy, and packaging applications.

In this chapter we discuss not only the scientific, engineering, and commercial aspects of polystyrene-foam products but also some experimental materials, as well as structural foam molding as applied to styrene polymers.

II. METHODS FOR PREPARING EXPANDABLE-STYRENE POLYMERS

This section covers the preparation of styrene-containing materials from which foam can be generated, usually by steaming. It does not discuss the formation of foams by means of extruders and injection-molding machines into which blowing agents are added separately from the polymer. These methods will be considered in Section VIII, IX, and X.

The basis for this section is a paper [3] presented in 1968. This paper has been revised, and parts of it are reprinted here by courtesy of the publisher, the American Chemical Society.

A. Diffusion of Blowing Agents into Polystyrene

1. In Contact with Anhydrous Organic Liquids

a. Particles Plus Excess Liquid. The first patent [4] on the preparation of expandable polystyrene containing volatile hydrocarbons, applied for in Great Britain in 1944, teaches the soaking of polystyrene particles, held in a cheesecloth bag, in a solution comprising a nonsolvent, petroleum ether (90-99 vol %), and a solvent (10-1%), such as acetone, ethyl acetate, or benzene. At 20 to 25° C impregnation is complete within 0.5 to 12 h, depending on particle size and composition. A mixture of diethyl ether and ethanol is also claimed [5] to render polystyrene expandable after 3 days of soaking. Nonvolatile additives [6, 7] may be incorporated in the polystyrene, hastening its impregnation by n-pentane and avoiding the use of solvents as disclosed above. Cited additives are rubbery isobutylene polymers, rubbery diene polymers (e.g., butadiene-styrene elastomers)

FIG. 3. Expandable-polystyrene package molded by General Foam Plastics, Los Angeles, Calif., for packaging Paul Masson wines. Reprinted by courtesy of the Sinclair-Koppers Co.

fatty acid salts or esters (e.g., stearates), organic insoluble pigments, and inorganic pigments smaller than 5 microns in diameter. Polystyrene (ground to 8-20 mesh) stirred with an equal weight of n-pentane for 24 h absorbed 4.7% by weight; for 48 h, 7.7%. However, the presence within the polystyrene of 1% of intimately mixed calcium carbonate of a particle size of 0.1 micron increased the 24-h absorption of n-pentane to 7.8% [6]. Although these modifications of the polymer effectively decrease absorption time, they also increase the rate at which n-pentane diffuses out of the polymer, thereby shortening shelf life.

A difficulty with the steeping processes is that the particles become softened and tend to agglomerate. Therefore they are stirred vigorously and kept at relatively low temperatures. The addition of finely divided solids [8] to the dispersion prevents agglomeration and permits the use of higher temperatures to hasten impregnation. Thus a slurry of 24% polystyrene in pentane was stabilized by 5% (polystyrene basis) calcium carbonate and 2% calcium phosphate. After stirring at 90° C for 3 h, the beads contained 10% pentane and were not stuck together. By another process [9] a 50% slurry in n-pentane was stabilized by 1% calcium silicate and stirred for 0.5 h at 40°C and 40 psig under nitrogen. This product contained 8% n-pentane.

b. Particles Plus the Minimum Amount of Liquid. A process without a dilution medium or antiagglomeration agent would be desirable. The first of such processes [10] teaches the continual rolling for 32 h at 70° C of a closed vessel containing 1000 parts of polystyrene particles with a solution comprising 65 parts of hexane, 5 parts of benzene, and 15 parts of methanol. Another process [11] teaches an impregnation under carbon dioxide well above the softening point of the polymer: 900 parts of polystyrene and 80 parts of hexane under carbon dioxide at 10 atm for 4 h at 140° C. This product was broken apart after cooling.

The absorption of butane into polystyrene proceeds more rapidly than does that of pentane or hexane. Thus the addition of 7% of butane, plus an excess only for air-purging, to polystyrene (or to rubber-modified polystyrene) particles in a sealed container provided useful expandable polystyrene after agitation for 24 h at room temperature [12]. If the butane was mixed with a noncombustible gas of lower density, the explosion hazard was avoided [13].

c. Molded Objects Immersed in Liquids. Molded, nonfoamed articles, such as drinking cups from rubber-modified polystyrene, are immersed in absorbable liquid blowing agents (e.g., trichlorofluoromethane). The molding is then heated, creating an insulating surface according to the location and amount of heat, as well as the extent of absorption of liquid [14-17].

d. Sheet Immersed in Liquids. Continuous processes for converting sheets of styrene polymer to foam comprise (a) immersing the sheet in a mixture of ketone penetrant and hydrocarbon blowing agent; (b) removing excess solvent; and (c) immersing in boiling water. Knobloch et al. [18] obtained a foam sheet of 1.8-lb/ft^3 density from an 80:20 copolymer of styrene and butadiene immersed in a 90:10 mixture of pentane and acetone. A similar process [19] for polystyrene sheet employs a mixture of petroleum ether and methyl ethyl ketone as the immersion bath.

2. In Contact with Organic Vapors

Finely divided polystyrene beads (22-48 mesh) supported on a screen suspended above the surface of pentane in a closed vessel absorbed as much as 9.2% in 2 days at 30° C. In the same way [20] a styrene-acrylonitrile copolymer was rendered expandable by exposure to vapors of a 90:10 mixture of pentane and methylene chloride.

A graft copolymer sheet of 5 parts of rubber (25:75 styrene-butadiene) and 95 parts styrene was kept for a week at -70° C in solid carbon dioxide in order to make it expandable [21].

3. In Aqueous Slurries

Patent applications on the diffusion of pentane fractions of petroleum into slurries of polystyrene beads in water at elevated temperatures were made in both the United States and Germany in 1953. In the U. S. application [22] the suspension was stabilized by a finely divided calcium phosphate and an anionic surfactant, and the impregnation was carried out at 90° C. In the German application [23] the suspension was stabilized by an emulsifying agent alone, and the impregnation was conducted at 80° C. To permit the introduction of butanes, rather than pentanes, as blowing agent without creating excessive pressure in the reactor, methylene chloride or ethyl acetate (strong solvents) were first added to the slurry; then the polystyrene beads readily absorbed butane within 5 h at room temperature [24]. Reactor pressure was 30 psig. The impregnation of polystyrene in aqueous suspension could be performed without a pressure vessel when a mixture of equal parts of petroleum ether and ethyl acetate was slowly added to the bead slurry [25]. Addition time was 6 h; soaking time, 18 h. The use of polyvinyl alcohol as the sole or partial suspending agent was claimed to permit the rapid impregnation by pentane of polystyrene slurries at rather high temperatures, 115 and 120° C [26, 27]. Bead suspensions at 105° C could be stabilized by a mixture of hydroxyethyl cellulose and Tamol SN [28].

Instead of charging beads to be impregnated with pentane, one may charge ground or chopped strands of polystyrene. When the impregnation is conducted at very high temperatures, 130 to 135° C, these particles become rounded and the suspension is stable provided suspending agents (e. g., a mixture of carboxymethyl cellulose and the ammonium salt of sulfonated polyvinyltoluene) are used. Pentane could be added under pressure at the elevated temperature, preferably in a slow stream or by several small additions [29]. In another modification [30] of this process the length and diameter dimensions of the cylindrical particles (e. g., about 0. 1 in. in length and 0. 02 in. in diameter) were such that they became spherical after being heated in a suspension stabilized by basic magnesium carbonate at 130 to 135° C. The temperature of the suspension was then reduced, and pentane was added at a lower temperature, 118° C.

In a process [31] for simultaneously obtaining a rapid rate of pentane diffusion and good foam-forming qualities, the autogenous pressure of the vessel was maintained constant by increasing the temperature in the range 90 to 120° C within 4 h.

B. Polymerization of Styrene Solutions of Volatile Hydrocarbons

1. Addition of Hydrocarbon Before Polymerization

a. Bulk Polymerization. Expandable polystyrene was inadvertently prepared [32] in 1945 in an attempt to bulk-copolymerize 10% isobutylene with styrene. The product formed a low-density foam when heated. An early method (1950) for rendering polystyrene expandable by petroleum ether was to dissolve 6 parts of petroleum ether in a 40% solution of polystyrene in benzoyl peroxide-catalyzed styrene and to hold the mass for 28 days at 32° C [33]. In a recent version [34] of this process the monomer (chlorostyrene) and blowing agent (trichlorofluoromethane) in a polyvinyl fluoride bag were irradiated with gamma rays.

b. Suspension Polymerization. The first patent application [35] on polymerizing a suspended styrene solution of petroleum ether (7%) was made in Germany in 1951. The suspending agent was polyvinylpyrrolidone-2; the initiator, benzoyl peroxide (0.7%); and the temperature of polymerization, 82° C. These products were used for the development of the popular process for steam-molding expanded beads. Improvements on this system involved the application of special initiators to reduce the residual monomer content, in one case [36] with a mixture of azobisisobutyronitrile and di-t-butyl peroxide, and in the other [37] by using a difunctional peroxide. An objectionable feature of these processes is the formation of blisters, craters, and pock marks by pentane escaping from the particles at a critical viscosity during the latter stage of polymerization. This effect can be greatly reduced by increasing the pressure in the polymerization vessel by at least one-fifth of the original pressure by introducing an inert gas, such as nitrogen, when the polymerization has reached the particle-identity point [38]. Another way [39] to avoid off-round and blemished beads is to withhold about one-fifth of the petroleum ether from the initial charge and then add it during the first 3 h of polymerization.

2. Addition of Blowing Agents to Styrene Solutions of Polystyrene

If the pentane is added to a suspension polymerization of styrene after the bead-identity point has been reached, the formation of blisters is avoided, but the diffusion of pentane into the bead is still quite rapid. Thus the two objections to the "pentane-in-monomer" and the "postpolymerization impregnation" processes are avoided [40-42]. The same system has also been employed for the introduction of normally gaseous blowing agents, such as butane, propane, sym-dichlorotetrafluoroethane, propylene, butene, and butadiene [43-45].

The styrene monomer may be first polymerized to about 65% conversion in the absence of water, and then the blowing agent (cyclopentane) is added with additional peroxide. This solution is suspended in water in the presence of potato starch, and the polymerization is finished [46]. In another modification [47] a styrene solution of "waste" polystyrene and peroxide was suspended by polyvinyl alcohol in water and then pentane was added to the suspension after the solids content of the oil phase was greater than 70%. Polystyrene particles could also be suspended in water by a mixture of polyvinyl alcohol and a phenyl sulfonate, a mixture of equal parts of pentane and catalyzed styrene (8% each on polymer) being then diffused into the polymer [48]. The temperature was raised, and the polymerization was finished.

C. Suspension-Polymerization Systems

Two variables may be manipulated in order to control the size and shape of particles obtained by suspension polymerization: (a) the suspending agent and (b) the agitation system.

1. Suspending Agents

Both types of suspension stabilizers--the finely divided water-insoluble solids and the soluble film-formers--have been extensively applied to the preparation of spherical styrene polymers with particle sizes of about 10 and 40 mesh. Examples of such materials are the following:

Finely divided solids:
- Tricalcium phosphate with anionic surfactants [49].
- Tricalcium phosphate with sodium β-napthalene sulfonate and sodium polyacrylate [50].
- Zinc oxide without or with ammonia [51].
- Bentonite plus gelatin [52].

Water-soluble, high-molecular-weight polymers:
- Polyvinylpyrrolidone-2 [53].
- Vinylpyrrolidone-2 copolymer with 6% methyl acrylate without [53] or with [54] a surfactant.
- Polyvinyl alcohol without [42] or with [55] a polymerized sodium salt of alkenylnaphthalenesuflonic acid.
- Methyl cellulose [56].
- Hydroxyethyl cellulose without [57] or with [58] a 94:6 copolymer of vinylpyrrolidone-2 and methyl acrylate.
- Ammonium polyvinyltoluence sulfonate plus anionic surfactant [59].

For preparing expandable polystyrene beads of extremely fine particle size (2-10 microns) a mixture of colloidal silica and a diethanolamine-adipic acid condensate was used as suspending agent for a styrene solution of neopentane [60].

2. Agitation Systems

The particle size of polystyrene made in a suspension polymerization is not only influenced by the type and amount of suspending agent but also is subject to mechanical factors, such as the speed of agitation and the shape and size of the vessel and the agitator. The effects of such mechanical factors in large reactors (50 and 500 liters) have been described by Yanishevskii and Voevodin [61].

D. Deposition of Expandable Polystyrene from Solution

Particles of polystyrene containing pentane were precipitated from a solution of polystyrene in 80:20 pentane-dimethylformamide by the addition of a nonsolvent (methanol) [62].

The polymer may be dissolved in a water-immiscible liquid and then suspended into droplets; the resulting suspension is partially stripped until a suspension of expandable particles is obtained. In an example of this process a solution of chlorinated polystyrene in methylene chloride was suspended by polyvinylpyrrolidone-2, and then the resulting suspension was simultaneously sparged and agitated by nitrogen [63].

Expandable films were prepared by evaporating a solution of polymer and blowing agent. The blowing agent was a decomposable solid [64] or a volatile liquid [65]. In the latter case a solution of one part of a 90:10 styrene-butadiene copolymer and 0.3 part of pentane in 3 parts of methylene chloride was evaporated. The resultant film was expanded in boiling water to a foam with a density of 1.2 lb/ft^3.

E. The Quenched-Pellet Process

The quenched-pellet process consists of extruding a mixture of polystyrene and pentanes, chilling the extrudate to avoid foaming, and chopping the strand into particles of the desired size. The following features were disclosed in the first patent applications [66]: (a) extrusion into a water bath at 50 psi; (b) extrusion into a water bath to cool the strand instantly to below 50° C at atmospheric pressure; and (c) holding the quenched pellets at an elevated temperature (50° C to the plastifying temperature) to relieve orientation strains.

The quenched-pellet process was developed primarily for the incorporation of colorants. Several improvements and modifications of the apparatus to make the quenched pellets have been made: extrusion through a die made of or lined with polytetrafluoroethylene to broaden the permissible temperature range of extrusion [67]; a conveyor system for continually passing the freshly chopped pellets through a normalizing bath to relieve orientation strains at 65 to 85° C [68]; an extruder to facilitate mixing of pentane with polystyrene [69]; and replacement of the water bath by a chilled roll containing grooves in which the extrudate was cooled [70]. Orientation strains were also relieved by very short exposure (0.5-15 sec) above the heat plastifying temperature (77-99° C) [71].

The cell size of foam made from quenched pellets was substantially reduced by instituting cyclic shock waves [72] or flexural stresses (i.e., through guides to form a zigzag path) [73] in the unfoamed extrudate. The pellets have been treated to reduce cell size by either storing them at -10° C or by agitating them until a cloudy appearance developed [74].

Various finely divided solids (ca. 2%) have been added [75] to the extruder charge as cell-nucleating agents for making butane-expandable pellets. The charge to the quenched-pellet extruder can be in the form of expandable beads already containing the blowing agent [76, 77]. However, because spherical beads tend to slip around the screw in the feed section, they do not extrude smoothly without "stuffing." This problem was overcome by feeding expandable disks obtained by a modified polymerization process [78]. Considerable quantities (2-40%) of a lightweight, finely divided, inexpensive filler, especially wollastonite, have been incorporated without substantially harming the foaming properties [79].

The quenched-pellet operation has been used as the finishing step in the bulk polymerization of a styrene solution of pentane [80].

F. Water-in-Monomer Polymerizations

Porous polymers or mixed polymers of styrene have been prepared [81-83] by polymerizing the monomer with water emulsified therein. The water was removed by heating at 50 to 80°C to give open-celled foams. The foams discussed heretofore were of the closed-cell type.

III. COMPOSITION OF EXPANDABLE STYRENE POLYMERS

Having presented many methods for preparing expandable styrene polymers, we now discuss variations in their composition. The basis for this discussion is a paper [3] presented in 1968. This paper has been revised, and parts are reprinted here by courtesy of the publisher, the American Chemical Society.

A. The Blowing Agent

The blowing agent for most commercial expandable polystyrenes has consisted of hydrocarbons obtained from the pentane boiling range of petroleum. Expandable polystyrene typically contains about 6.3 to 7.3% pentane. For each composition there is an optimum amount of blowing agent for complete foaming. Only slightly more (i.e., to 1 part per 100 of polymer) than this amount leads to such undesirable properties as exudation, foam collapse, and excessive cell size [84]. The tolerable amount of blowing agent decreases with the molecular weight of the polymer. A practicable volume of n-pentane (6.5% by weight) for foaming is identical with the free volume determined by the compressibility of amorphous polystyrene. This aspect will be discussed later in Section IV.

Expandable beads made in water suspensions must be dried at relatively low temperatures (e.g., 50-60° C maximum) to avoid excessive evaporation of n-pentane (bp 36° C). The loss of n-pentane in 1 h is related to temperature and the n-pentane content of the beads as shown in Table 2 [84].

TABLE 2

Loss of n-Pentane in1 h from Expandable Beads at Varying Temperature[a]

Temperature (° C)	Loss (wt%) from beads with an n-pentane content of:			
	4.0%	5.0%	6.0%	7.0%
35	0.0	0.0	0.0	0.0
45	0.0	<0.1	<0.1	0.1
52.5	0.0	<0.1	<0.1	0.1
65	0.0	0.1	0.1	0.9

[a]Data from Ref. [84].

To reduce pentane losses in shipping, expandable beads are ordinarily transported in 50-gal drums constructed from fiberboard with an inner liner of aluminum foil. In these containers the beads meet most expansion requirements for a few months, depending on storage conditions. When expandable beads (screened through a 10-mesh sieve and retained on a 25-mesh sieve) were exposed in open containers at 23° C, their n-pentane content decreased from 7.0 to 6.0% in 6 to 7 days.

Several other blowing agents have been disclosed, and many of these have particular advantages. Neopentane (tetramethylmethane, bp 9.5° C) is retained well by polystyrene and generates low-density foam of extremely fine cell size [56]. Isopentane (bp 28° C) also provides a smaller celled foam and in spite of its lower boiling point is held more tenaciously than is n-pentane (bp 36° C) [84]. In combination with selected plasticizers, isopentane provides expandability to very low densities without excessive collapse [85]. As already pointed out, normally gaseous materials, such as propane, butane, butene, and sym-dichlorotetrafluoroethane, have been incorporated into polystyrene during its suspension polymerization near the bead-identity point. Longer retention of butane and enhanced expandability in butane-expandable beads are claimed [86] to result from adding sodium nitrite to the initial suspension charge, thereby leaving styrene monomer in the beads.

Solid gas-releasing blowing agents can also be used to expand polystyrene. Azobisisobutyronitrile may be the initiator of polymerization, and the unused portion remaining in the polymer serves as a blowing agent when the polymer is heated above its plastification temperature [87]. Nitrogen-releasing blowing agents have been encapsulated in polystyrene by polymerizing the monomer on the surface of the blowing agent [88]. Azodicarbonamide has been incorporated into polystyrene pellets by extruding them at a temperature below the decomposition point [89]. Carbonate esters that release carbon dioxide have also been disclosed, employing an activator to promote the release of the gas (e.g., sodium carbonate with succinyl monoglyceryl carbonate [90]).

B. The Polymer

1. Amorphous Polystyrene

All of the expandable polystyrene we have been discussing is of the amorphous type, which is obtained by free-radical initiation. This polymer is completely noncrystalline and in the absence of impurities, such as monomer and blowing agent, has a glass-transition temperature of about 100° C. Both the rate of expansion and the extent of expansion are enhanced by reducing the molecular weight, but the foam becomes less resistant to collapse from prolonged steaming [84].

2. Isotactic Polystyrene

Isotactic polystyrene is obtained by stereospecific polymerization methods and is characterized by a high degree of crystallinity and a crystalline melting point in excess of 200° C. Thus the processes for expandable amorphous polystyrene are not directly applicable. Foams of outstanding resistance to heat (i.e., dimensionally stable for 4 days at 175° C) and solvents were obtained by using aromatic hydrocarbons as blowing agents for a completely crystalline acetone-insoluble polymer [91]. In another method [92] a heated mixture of molten polymer and methyl chloride, propane, or butane was suddenly depressurized. If the polymer particles are obtained from a quenched (temporarily amorphous) sheet, they can be foamed by depressuring a mixture with n-pentane heated to 170° C, a temperature below the crystalline melting point [93]. Isotactic polystyrene pellets have been impregnated with a 30:70 mixture of n-pentane and 2,2-dimethylbutane at 220° C in an aqueous suspension stabilized by magnesium carbonate and then cooled to below 100° C at a rate greater than that of polymer crystallization [94]. Isotactic foam has been generated continuously from a butyllithium-initiated polymerization conducted in the presence of a 4:1 mixture of benzene and petroleum ether [95].

3. Irradiated Polystyrene

The radiation-induced crosslinking of expandable polystyrene with ionizing radiation in the range 200 to 300 megarep raises the serviceability temperature of the foam in proportion to the amount of radiation. Foams stable to 166° C have been made in this manner [96].

4. Mixed Polymers

A mixture of 5 parts of polyethylene and 95 parts of expandable polystyrene was extruded and converted to quenched pellets. The pellets were expanded and converted into molded foams with improved resistance to solvents and water-vapor transmission [97]. There was a polyethylene skin on the surface.

5. Styrene Copolymers

a. Noncrosslinked. Foams with improved resistance to gasoline and lubricating oils are obtained at reasonable cost by substituting acrylonitrile for 20 to 30% of the styrene in the polymer. Expandable copolymers have been made by the pentane-in-monomer process and by the impregnation of already formed pellets [30, 53]. A difficulty with the former process is the formation of yellow beads with high residual acrylonitrile content when one attempts to prepare copolymers of over 25% acrylonitrile. A difficulty with the latter process is the need to have a more expensive expanding agent, trichlorofluoromethane, to assist the impregnation of pentane. A pentane-in-monomer process to overcome these objections was realized by adding alkaline reagents at the bead-identity point to hydrolyze the acrylonitrile in the water phase [98].

α-Methylstyrene (e.g., 20%) in the copolymer improves the heat resistance of the molded foams [99]. Copolymers of 30 to 60% methyl methacrylate have been prepared by a partial conversion by bulk polymerization followed by suspension and addition of petroleum ether blowing agent [100].

Polymers or copolymers of chlorostyrene and dichlorostyrene were prepared by suspension polymerization in the presence of a volatile blowing agent [101].

b. Crosslinked Polystyrene. Polystyrene lightly crosslinked during polymerization by 0.01 to 0.25% divinylbenzene imparts to foams, expanded by carbon dioxide or other gases, greater expandability and resistance to collapse at elevated temperatures [102]. Other divinylbenzene copolymers have been reported, employing as blowing agent saturated aliphatic or fluoroaliphatic volatile liquids [103], tetramethylsilane [104], and azobisisobutyronitrile [105].

Molded-foam copolymers with 10 to 30% glycidyl acrylate can be crosslinked by heating in the presence of polyamines. Heat resistance is increased to as high as 150° C by this treatment [106]. The use of certain allyl or diallyl esters to control the crosslinking of polystyrene to the desired very slight extent has been disclosed as a means of obtaining foams of improved and regular cell structure [107]. Block copolymers of polystyrene, polydiene (particularly polyisoprene), and polystyrene have been foamed by sudden depressurizing of a solution in isopentane. These foams are particularly adaptable to crosslinking by irradiation [108].

c. Postreacted. The serviceability temperature of polystyrene is significantly increased by the hydrogenation of polystyrene before foaming [109]. Polystyrene foams have been rendered infusible by alkylating the polymer with mixtures of monohalomethylated and dihalomethylated compounds in the presence of dehydrochlorination catalysts [110].

C. Additives

1. Self-extinguishing Agents

Polystyrene foams ignite readily in contact with a flame. Therefore, where large quantities of foams are exposed, such as in the warehousing of insulation board, a self-extinguishing grade is desirable. Although present self-extinguishing grades can burn in contact with a flame, they are quickly self-extinguishing and have little or no tendency toward flame spread. The state of the art in this field was recently reviewed by Lindemann [111], who listed 204 references.

The self-extinguishing agent for expandable polystyrene is usually a mixture of an aliphatic bromine compound and a relatively stable peroxide [112, 113]. For example, the self-extinguishing action of 5% acteylene tetrabromide alone with polystyrene is duplicated by only 0.5% acteylene tetrabromide in combination with 0.5% dicumyl peroxide. Other bromine compounds cited as self-extinguishing agents for polystyrene are 1,2-dibromotetrachloroethane [114], tris (2, 3-dibromopropyl) phosate [115], pentabromomonochlorocyclohexane [116], hexabromocyclododecane [117], 2,2,3,3-tetrabromobutanedioldinitrate [118], and the bromine adducts of vegetable oils [119].

Many free-radical-generating compounds have been disclosed as synergists that can be used in place of peroxides to enhance the self-extinguishing action of bromine compounds. Examples of some of these materials are the disulfides [120], sulfenamides [121], N-nitroso-N-methylaniline [122], tetraphenylhydrazine [123], pentaphenylphosphorane [124], α,α'diphenyl-α-methoxybibenzyl [125], 2,3-dimethyl-2,3-diphenylbutane [126], polymers of diisopropylbenzene [127], tetraethyllead [128], and heavy-metal salts [129] or chelates [130].

2. Bead-Expansion Antilumping Agents

The tendency of expandable polystyrene to form agglomerates, or lumps of expanded beads during the preexpansion step is a well recognized problem. Lumping is overcome with varying degrees of success by moistening the beads with and without surfactants, by adding lubricants like zinc stearate, and by increasing the intensity of agitation in the pre-expander. Special additives have also been disclosed for pretreating the beads to prevent lumping. For example, beads have been contacted with a solution of perfluorooctanoic acid in water [131]; an emulsion of polysiloxane oil and water [132]; a solution of sodium stearate and citric acid [133]; an alcohol solution of hydroxyoleamide [134]; kaolin coated with hydroxyethylated or hydroxypropylated amides of fatty acids [135]; water saturated with ozone [136]; emulsified unsaturated polyester resins [137]; ethylenebisstearamide added during polymerization [138]; amorphous hydrated calcium aluminosilicate powder [139]; an alcohol solution of an amide derived from cocoanut-oil fatty acids and diethylenetriamine [140]; a dispersion of magnesium carbonate and a fatty, hydroxyl-containing amide in a solution of polyvinylpyrrolidone in methanol [141]; and a finely divided thermoplastic powder [142].

3. Colorants

The quenched-pellet process (see Section II. E) is suitable for the incorporation of dyes or pigments into expandable polystyrene. Oil-soluble dyes can be transferred to the surface of expandable beads by immersion in an aqueous dye bath containing a carrier like dimethylformamide [143] or a nonionic surfactant [144] . The diffusion of oil- soluble dyes into expandable polystyrene has been accomplished during the impregnation of polystyrene beads in an aqueous suspension [145]. Insoluble pigments can be incorporated into polystyrene by first wetting them with a styrene-ethyl acrylate copolymer and then polymerizing a styrene dispersion of the pigment in an aqueous suspension stabilized by polyvinyl alcohol [146]. In another method expandable polystyrene was coated with pigment dispersed in a solution of a polymer in gasoline, thereby leaving on the surface of expandable polystyrene a coating of pigment dispersed in polymer [147]. Ordinary oil-soluble dyes are too reactive with the peroxides used in the polymerization of styrene, but certain dyes are claimed [148]to be satisfactory when added before or during suspension polymerization.

4. Regulation of Cell Size

a. Steam-Expanded Foam. Several finely divided polymers have been mentioned as nucleating agents to reduce the cell diameter of polystyrene foam from the range 50 to 20 mils to less than 5 or preferably less than 2 mils. For this purpose chlorinated wax [149] or polyvinyl propionate [150] can be dissolved in monomer. A dispersion of polyvinyl chloride in the beads is obtained by polymerizing styrene with either vinyl chloride monomer [151] or powdered polyvinyl chloride [152]added thereto.

Polystyrene has been pelletized with polyethylene glycol or derivatives [153] and impregnated in an aqueous suspension. Expandable beads have been soaked in aqueous solutions of surfactants [154]. Mechanical means have been employed to cause a reduction in cell size: for example, cooling to below -18°C [55] or by crushing (e.g., from 1/4 in. to 3/16 in. thickness) [155]. Cooling a slurry of expandable beads from 100 to below 25°C in less than 15 min also reduces cell size [156].

IV. THE FORMATION AND EXAMINATION OF CELLS IN EXPANDABLE POLYSTYRENE

A. Cell Formation

Investigations of the expansion of impregnated polystyrene beads have been reported in detail by Ingram [157] and by Skinner and co-workers [158, 159]. This information is summarized here, with excerpts from the first-named paper by permission of Wayne State University.

Directions for making and evaluating small quantities of expandable polystyrene in the laboratory have been reported in the literature [84, 160]. Beads are conveniently expanded while supported on a 100-mesh screen suspended in a loosely closed space over steam generated by boiling water or a line source. Typical expansion rates for the three types of polystyrene expanded by n-pentane are shown in Fig. 4, in which the bulk density of

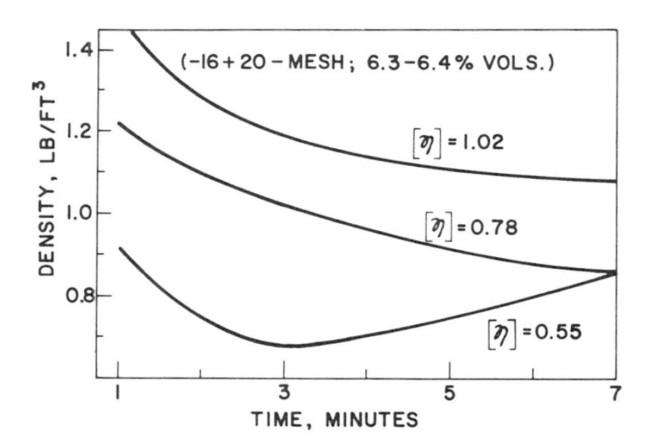

FIG. 4. Expandability versus molecular weight. Reprinted from Ref. [84], p. 153, by courtesy of McGraw-Hill.

the foam particles is plotted against time in minutes. The only variable here is the molecular weight of the polymer, as indicated by the intrinsic viscosity (η) values on the three curves. Expansion rate, ultimate expansion, and collapse of the foam in steam at 100° C are greatly influenced by molecular weight. The curve at the top represents that of a high-strength, injection-molding grade of polystyrene. Though expansion is relatively sluggish, this type of foam has a high resistance to shrinkage from prolonged steaming at elevated pressures as well as atmospheric pressure. The bottom curve is obtained from a rather brittle polystyrene with barely adequate physical properties. Although this material expands very rapidly, the foam collapses at about 3 min of exposure to steam at atmospheric pressure. The commercial expandable polystyrenes exhibit curves that fall between these two limits.

As expandable beads are heated, the blowing agent vaporizes to form many small pockets of gas (called cells) within the softened polystyrene. By measuring cell size as a function of foam density, Pogany [161] concluded that all cells form simultaneously at the beginning of expansion and then increase in size without coalescence or rupture at least within normal conditions of expansion in steam at 100° C.

As heating is continued, more blowing agent vaporizes from the polystyrene into the cells, thus increasing the pressure in the cells (up to 8 atm) and pushing the thin cell walls farther apart. At the same time that the cells are expanding from the pressure of the vaporizing agent, some of the vapors permeate out of the cells and out of the bead itself. Eventually the rate of vaporization into the cells slows down enough to equal the rate of permeation out of the beads. At that point expansion stops. On continued heating the blowing agent is lost faster than it vaporizes into the cells, and contraction sets in. In commercial practice, heating is not continued nearly long enough to permit the beads to shrink. (Compositions prone to shrinkage are usually avoided.)

The question may be asked, How do these expansions compare with the theoretical value one might obtain from pentane alone, assuming no loss of vapor and no restriction by the polymer to expansion? The theoretical densities, about 1.4 lb/ft^3 in bulk, are actually higher than those realized in practice, and densities as low as 0.50 lb/ft^3 have been attained by continued steaming. The explanation of this phenomenon is that air and steam permeate into the expanding beads and act as auxilliary blowing agents. The internal pressure of pentane offers no restriction to the incoming permeation of air and steam, for their permeation is affected only by the partial pressure of air and steam inside the cells.

Excellent experimental proof of this effect is shown in Fig. 5 [158]. The expansion at 100° C of a single bead impaled on a needle was observed. Expansion in air was permitted for 15 min to a density of about 3 lb/ft^3; then the atmosphere was changed completely. In the case of steam the

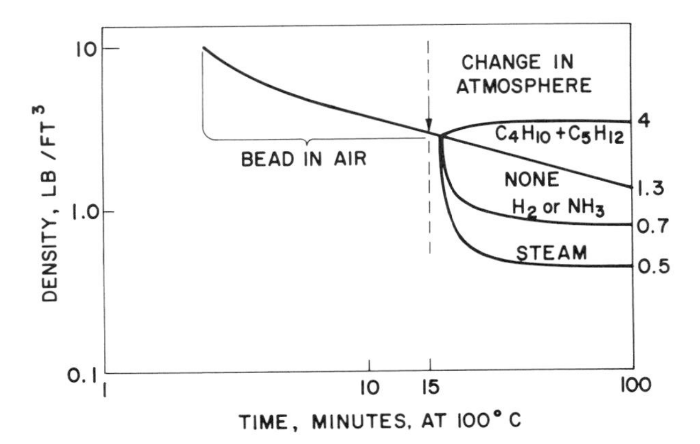

FIG. 5. Expandability versus atmosphere [158].

then the atmosphere was changed completely. In the case of steam the density dropped abruptly to below 1 lb/ft^3 and ultimately to 0.5 lb/ft^3 in 100 min. Without changing the atmosphere, the density reached only 1.3 lb/ft^3. When the atmosphere was changed to hydrogen or ammonia, the ultimate density was about 0.7 lb/ft^3, whereas a change to an atmosphere of a mixture of butane and pentane produced essentially no further expansion beyond 3 lb/ft^3. Thus the cell walls act as semipermeable membranes, and expansion continues until the pressure of the entering gas equals its external pressure.

It seems obvious that expandability should increase with an increase in blowing agent. However, Fig. 6 shows that the amount of n-pentane that can be utilized for expansion is limited. Since the 3-, 5-, and 7-min curves of foam density versus n-pentane content do not parallel each other, we can assume that excess pentane is being wasted by evaporation early in the expansion process. Also, at 7 min there is little difference in the density of foam generated from 6 to 8% of n-pentane. We see in Fig. 7 that a steep slope exists at 7% pentane in the weight-loss curve of exposed beads of either board- or cup-molding size. Also, the smaller cup-molding beads lose the blowing agent much faster than the bigger board-molding beads unless a pentane pressure is maintained in the container, as shown in the top curve. The curve at the left is especially interesting for theoretical reasons because it represents a true solution of pentane in liquid styrene dimer. This behavior of extremely rapid loss of pentane from the polystyrene oil suggests that in the solid polymer there are much stronger retentive forces in operation than those acting between two liquids.

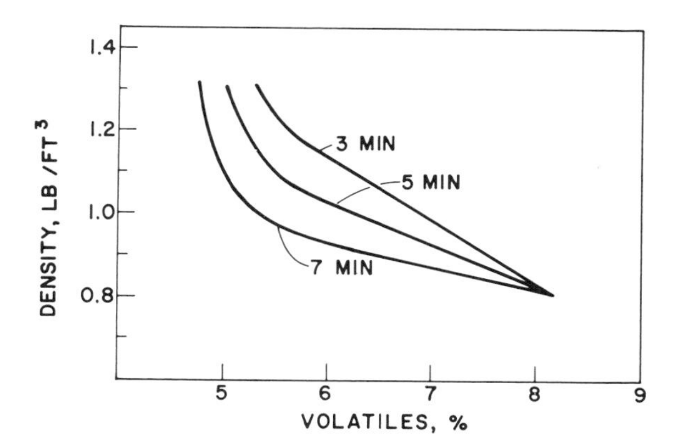

FIG. 6. Expandability versus volatiles. Reprinted from Ref. [84], p. 153, by courtesy of McGraw-Hill.

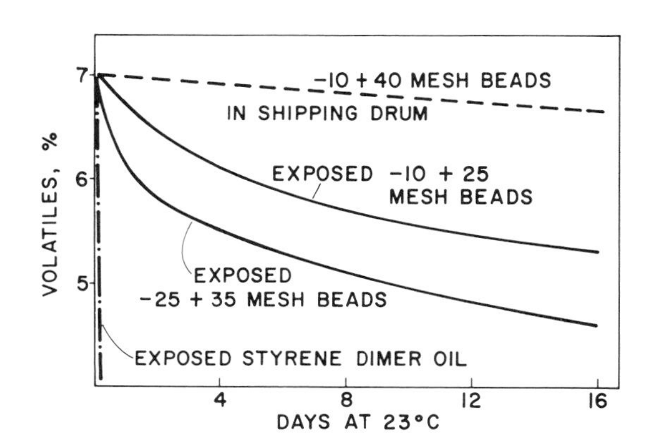

FIG. 7. n-Pentane loss from polystyrene. Reprinted from Ref. [84], p. 153, by courtesy of McGraw-Hill.

Figure 8 shows another reason why excess pentane should be avoided, in that objectionably large cells are obtained. The problem persists also in aged beads. Apparently the excess of pentane did irreparable harm to the cell structure. It seems that above a certain concentration the blowing agent can enter the polystyrene only by pushing the chains apart, thus

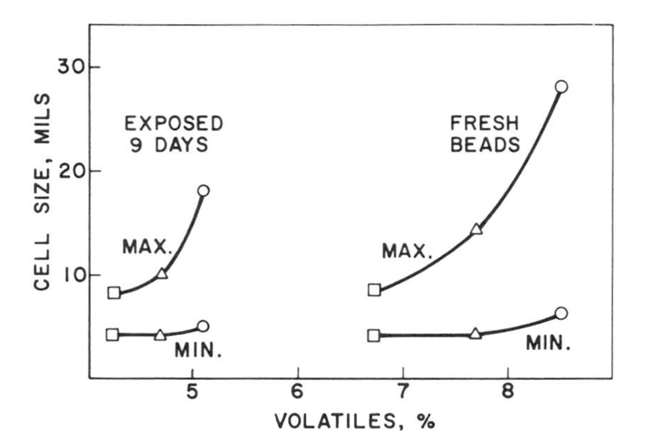

FIG. 8. Cell size versus volitiles and age. Reprinted from Ref. [84], p. 153, by courtesy of McGraw-Hill.

leading to the damage. If even more blowing agent is used, enough damage is done to form channels that permit the blowing agent to permeate rapidly from the polymer, thus lessening expansion.

If we place isopentane in polystyrene beads, we find, as seen in Fig. 9, that it is retained more strongly than n-pentane. This retention occurs in spite of the fact that isopentane is much more volatile, boiling at 28° C,

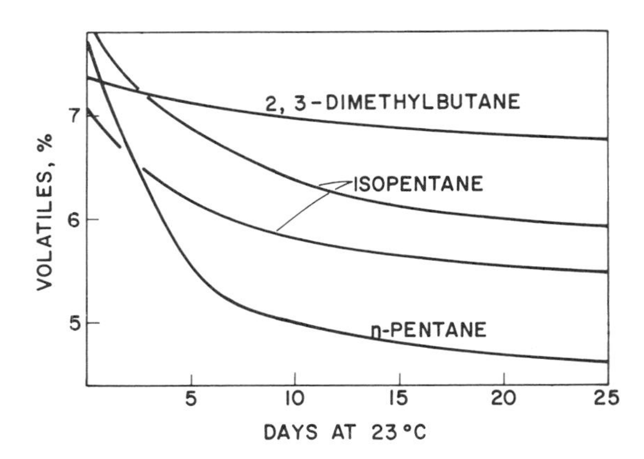

FIG. 9. Loss of 2, 3-dimethylbutane (bp 58° C), isopentane (bp 28° C), and n-pentane (bp 36° C) from polystyrene beads. Reprinted from Ref. [84], p. 154, by courtesy of McGraw-Hill.

whereas n-pentane boils at 36° C. Also, the cells of isopentane-expanded foams are smaller than those of n-pentane-expanded foams when the expanded beads are made in an identical manner. These observations are believed to be the result of the branched molecular structure of isopentane and the linear structure of n-pentane.

Let us consider how pentane is distributed among the chains of polystyrene. Does pentane occupy cages that already exist in a polystyrene matrix or does it push the chains apart? In other words, is polystyrene "packed" with pentane or is it swollen by pentane? Excellent support for the cage filling structure of expandable polystyrene has been published. According to Haward [162], who extrapolated to zero internal pressure the polystyrene-compressibility data of Breuer and Rehage, the free volume frozen into polystyrene at its glass-transition temperature is 13%, not 2.5% as had been taken for granted for all thermoplastics for many years. In Table 3 are compared three different methods of deriving the free volume in polystyrene.

TABLE 3

Free Volume and Pentane in Polystyrene[a]

Method	T (° C)	Free volume (%)	Maximum density (g/cm^3)	n-Pentane capacity (wt %)[b]
Compressibility [162]	100	13	1.18	6.5
Molecular structure [163]	100	17	1.22	8.5
Viscosity; ΔV versus T [164]	100	2.5	1.07	1.3
Expandable polystyrene	25	14[c]	1.22[d]	7.5[e]

[a]Reprinted from Ref. [157] by courtesy of Wayne State University.

[b]Estimated values unless otherwise indicated.

[c]Calculated by the following formula:

$$\frac{0.075}{d_{C_5H_{12}}} \times \left[(100)(d_{ps}) + (\%\,FV)(d_{C_5H_{12}})\right],$$ where d is density, FV is free volume, and the subscript "ps" stands for polystyrene.

[d]Calculated by the following formula:

$$\frac{d_{ps}}{1.00 - \%\,FV/100} = \frac{1.05}{1.00 - 0.14}$$

[e]Observed value.

A free-volume of 17% was obtained by Biltz [163], who calculated the closest packing of molecules, based on a study of liquids. The method generally accepted for thermoplastic polymers is the third one, originally proposed by Williams, Landel, and Ferry in 1955 [164]. This method, based on viscosity and the thermal coeeficient of expansion, gave a value of about 2.5% for a variety of cellulosics, rubbers, and methacrylates. If we calculate the amount of n-pentane that can be packed into polystyrene at the glass temperature, a free volume of 13%, according to Haward, gives a limit of 6.5% by weight (extrapolating the density of n-pentane to be 0.55 g/cm^3 at 100° C). The free volume of 17% from Biltz gives 8.5% n-pentane, whereas the classical value of 2.5% gives only 1.3% n-pentane. Actually the limiting amount of n-pentane in high-molecular-weight polystyrene is about 7.5%, since the presence of higher amounts leads to excessive evaporation rates, very soft polymer, and blistered, shrinking foam. If 7.5% n-pentane does occupy the limit of the free volume, then the authentic free volume should be at 14% at room temperature. This is in good agreement with Haward's calculation for the glass-transition point. There is supposed to be little or no change in free volume with temperature below the glass temperature, provided there is no crystallization.

Having seen the effect of changing blowing agent in polystyrene, three different polymers were investigated with n-pentane as a common blowing agent. In Fig. 10 are presented weight-loss curves for polyethylene, polystyrene, and a styrene-acrylonitrile copolymer. The polymers that

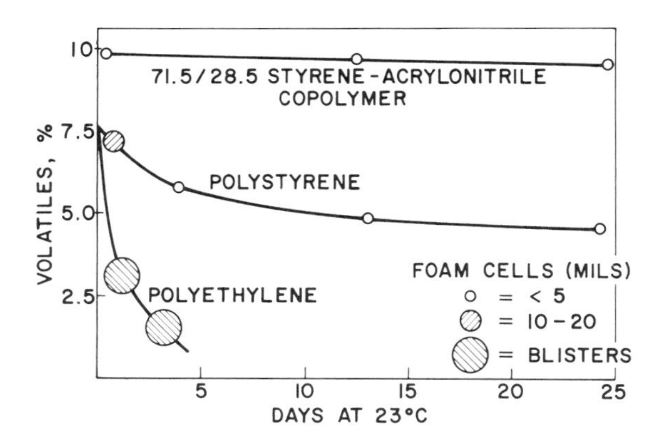

FIG. 10. n-Pentane loss versus polymer. Reprinted from Ref. [84], p. 154, by courtesy of McGraw-Hill.

exhibit the fastest weight loss also have the largest cell size in their foams. An interesting observation with polyethylene is the fact that it does not hold pentane any better than mineral oil does. This is not too surprising, since the molecules of polyethylene in the amorphous regions are free to move, whereas polystyrene the molecules are frozen up to about 100° C, the glass-transition temperature. In the copolymer the molecules are not only frozen but are also attracted by strong intermolecular polar forces between the acrylonitrile portions, thus accommodating much higher amounts of pentane without harming shelf life or cell structure.

Consider the various phenomena that occur on heating expandable beads. Figure 11 shows the thermal, volume, and weight changes observed [165] on heating Sinclair-Koppers Dylite F-40 expandable-polystyrene beads. The upper curve is the differential thermal analysis (DTA) curve, comparing the difference in temperature between 0.3-g samples of 40-mesh glass beads and expandable-polystyrene beads of similar size. Both samples are heated simultaneously in an aluminum block, the temperature of which was raised at a rate of 2° C/min. If no heat were absorbed or evolved by the expandable polystyrene, a straight line would result. However, if the curve dips down, heat is being absorbed and an endothermic process occurs. The upright curve is the expansion curve obtained on a hot-stage microscope, and the bottom curve is the thermogravimetric analysis (TGA), which is a weight-loss curve obtained by supporting the heated sample from the beam of an analytical balance.

There are three distinct endotherms (A, B, and C) in Fig. 11. At 50° C (A) pentane starts to evaporate at a noticeable rate, presumably

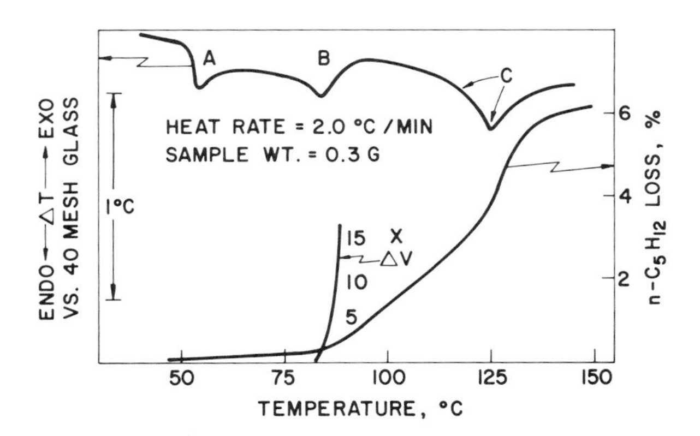

FIG. 11. Differential thermal analysis and thermogravimetric analysis of expandable polystrene. Reprinted from Ref. [165], p. 73, by courtesy of Technomic Publishing.

because the motion of some phenyl groups begins at this point [166, 167]. At 85° C (B) expansion starts, and the loss of pentane is accelerated. We believe B is the true glass-transition point of pentane-softened polystyrene. If the pentane is removed by evaporation, a value of 98° C is obtained. The starting polymer gives a value of 100° C. The final endotherm at C occurs over a range of temperature, 115 to 125° C, which happens to bracket the steam-molding temperature, 119 to 124° C, of the foam. The final absorption of heat may be caused by flow of polymer when the foam collapses.

Rising temperature affects expansion three ways:

1. The expansion rate for the blowing agent and the permeation rate for for external gas increase so that expansion is more rapid.

2. The polystyrene becomes softer, thus expanding faster but also becoming more sensitive to shrinkage and rupture of cell walls. (The latter is from too rapid an increase of pressure in the cells.)

3. The blowing agent is lost at an increasingly fast rate.

In practical terms this means that (a) below 100° C expansion is too sluggish because the polystyrene is too stiff; (b) above 120° C the polystyrene is too soft and the blowing agent evaporates too rapidly; (c) at 110 to 120° C the most efficient use of blowing agent and external steam is obtained, but sensitivity to shrinkage after expansion is high; and (d) the range 100 to 110° C is probably the best compromise.

Finally, the lower the external pressure the greater will be the push of the internal gases to expand the cells. At the same time, with lower external pressure the diffusion of gases into the expanding cells would be less. In the work of Skinner [158] the lowest pressure tested (135 mm Hg absolute) gave the greatest expansion at 110° C. However, these materials were most subject to shrinkage on restoring atmospheric pressure.

B. Examination of Cell Structure

1. Light Microscopy

A study by Ingram, Cobbs, and Couchot [168] of polystyrene foams was published in 1967. The ensuing section is based on this paper, which has been revised and duplicated in part by courtesy of the publisher, the American Society for Testing and Materials.

a. Molded or Extruded Boards. With foams of relatively large cell diameter (5 to 10 miles or more), suitable specimens for microscopic examination are cut with a razor blade, microtome, or automatic meat slicer. Such specimens are thin enough for light to pass without obstruction between most of the cell walls. As the cells become smaller than 5 mils in diameter, it becomes increasingly difficult to cut specimens thin enough to see between the cell walls yet thick enough to avoid rips and ridges. A method was therefore devised [168] to prepare specimens of 2-mil thickness. The method consists of mechanically advancing a cylindrical plug of foam through an open-ended cylinder so that slices by a razor blade, in a special holder, can be made against a machined-flat surface on the end of the cylinder. The foam slices, held between 1 x 3-in. glass microscope slides, are placed in a microscope attached to a camera, and photographs are taken at magnifications of X30 and X170. At a lower magnification, such as X5, which is desired for showing many foam particles, the specimen is placed in a photographic enlarger and projected onto either film or paper.

In this section we present a series of photomicrographs in order to compare the structures of several polystyrene foams with a density of approximately 1 lb/ft^3. In Fig. 12 (top) a 1.1-lb/ft^3-density foam (foam A) is first shown at X5 magnification. Several heterogeneities are noted in the composite structure. First, the foam particles, which are referred to as prepuff, vary somewhat among themselves in size, shape, cell structure, and density. The variation in particle size is dependent on the method of formation and the extent of classification. The variation in cell structure is dependent on the formulation, the amount of blowing agent, and the age of the raw material. In this specimen the larger particles, having a greater amount of blowing agent, exhibit the larger cells (10-30 mils), whereas most of the particles consist of cells whose diameters are in the range 4 to 8 mils. The shells of the prepuff appear to be denser where the particles have welded together. (In photomicrographs of other foams to be shown later this densification of the shell of each prepuff is much more pronounced.) At the surface of the foam the particles are squeezed by compression of the soft expanding mass, thus creating a higher density at the molded surface.

The same foam is shown at X30 magnification of a region of the smaller cells in order to examine more closely the particle-to-particle boundaries. Although the boundaries are discernible, they show very little densification of foam beyond that of a slight compression of one layer of the fused cells. When such a foam is torn, it exhibits a random pattern of rupture through bead boundaries as well as along them.

This foam specimen is also magnified to X170 to show the cell structure in detail. From this view one can appreciate the fact that air occupies 98.6% of the volume of a foam of 1-lb/ft^3 density. The photomicrograph shows cells with both top and bottom sheared away, suggesting a strut-and-window pattern. (We forewarn the reader that insufficient depth of focus in the optical microscope causes the appearance of struts. Subsequent electron-microscopy studies reported in this chapter show no heavy struts in such foams.)

In Fig. 12 (bottom) is shown, in X5 magnification, foam B, whose largest cells are about 12 mils in diameter and the smallest about 4 mils.

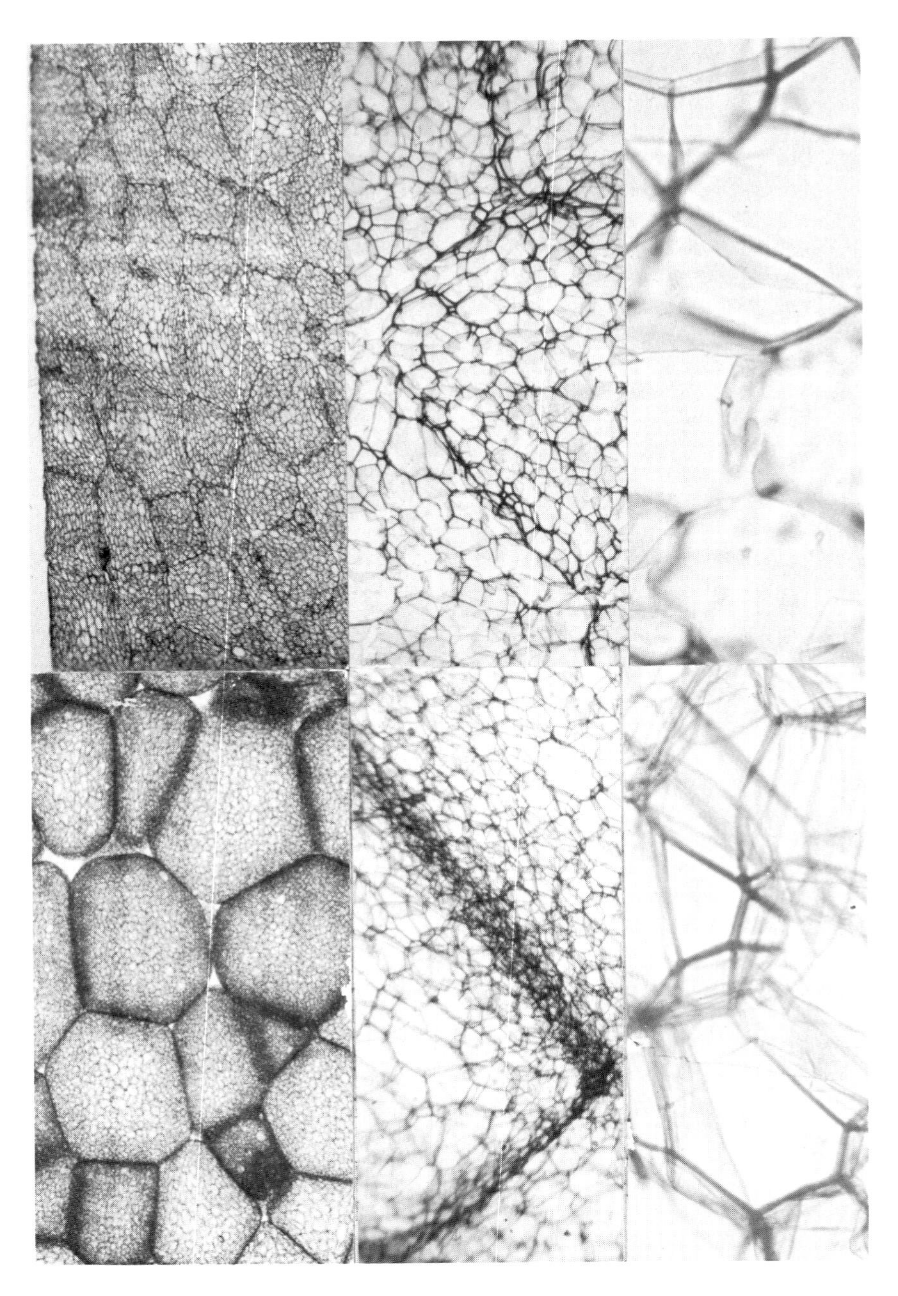

FIG. 12. Appearance of the internal cells of molded bead foams under the optical microscope. Top: foam A, density 1.1 lb/ft^3; bottom: foam B, density 1 lb/ft^3. Magnification: left, X5 (the molded surface is seen at top left); center, X30; right, X170. Reprinted from Ref. [168], p. 56, by courtesy of the American Society for Testing and Materials.

Although there is greater uniformity in cell structure than that in foam A, the particle boundaries, both fused and unfused to each other, are more apparent because of consistently small cells at the surface. This is also apparent in the boundaries shown in the X30 photomicrograph. The X170 view again shows the strut-and-window pattern.

In Fig. 13 (top) is shown the X5 view of a foam structure (foam C) that, for polystyrene, is exceptionally small and uniform, with cell diameters in the 2- to 4-mil range. However, other heterogeneities are apparent in this photomicrograph: first, an imperfect welding of several particles to each other, and second, an erratic occurrence of up to four voids of 10- to 25-mil diameter. These voids are believed to originate from water droplets. The X30 view depicts the extremely fine surface cells. In the X170 view the struts of this very-small-celled foam appear to be thinner than those of the large-celled foams of Fig. 12.

In Fig. 13 (bottom) is the X5 view of a cell structure (foam D) showing the heterogeneities that can occur when a certain raw material has not matured sufficiently before expanding. Whereas one particle appears to be a hollow sphere and some particles have very large cells (to 38 mils in diameter), others exhibit cells of 4- to 8-mil diameter. Many particles have shells of very fine cells. The shell effect is especially apparent in the X30 view, showing a shell of 1- to 2-mil cells covering a foam of about 8- to 17-mil cells. In the X170 view is a cell of about 12- to 16-mil dimension. The struts appear much thicker than those of the smaller celled foams A, B, and C.

Figure 14 (top) shows first the X5 magnification of a foam (foam E) with 2- to 4-mil cells in all particles, some of which have interiors with 6- to 8-mil cells, and some of which have no, one, or two voids in the 25- to 38-mil range. Of particular interest is the fact that the particle boundaries show very little densification or smaller cell structure -- an observation supported by the X30 magnification. The boundaries appear to be formed from the fusion of a single layer of 2- to 4-mil cells at the surface. Such foams exhibit a completely random pattern of rupture when torn.

Figure 14 (bottom) shows a material consisting of small-celled (2- to 6-mil diameter) foam "peppered" with voids of about 8 to 10 mils. Nearly every particle also contains a crater, apparently left by a bubble of escaping vapor as the particle was hardening during manufacture. (It is not possible in a cross section to show the crater in every particle.)

Figure 15 shows an extruded-board foam. The density of this product is 1.8 lb/ft^3, whereas the foams shown in Figs. 12, 13, and 14 were of 0.9- to 1.3-lb/ft^3 density. The cells are larger, 25 to 50 mils, than those of the bead foams, but because of the lack of particle boundaries, the extruded-foam cells present a uniform appearance. At the higher magnifications, X30 and X170, the strut-and-window patterns are apparent, but not as pronounced as in the bead foams.

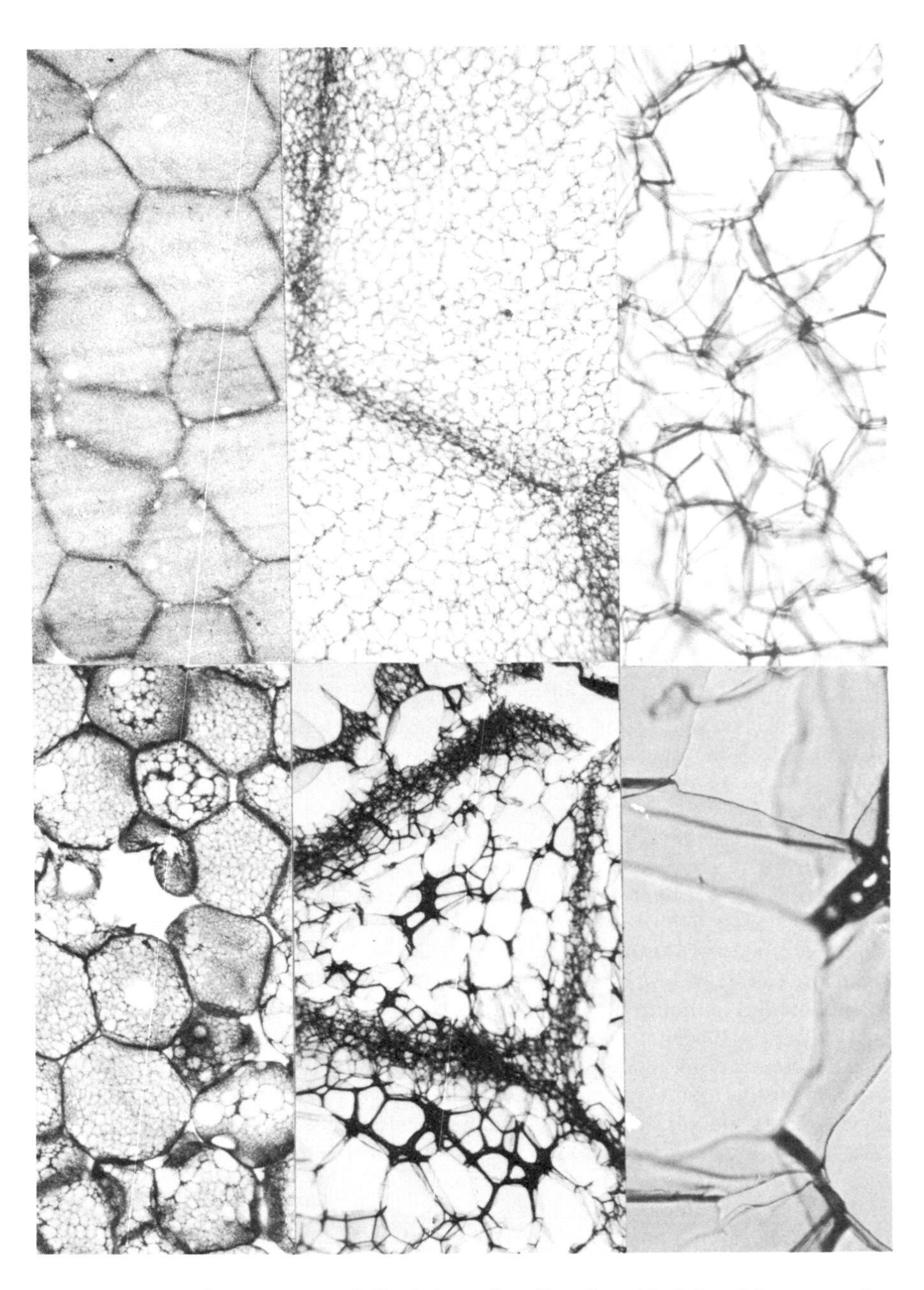

FIG. 13. Appearance of the internal cells of molded-bead foams under the optical microscope. Top: foam C, density 0.9 lb/ft^3; bottom, foam D, density 1 lb/ft^3. Magnification: left, X5; center, X30; right, X170. Reprinted from Ref. [168], p. 58, by courtesy of the American Society for Testing and Materials.

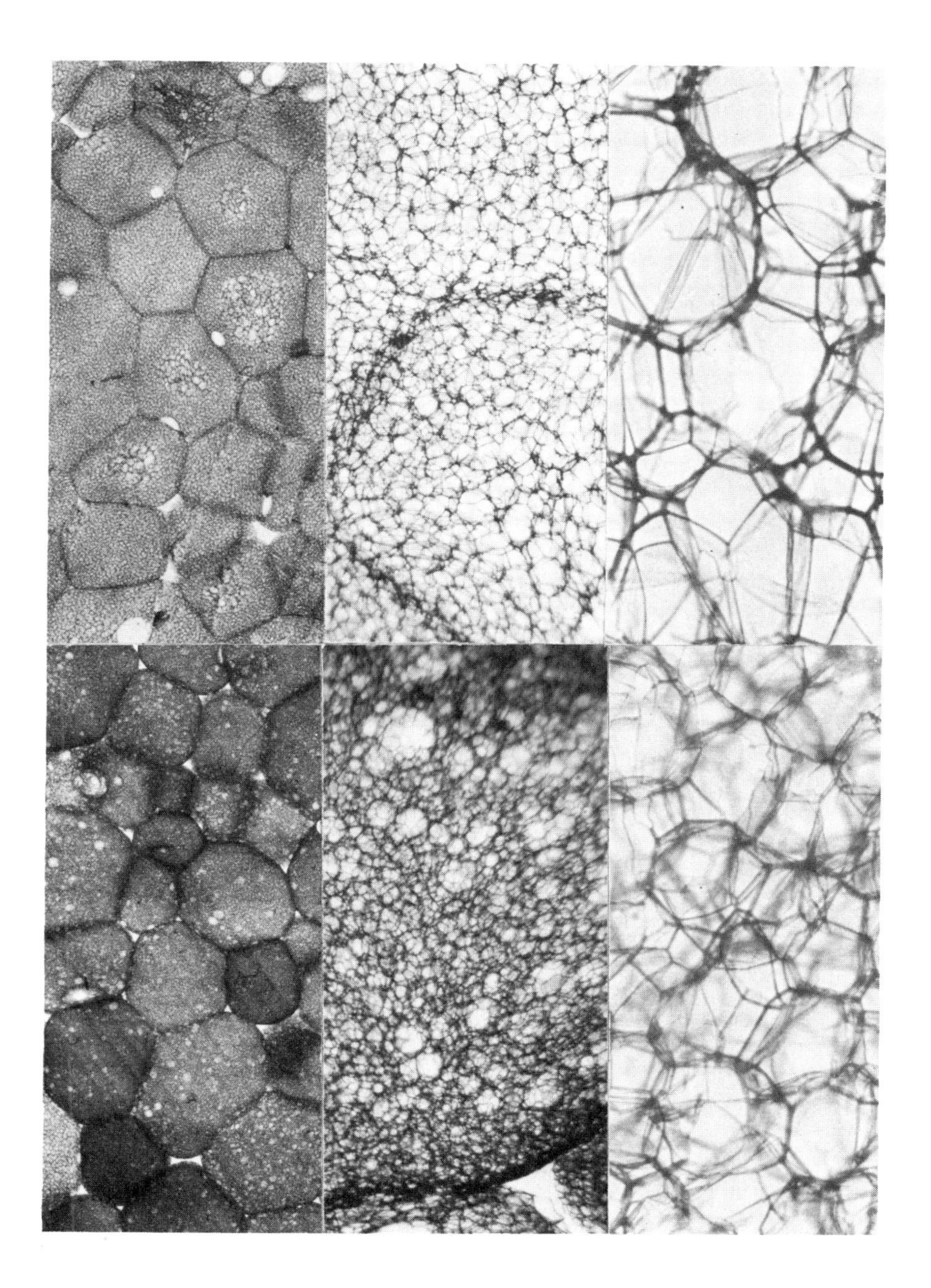

FIG. 14. Appearance of the internal cells of molded-bead foams under the optical microscope. Top: foam E, density 0.9 lb/ft^3; bottom, foam F, density 1 lb/ft^3. Magnification: left, X5; center, X30; right, X170. Reprinted from Ref. [168], p. 60, by courtesy of the American Society for Testing and Materials.

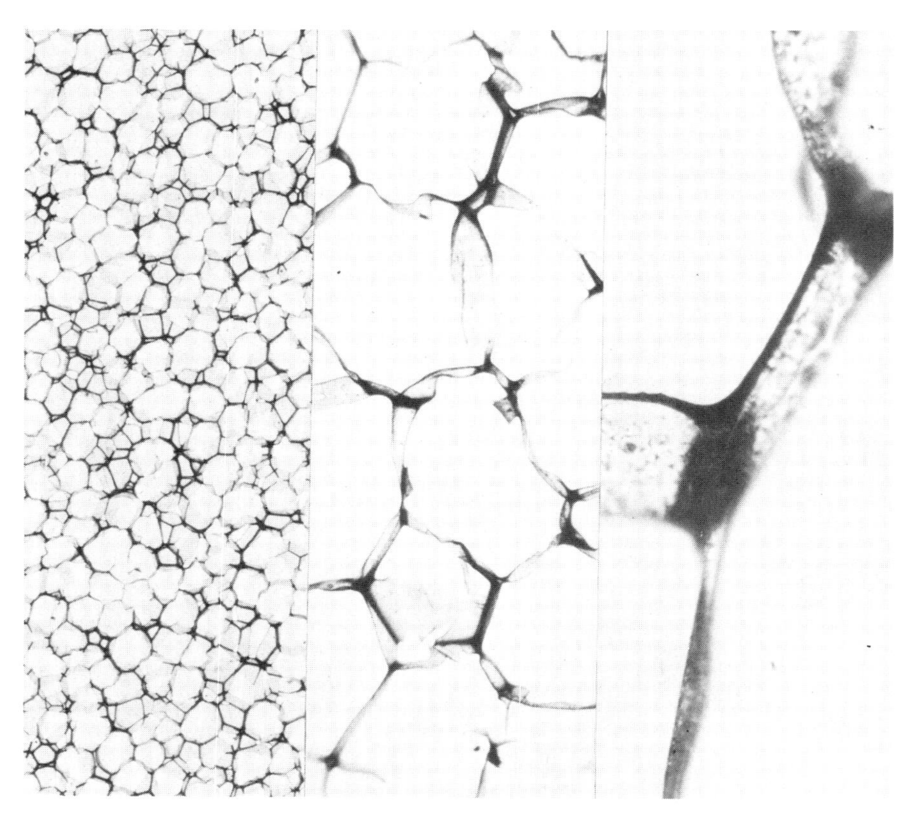

FIG. 15. Appearance of the internal cells of extruded-board foam of 1.8-lb/ft^3 density under the optical microscope. Magnification: left, X5; center, X30; right, X170. Reprinted from Ref. [168], p. 61, by courtesy of the American Society for Testing and Materials.

b. Extruded Sheet. Two types of extruded polystyrene sheet are considered: the so-called low-density and medium-density types of about 4 to 6 and 12 to 20 lb/ft^3, respectively. Both varieties are produced by extrusion of a hot expandable mass through a tubular die. Sheets of polystyrene foam are thermoformed into such consumer products as meat trays and egg cartons from the low-density foam, and beverage cups from the medium-density foam. The formation and application of extruded film and sheet are discussed in Section XIII.

In examining the cell structure of extruded foams it is important to observe the skin as well as the extent of flattening of the foam cells by the air pressure inside the tubular extrudate. To study this effect, one obtains a so-called cross-machine specimen from which to view the cut edge of a cylinder of extrudate, as well as an in-plane view by looking through the sheet. This view, perpendicular to that of the cross-machine specimen, permits one to determine the extent of orientation introduced by pulling the foam.

Specimens of extruded foams were cut with a razor blade until one of the desired size, uniformity, and thickness was obtained. For the in-plane specimens the skin of a small area, about 1 cm^2, was first cut away. The sections were mounted between two glass microscope slides, caution being taken not to crush the sections. Photographs were taken at two magnifications: X28 with a 35-mm photographic enlarger, using the mounted section as a negative, and X170 with a microscope.

Figures 16a and 16b depict X28 and X170 cross-machine views of a relatively fine-celled, satin-textured, nonbrittle sheet of 5-lb/ft^3 density and 80-mil thickness. Figure 16a shows both edges of the sheet, indicating no densification of one surface and a slight densification of the other. Figure 16b plainly shows the flattening effect of air pressure on the cells, since many of the "holes" are 3 mils wide and 9 to 10 mils long. The in-plane view (X28) of Fig. 16c shows very little distortion from the perpendicular forces of air compression or lateral force of windup tension. The windows are about 7 mils wide. Figure 16d shows a nearly complete cell, measuring about 12.5 mils at its greatest diameter.

Figures 16c and 16f show the cross-machine and in-plane views X28 of a coarse-celled, brittle polystyrene-foam sheet. Such coarse foam can result from a deficiency of a cell-nucleating agent or an excessively high extrusion temperature or both. In the cross-machine view the cells measure 11 to 14 mils in width and 18 to 25 mils in length, and in the in-plane view the wall-to-wall distance is 14 to 18 mils.

A foam sheet in the so-called medium-density range, in this case 20 lb/ft^3, and 105 mils thick is shown in the cross-machine view at X28 and X170 in Figs. 17a and 17b, respectively. Such foams are made with a lower quantity of blowing agent than the 5-lb/ft^3 foams shown in Fig. 16. At the surface of the foam is seen a dense skin, which was created deliberately to enhance the utility of the material for making cups that can be stacked in a vending machine. Figures 17c and 17d represent the interior of the same medium-density sheet. This foam has a "Swiss-cheese" structure, rather than the strut-and-window structure of all the other foams.

2. Scanning Electron Microscopy

Earlier electron-microscopy studies [168] of the surface and interior cell windows showed considerable difference between slow-cooling (i.e., slow-molding) and fast-cooling types of bead foam. Since the windows of the fast-cooling foams were much more heterogeneous, they were assumed to permit outgassing more rapidly. (The molding process is discussed in

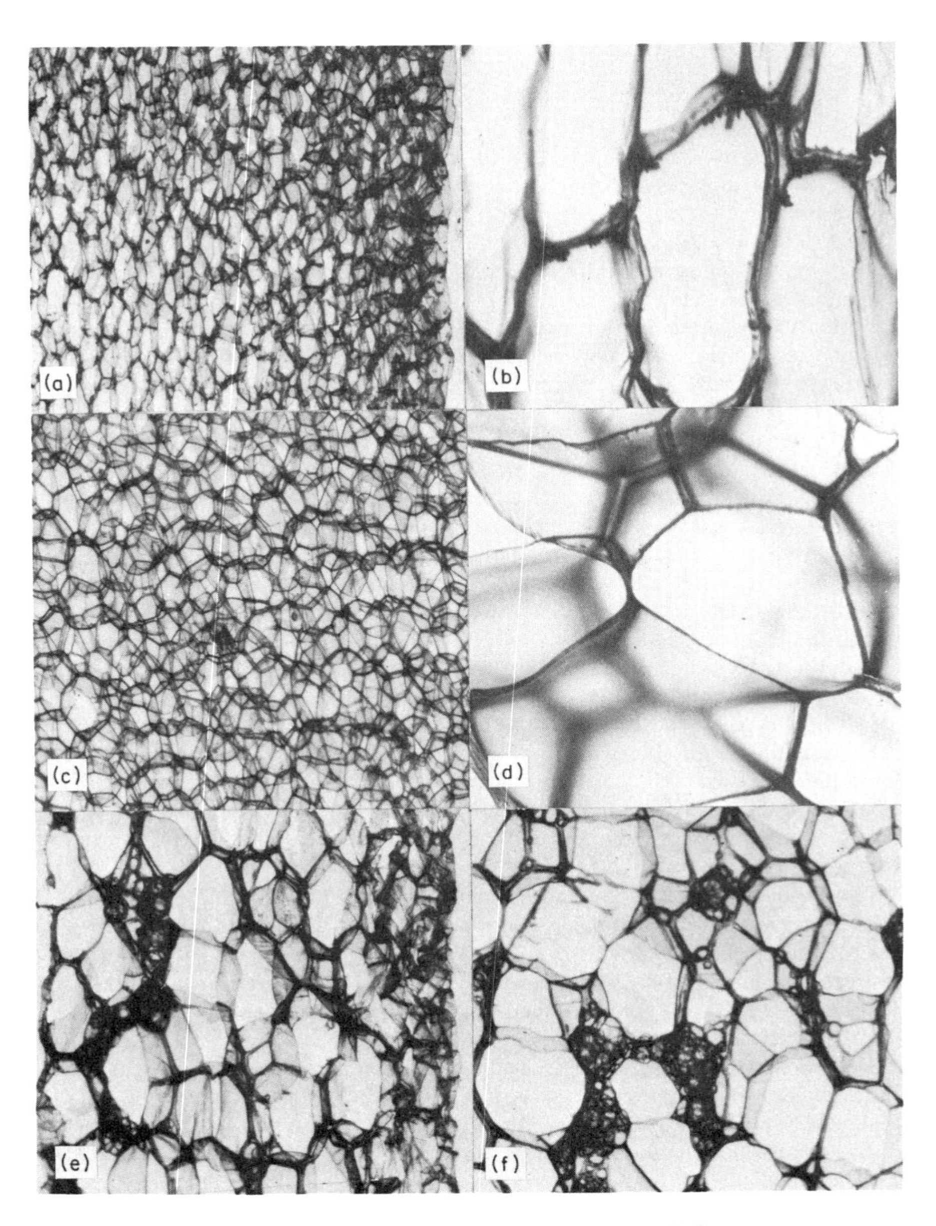

FIG. 16. Appearance of the internal cells of 5-lb/ft^3 extruded foam sheet under the optical microscope: (a) fine-celled, cross-machine view (X28); (b) fine-celled, cross-machine view (X170); (c) fine-celled, in-plane view (X28); (d) fine-celled, in-plane view (X170); (e) coarse-celled, cross-machine view (X28); (f) coarse-celled, in-plane view (X28). Reprinted from Ref. [168], p. 62, by courtesy of the American Society for Testing and Materials.

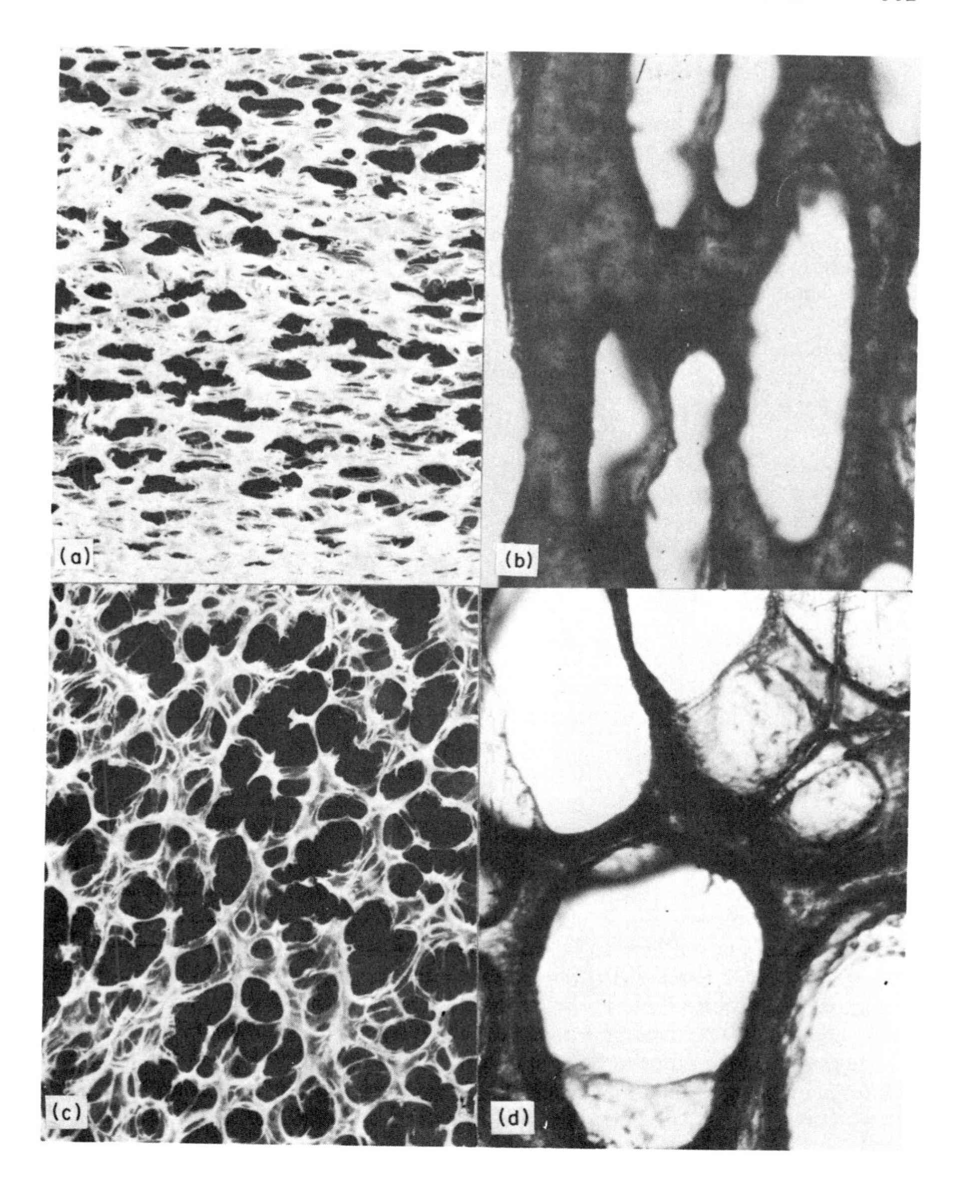

FIG. 17. Appearance of the internal cells of 20-lb/ft^3 extruded foam sheet under the optical microscope: (a) cross-machine view, X28; (b) cross-machine view, X170; (c) in-plane view, X28; (d) in-plane view, X170. Reprinted from Ref. [168], p. 63, by courtesy of the American Society for Testing and Materials.

detail in Section VI.) A recent investigation of similar foams was conducted with a scanning electron microscope by Angeloni and Kifer [169]. The scanning electron microscope has a great depth of focus, thus making specimen preparation much easier and giving photomicrographs with much better definition than heretofore possible. Figure 18 shows the interior cells of a slow-cooling type of preexpanded bead (Sinclair-Koppers Dylite F-40) at X75, X225, and X750. The windows are quite uniform. There is no buildup of polymer where the windows join. What had appeared as heavy struts under the optical microscope were actually folds or waves viewed edgewise. Scanning-electron-microscope views of the interior cells of a fast-cooling type of preexpanded bead (Sinclair-Koppers Dylite KFP-164) at X75, X225, and X750 are shown in Fig. 19. No difference in internal cell structure is apparent between these two materials.

The surfaces of both these foams, however, do exhibit differences between the two materials. Figure 20 shows the surface of the slow-cooling preexpanded beads at X75, X225, and X750. In each view the foam surface is transparent enough to show some details in the cell structure below. Figure 21 depicts X75 and X225 views of the surface of the fast-cooling material at about the same density, 1 lb/ft^3. In comparison with Fig. 20, we observe the following:

1. A much more opaque surface since the subsurface cell structure cannot be seen.

2. Much smaller and more numerous surface cells.

3. A pebble-grain effect in that the surface is depressed at the cell boundaries.

Examined next are the surfaces of foams molded from these preexpanded beads. The surface of the slow-cooling foam (Fig. 22) at X75, X225, and X750 appears transparent by scanning electron microscopy, permitting a view of some of the substructure. However, the surface of the fast-cooling foam in Fig. 23 not only obscures the substructure but is also very heterogeneous. The surface appears fibrous and wavy, which suggests that cells have collapsed.

The scanning electron microscope, because of the greater depth of focus, has given a better understanding of foam-cell structure. Earlier interpretations of thick struts and thin windows from optical microscopy were erroneous.

V. PREEXPANSION METHODS AND PREMOLDING TREATMENTS

A. One-Step Preexpansion

The most widely used method of preexpansion in the United States involves continuous preexpansion in a Rodman steam preexpander [170],

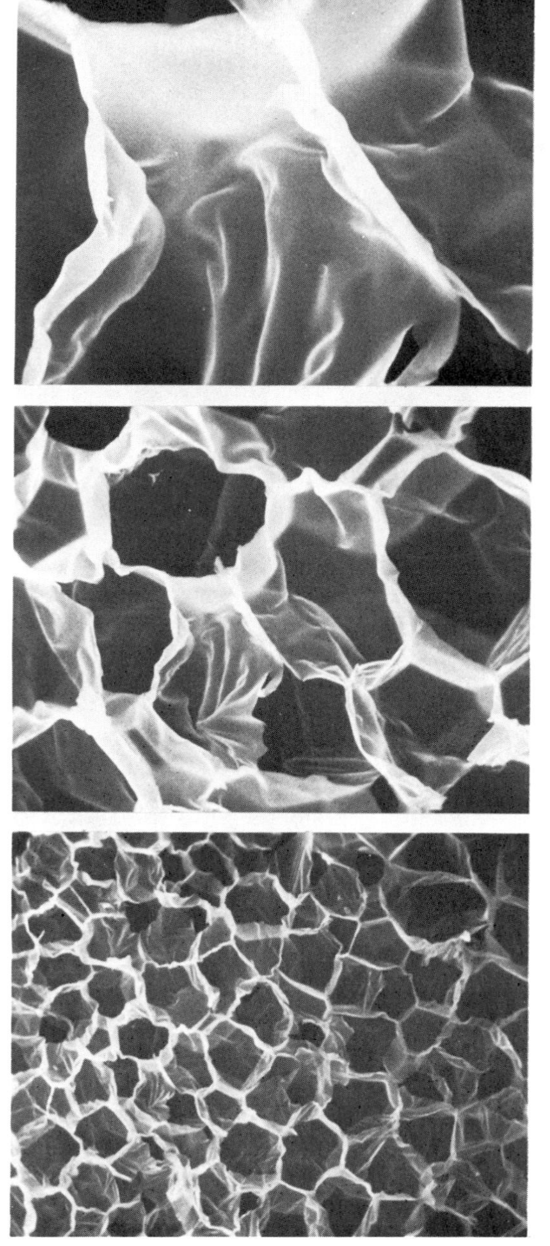

FIG. 18. Scanning-electron-microscope views of the internal cells of expanded beads of a slow-cooling type of foam: left, X75; center, X225; right, X750.

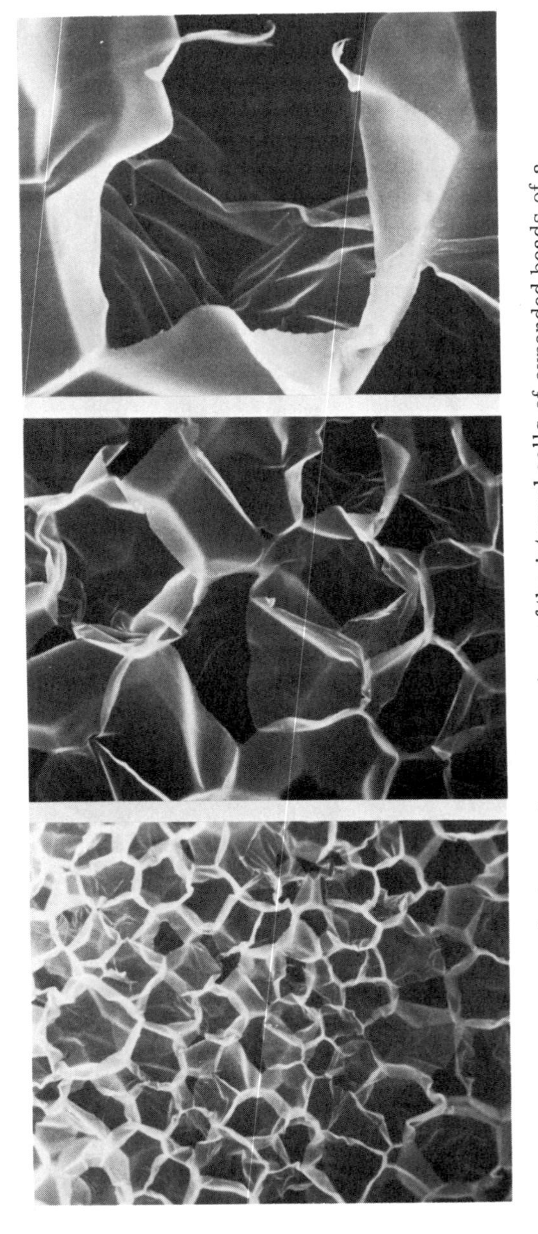

FIG. 19. Scanning-electron-microscope views of the internal cells of expanded beads of a fast-cooling type of foam: left, X75; center, X225; right, X750.

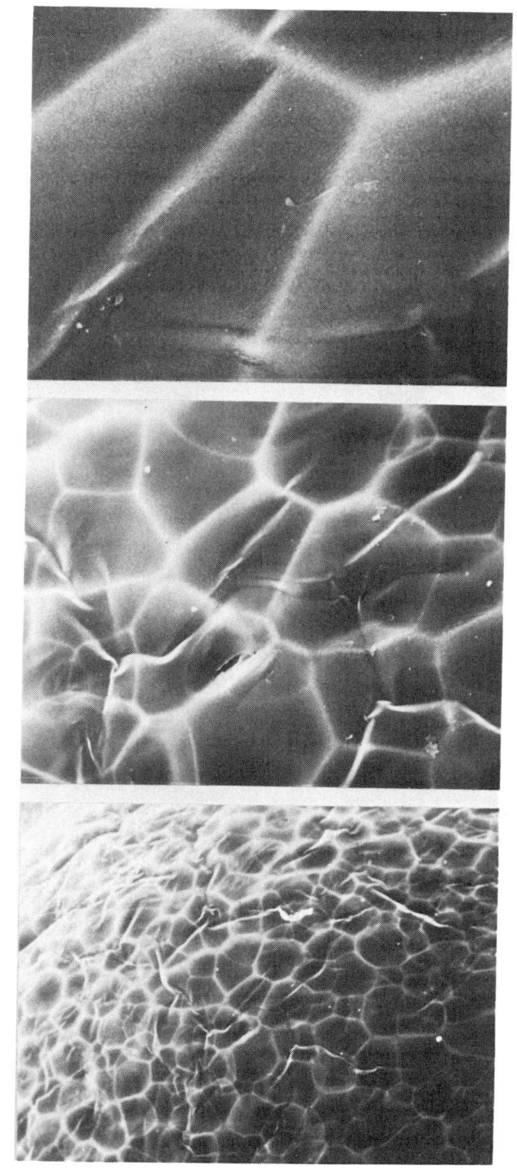

FIG. 20. Scanning-electron-microscope views of the external cells of expanded beads of a slow-cooling type of foam: left, X75; center, X225; right, X750.

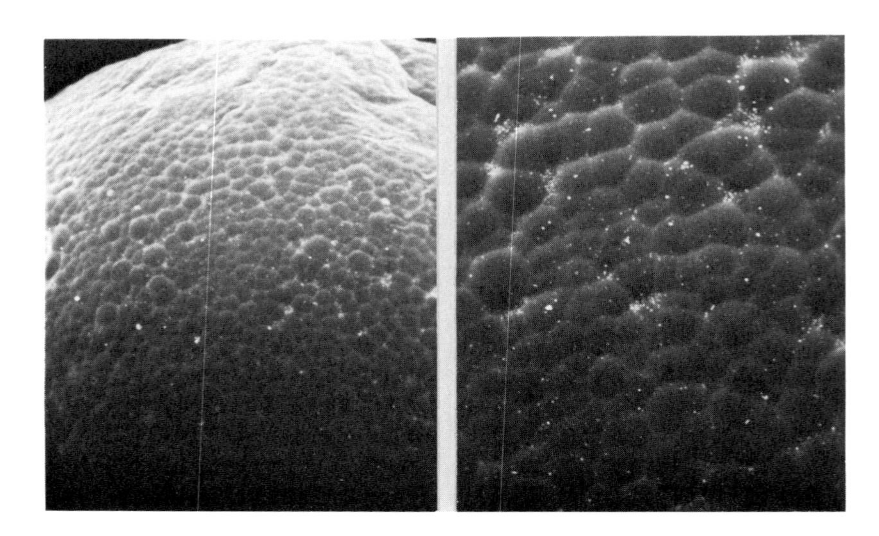

FIG. 21. Scanning-electron-microscope views of the external cells of expanded beads of a fast-cooling type of foam: left, X75; right, X225.

which is depicted in Figs. 24 and 25. The preexpander is a tank equipped with a series of alternating stirring blades and baffles. Steam, air if necessary, and expandable beads are fed in at the bottom; the steam acts through a venturi tube to draw in the beads. As they expand, the beads rise to the top and overflow to a storage chamber, usually with the assistance of a conveyor, such as an aiiveying system. A 200-gal-capacity preexpander expands from 200 to over 600 lb/hr of beads, depending on the material used and the density desired. With fast-expanding beads at least 500 lb/hr of 1-lb/ft^3-density foam can be produced. The density can be increased by raising the feed rate or by diluting the steam with air. The purpose of stirring is to prevent fusion of the expanding beads. However, some lumps of fused beads may form, and they can be removed by a screen in the conveying system. Continuous preexpansion will give a product of fairly uniform density, but the individual particles will vary in density because of differences in residence time. An auger or screw feed provides a more uniform product than a venturi feed. Couchman [171] described a modification whereby the outlet can be moved up or down to change the residence time, and thus density and uniformity. Variation in particles residence time decreases with increasing height of the outlet.

In Europe preexpansion by steam is commonly done in automatically operated stirred tanks fed batchwise from the top and discharged from the bottom. An advantage of this system is the ease of operation at slightly

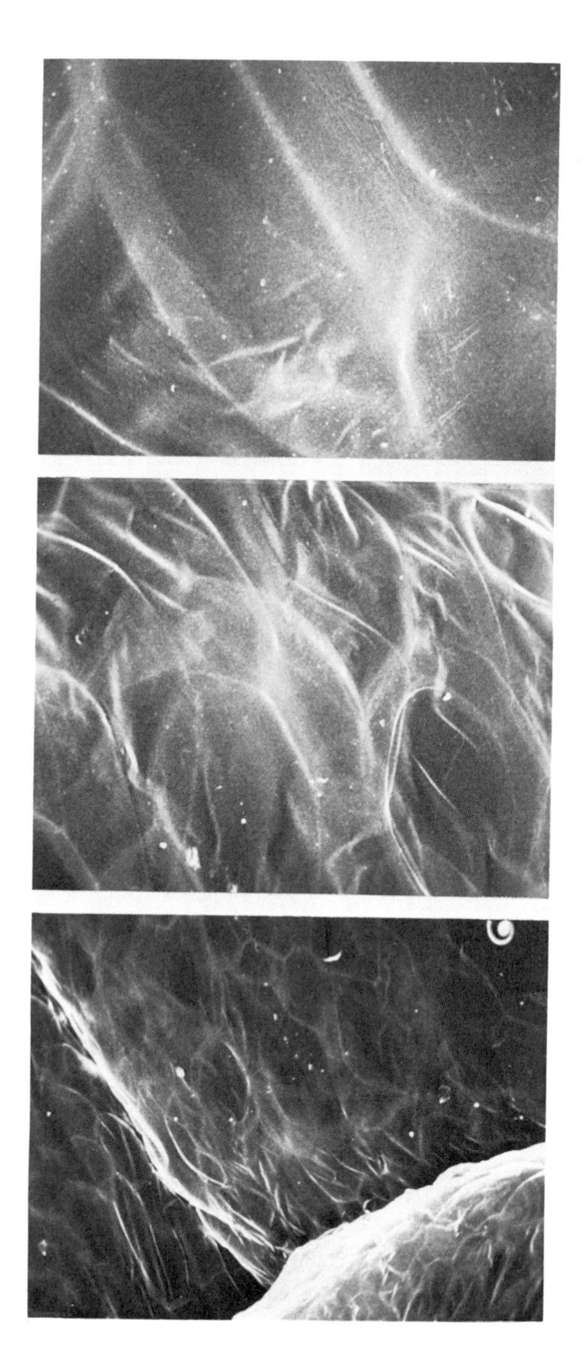

FIG. 22. Scanning-electron-microscope views of the molded skin of a slow cooling type of foam: left; X75 center, X225; right, X750.

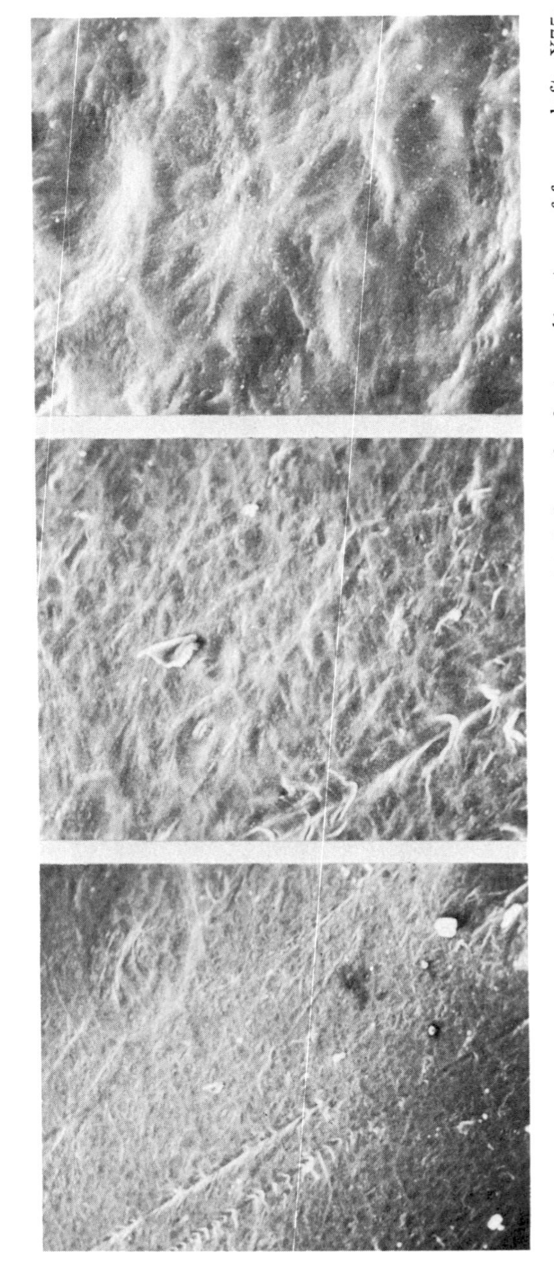

FIG. 23. Scanning-electron-microscope views of the molded skin of a fast-cooling type of foam: left, X75; center, X225; right, X750.

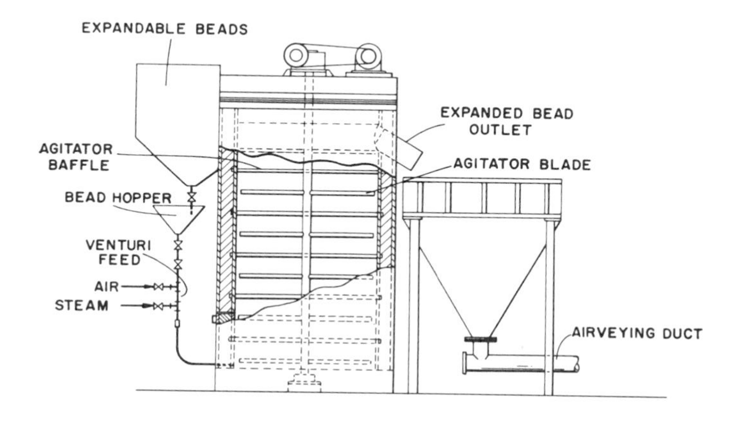

FIG. 24. Apparatus for continously expanding polystyrene beads in steam [170].

elevated pressures (to 2.5 lb/ft^2 gauge) to obtain foam of lower density, 0.8 to 0.95 lb/ft^3, for block molding.

Brockhues and Muhm [172] described a bath steam preexpander in which a cylindrical sieve rotates within a jacket as steam is fed to the space between the drum and jacket. Oxel [173] described a perforated moving belt through which steam is passed to expand beads on the belt. Another process [174] has the beads expanding while tumbling or free-falling in hot air or steam. In one case they tumble in an inclined rotating kiln, whereas in the other they fall through a steam or air tower. Fischer et al. [175] used a horizontal cylindrical batch preexpander with a piston for sealing the chamber and for discharging. Poron Insulation Ltd. [176] over-expands the beads, thus permitting most of the blowing agent to escape, so that subsequent molding is much quicker.

The Koppers Co. [177] has described a process for obtaining low densities by expanding with steam at 5 to 30 lb/in^2 gauge. Similarly by preexpanding at 15 lb/in^2 gauge and very slowly lowering the pressure to atmospheric, 0.3-lb/ft^3-density foam was obtained [178]. A continuous-pressure preexpander [179] involved an endless conveyor passing through a closed vessel supplied with pressurized steam. Denslow [180] described a similar apparatus designed particularly for expanding higher- softening materials, such as polychlorostyrene. Attanasio and Lambert [181] described a process wherein they heated the pellets under pressure high enough to prevent expansion. They then discharged the hot pellets directly

FIG. 25. Rodman preexpander. Reprinted by courtesy of Tri Manufacturing and Sales Co.

into an evacuated mold, providing expansion and molding in one step. Columba [182] described a process for preexpanding in a partial vacuum.

An early commercial method [183] involved heating the beads with infrared lamps as they were carried past on a belt. This method is expensive and does not lead to densities below 1.5 lb/ft^3. Several hot-air preexpanders have been described, but they also are of limited utility because of their explosion hazard and inefficient expansion rates in comparison with steam. Stanchel [184] used a Rodman-like preexpander that employed hot air instead of steam. Another steamless apparatus [185] used a rotating tiltable pan to hold the beads. The prepuff density was varied by changing the tilt of the pan, the speed of rotation, or the amount of heat supplied. A dry-heat batch preexpander [186] was provided with a trip mechanism that opened the stirred vessel when the expanded beads reached a predetermined height.

Boot and Anderson [187] coated expandable beads with an aromatic sulfonamide and then dielectrically heated them for expansion with radio-frequency waves.

B. Multistep Preexpansion

A number of preexpansion methods involve two or more steps, with an intermediate recovery period to permit air to diffuse into the cells so that greater expansion can be obtained in the next heating cycle [188]. Schuur [189] pressure-aged the beads between preexpansion cycles to hasten the diffusion of air. Compagnie de Saint-Gobain [190, 191] was issued patents describing two-step preexpanders. One process employed hot air (110-115°C) for the first step and steam at 115 to 117°C for the second. In another process they expanded with steam at atmospheric pressure, then after aging further expanded with steam at 8 lb/in^2 gauge. With these processes they obtained densities as low as 0.2 lb/ft^3. At Badische Anilin [192] foam densities as low as 0.12 lb/ft^3 were obtained by expanding with steam at atmospheric pressure (to 1.3 lb/ft^3) and then applying a vacuum while the foam was still hot. Dow [193] developed processes in which the foam particles (from atmospheric, vacuum, or pressure preexpansion) were partly deflated by rapid cooling. On aging, they reexpanded to the original low density. These reinflatable particles are proposed for insulation, filling cavities, and packaging since they are capable of completely filling a confined space.

C. Preexpansion of Strands

Polystyrene strands, filaments, and the like are expanded for use in loose-fill packaging. However, when expanded in conventional stirred-tank preexpanders, they become intermeshed and then will not flow. To overcome this problem Oswald and Kinsey [194] used a continuous porous belt that carried the strands from the loading station through a steaming station where the steam rose through the belt to form pieces of spaghetti like foam. A cylindrical screen deposited the strands individually on the belt so they would not fuse together. A Rodman-type continuous preexpander with special agitation has also been used [195]. A batch steam preexpander was described by Lowry [196].

In other processes extruded foam rod, after being cut into short pieces, was further expanded. Kienzle [197] coated pieces with magnesium stearate to prevent fusion and then steamed them at elevated pressure in a batch operation. Sare, Ropiequet, and Bergeron [198] sliced extruded rods into disks and then steamed them at atmospheric pressure to form saddle-shaped particles useful for packaging.

D. Maturing and Premolding Treatments

When the beads cool after preexpansion, the gases in the cells condense, creating a vacuum. If the foam is crushed at this time, it remains crushed and does not recover its shape. If the foam is crushed several hours after expansion, it is sufficiently resilient to recover from crushing. The gain in recovery is due to the process of maturing in which air permeates into the cells until the internal pressure approaches atmospheric pressure. If the foam is now heated, as in molding, it can expand further.

Miller's [199] studies on the maturing process showed that maturing temperatures of about 25° C were favored to maintain enough blowing agent for subsequent molding while permitting a good rate of air permeation. At higher maturing temperatures more blowing agent was lost; at lower temperatures air diffused into the expanded beads too slowly. As the density of the preexpanded beads was increased, maturing time decreased, so that beads of 2-lb/ft^3 density matured in only a few hours. Maturing time had little effect on mechanical properties. Miller's results are at variance with some of the patents cited in this section.

In commercial practice the expanded beads are conveyed (usually by an airveying system) to appropriate containers and permitted to age, usually overnight. Two of the most popular containers are bags and bins. The bags are usually of muslin, and the bins are made of screening material. In any case the container permits free passage of air. The use of open drums or other solid containers leads to uneven aging and resultant poor moldings.

The treatments that have been given to expanded beads before or after the maturing process are heating, pressure, vacuum, or crushing. Schuur [200] treated aged foamed beads at 2- to 6-atm air pressure followed by molding within the hour. The treatment retarded shrinkage during the molding of low-density material (less than 1 lb/ft^3). Treating freshly expanded beads with air pressure served to hasten the maturing process, so that the beads could be immediately molded [201].

When aged foamed beads were heated to 60° C before being charged to the mold, shorter cooling times for the final foam were needed [202]. A patent [203] described heating the aged foamed beads to 110 to 115° C before introducing to the mold, where final heat was applied with steam. This served to give less of a density gradient between the center and the outside of a foamed object. It would also serve to reduce the shrinkage of the final foam. Landon [204] superheated expanded beads before he shot them into a mold. This served to fill voids around an object and cocoon the object without heating the mold. Freshly expanded beads were dried with hot compressed air and then stored warm [205]. This led to a moisture-free foam of improved flexibility. Similarly Russell [206] obtained a low-density (0.4 lb/ft^3) styrene-acrylonitrile foam. The aging time of expanded beads was reduced from 2 h to 5 min by heating at 105° C in air [207]. Likewise, when the freshly expanded beads were conveyed in a heated atmosphere, they were dry and ready for immediate molding [208]. Bracht [209] permitted freshly expanded beads to cool slowly in a vacuum while simultaneously admitting air until atmospheric temperature and pressure were reached.

Fronko [210] disclosed that partly crushed freshly expanded beads would reexpand and give foams that were quicker cooling during molding. Chaumeton [211] reported that expanded beads could be compressed to 5 to 20 lb/ft^3 and then later reexpanded in hot water. They could be more readily shipped in the compressed state.

VI. MOLDING AND POSTTREATMENT OF FOAMS FROM EXPANDABLE POLYSTYRENE

The basis for the ensuing section is a lecture by Nelson and Immel [212]. The text has been revised, and parts of it are reprinted by courtesy of Wayne State University.

A. Molding of Preexpanded Polystyrene

1. Simultaneous Heating and Molding

a. Steam Injection through the Mold Cavity. The most significant property responsible for the remarkable commercial growth of expandable polystyrene is its ability to be steam-molded into such useful, lightweight, low-cost, closed-cell foams as beverage cups, packages, picnic chests, and ice buckets, as well as insulation board. The foam-molding method in current practice was disclosed originally in a patent by Stastny [213]. It involves charging a mold cavity with aged preexpanded particles (prepared by expansion in steam as described in the preceding section), injection of steam through perforations in the cavity walls until the particles expand to fill the empty spaces and fuse together, and cooling the molded foam until it can be removed without distortion.

The molding process can be defined in terms of temperature and pressure in the cavity versus time. Figure 26 depicts these changes for two different types of preexpanded beads molded into 4-in.-thick by 8-in. diameter foam cylinders of 1-lb/ft^3 density. The apparatus for obtaining these data is represented by the sketch in Fig. 27. The molding cycle consists of the following steps:

1. Closing the cavity to a slightly vented position.

2. Passing steam through the mold until heated.

3. With the aid of a vacuum, filling the mold with aged particles expanded to the desired density.

4. Into the closed mold, injection of steam for about 14 sec through perforated ports 0.5 in. apart on the face plates. The particles undergo a very rapid expansion of about 70%, expelling air and water through the

parting line of the mold, filling the voids, and welding together. Steam injection is stopped when the foam exerts the desired maximum pressure against the mold.

5. Cooling the foam by circulating water through the steam chests until the original charging pressure is reached.

As seen in Fig. 26, the foam-cooling time takes up most of the molding cycle. However, considerable differences exist between types of expandable polystyrene since a conventional (i. e., slow-cooling) foam required 4.5 min to cool, and a "fast-cooling" type required only 0.75 min. The scanning electron micrographs (Figs. 20-23) in Section IV. B. 2 show large differences between the external cells of slow- and fast-cooling foams. The surface cell windows of the slow-cooling foams appear relatively thick and glassy, whereas those of the fast-cooling foams are relatively thin and irregular.

Pogany [161] and Skinner and Eagleton [214] compared the internal and external temperatures of slow- and fast-cooling foams. The slow-cooling foam exhibited the expected effect that the interior was hotter than

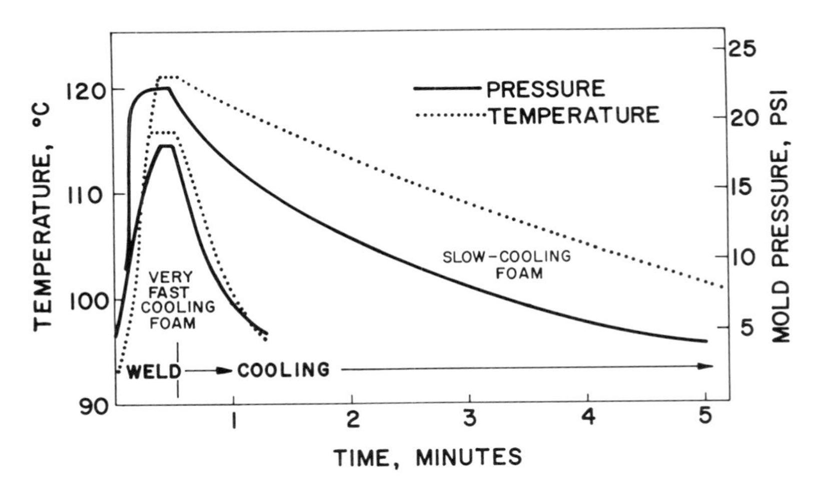

FIG. 26. Pressure and temperature changes in 4-in.-thick 1-lb/ft^3 foams during molding. Reprinted from Ref. [157] by courtesy of Wayne State University.

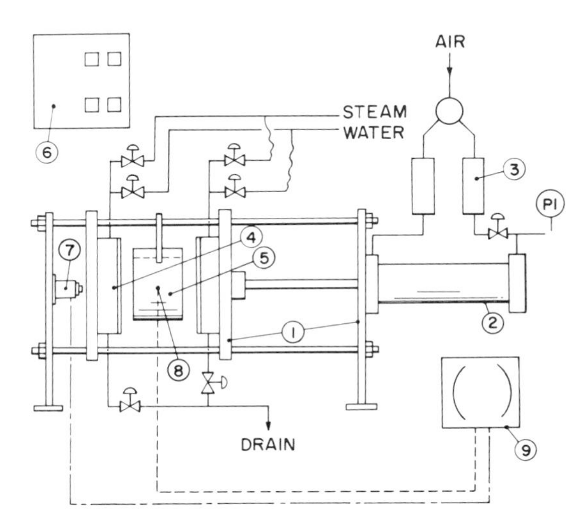

FIG. 27. Moldability test device. Key: 1, mounting frame with four tie bars and two movable steel platens; 2, hydraulic cylinder, 4 in. diameter; 3, oil reservoirs; 4, aluminum steam chests with perforated face plates and internal water-spray manifold; 5, 8-in.-diameter steam-jacketed ring mold (1-, 2-, 4-, and 6-in. thicknesses); 6, molding-cycle controller; 7, load cell; 8, thermocouple; 9, temperature-pressure recorder. Reprinted from Ref. [212] by courtesy of Wayne State University.

the exterior, assuming the removal of all heat by conduction to the cold mold. However, in fast-cooling foams the interior was actucally cooler than the surface. The quick-cooling effect is believed to be caused by diffusion of pressure-generating gases (steam, air, and pentane) out of the block, thereby lowering pressure and temperature.

The quick-cooling characteristic is imparted to bead foams either by incorporating certain additives [138, 152, 215-219] in expandable polystyrene, by quenching the hot bead slurry [220], or by treating the preexpanded particles, for example, by crushing [210] before molding or by injecting pentane [221] with steam into the mold. For any type of product the mold-cooling time of the foam decreases with increasing age of the preexpanded particles (after equilibration of air pressure) and with decreasing density and thickness of the foam.

The molding equipment varies with the application.

(1) Insulation Board. Polystyrene "bead board" is normally molded in the form of rectangular billets and then cut into boards of standard construction dimensions [222]. Billet faces may be as large as 4 x 16 ft and as much as 22 in. thick. A typical plant will have one or possibly two molds mounted on edge with the 4-ft dimension vertical. The top edge of the mold is usually hinged or otherwise removable to permit loading preexpanded beads into the mold by gravity from an overhead bin. The major faces of the mold cavity are steam chests from which steam is introduced into the cavity through perforations. The mold is usually constructed of aluminum or stainless steel with structural steel reinforcement.

The molding cycle for a billet of 1-lb/ft^3 density will range from 10 to 30 min, depending on product type and age of preexpanded particles. Large beads (through 10-on-20 mesh) of the fast-cooling types of both self-extinguishing and regular grades of expandable beads are commonly used to produce insulation board.

Molding is also done continuously, whereby preexpanded beads, fed to a channel of endless belts with perforations to admit steam, are fused and cooled while being restrained by the belts [223]. An advantage of the continuous machine is adjustability to making infinite lengths of boards as thin as 3/8 in. and as wide as 48 in. [224, 225],

Most insulation billets are reduced to standard construction dimensions, after a stabilization period of 3 to 4 weeks, with either hot-wire (nichrome, ca. 700° C) cutting equipment or power saws.

(2) Thin-Wall Containers. Thin-wall containers are molded commercially in self-contained, steam-chested, metal molds that employ a combination of steam injection and conduction heating systems. In nearly all cases the molding equipment is proprietary to the user [226-230]. Most thin-wall containers are molded at a density of 4.5 to 5.0 lb/ft^3, with a uniform wall thickness ranging from 0.085 to 0.110 in. Expandable beads of 25 to 45 mesh are employed. A typical molding cycle runs to 16 sec overall. Efficient grading of the preexpanded particles is absolutely essential for molding thin-wall containers. Therefore container molding plants have, in addition to the standard preexpansion and storage facilities, screening equipment for removing any oversize particles or small agglomerates that may come from the preexpander. Material losses from this screening operation are usually very low.

A wide variety of container shapes and sizes are available, from as small as 4 oz to as large as 16 oz. Stock and custom printing is also available.

(3) Custom and Proprietary Molding. This classification includes the molding of packaging parts, refrigeration and air-conditioning components, picnic coolers, ice buckets, and the like. Regular grade expandable polystyrene of 16 to 25 mesh is expanded to a density of 1.25 to 1.75 lb/ft^3. This type of molding is usually performed in cast aluminum molds mounted

in automatically operated holding presses. Steam chests surround the cavities. Automatic molding presses are available from numerous commercial sources in platen sizes ranging from 19 x 19 to 52 x 64 in. Utility valving, control timers, pressure regulators, and the like are generally an integral part of the molding press. Heating the expanded beads within the mold is accomplished by steam passing from the chest into the cavity wall. Multicavity molds are in common use.

(4) Typical Equipment Requirements. The sizes of the many expandable polystyrene molding plants in the United States vary considerably. A typical size plant for each of the three major end-use classifications is presented in Table 4 in terms of either pounds of product per year or number of units produced per year. The equipment requirements for plants of these sizes are listed, along with an estimate of utility requirements.

b. Steam Injection through Probes or Needles. When the foam shape is regular, as in the core of a building panel, steam can be injected through perforated tubes (i. e., probes) [231], which are extracted just before the completion of the molding process. Probe-molded billets can be produced at greater thicknesses and with more welding of the interior beads than by the conventional process of injecting steam through the wall of the mold. However, the probes, if not extracted soon enough, are difficult to remove from the block. This problem has been overcome by the use of hollow needles [232, 233] 1 to 2 mm in diameter.

c. Injection of Hot Gases. This so-called counterpressure process [234] consists of applying to the preexpanded beads in the mold an external pressure of hot nitrogen which is just equal to the expanding pressure the beads exert when they reach the fusion temperature. When the beads reach fusion temperature, the pressure is released, giving a block that is uniformly well fused throughout. This process has been applied to the molding of high-density (e.g., to 6 lb/ft^3) polystyrene foams as thick as 52 in.

d. Conduction of Heat through Molds. Where the cavity thickness is uniform and less than 0.25 in., the preexpanded beads can be fused by heat conducted into the cavity through the mold. Foam drinking cups were originally molded in this manner [235-238]. Conduction molding cycles can be shortened if sufficient water is introduced with the foam particles so that steam is generated inside the mold.

e. Dielectric Heating. High-frequency dielectric heating of preexpanded beads coated with an aqueous ionic layer in a nonconductive mold like plastic or cement permits the rapid fabrication of polystyrene-foam parts that are practically dry [239-242]. The minimum amount of steam and heat needed to fuse the beads is generated inside the mold. Savings in mold costs and steam-generating equipment are possible. Commercial applications are still under development [243, 244].

TABLE 4

Typical Equipment and Utility Requirements for Expandable-Polystyrene-Molding Plants

	Board	Custom	Cups
Raw-material usage, lb/month	100,000	60-100,000	180,000
Output per month	1,000,000 board ft	95,000 lb of product	30,000,000 6-oz cups
Plant area, ft^2	30,000	50,000	30,000
Utilities:			
Steam boiler, hp	100	450	500
Water, gal/min	60	500	640
Air compressor, hp	10	250	125
Electricity, A/phase	150	900	300
Equipment:			
Preexpanders:			
Number	1	2	2
Size, gal	200	55	55
Storage-bin capacity, ft^3	10,000	5000	1000
Screening equipment	No	Yes	Yes
Molds:			
Number	1	8	172
Size	48 x 144 x 20 in.	48 x 48 in.	6 oz
Presses:			
Number	--	8	40
Platen area	--	52 x 52 in.	--
Cavities	1	--	160
Finishing equipment	Hot wires and saws	Dryer	Printing and automatic packaging

f. Treatment of Preexpanded Particles Before Molding. Specific effects can be obtained by treatment of the expanded particles before molding. Examples are the following:

1. Mixing with aluminum slivers [245] or a dispersion of titanium dioxide in a polyvinyl acetate emulsion [246] to produce a dielectric lens for use as a radar reflector.

2. Coating of dichlorodifluoromethane-expanded particles with a 90:10 vinylidene chloride-vinyl chloride copolymer latex to produce foams with better thermal resistance [247].

3. A light coating of certain nonsolvent oils to reduce the mold-cooling time [248].

2. Molding of Preheated Foam Particles

a. Compressing. Expanded polystyrene beads of very low density (below 0.7 lb/ft^3) cannot be molded by steam-injection methods because the moldings are badly distorted. This problem has been overcome by (a) heating foam particles of 0.30- to 0.35-lb/ft^3 density (obtained by a two-stage expansion process) in air for a specified period (e.g., 60 sec at 117° C) and (b) compressing the particles to half their volume (i.e., to 0.6-0.7 lb/ft^3) [249]. Another method teaches the compression of fully expanded pentane-depleted beads (i.e., 20 min in steam at 115° C) by a piston in a hot-air atmosphere [250]. The expanded beads may be rapidly heated (e.g., by an infrared-energy source), so that only the surface melts, and then quickly compressed to form moldings with a high-density skin [251, 252].

b. Evacuating. If the preheated particles of low desnity (e.g., 0.7-0.8 lb/ft^3) are charged into a warm mold that is then quickly evacuated, they form well-fused dimensionally stable, dry foams at molding cycles substantially shorter than those of the conventional steam-injection process [253, 254].

B. Simultaneous Expansion and Molding

1. By External Heating

The direct molding of expandable beads without the intermediate pre-expansion step is unsatisfactory because of the excessive density gradient in the foams. Spherical beads, with a bulk density of 38 lb/ft^3, only lie on the bottom of the mold. To overcome this problem the quenched-pellet process (Section II. E) has been modified to extrude and quench expandable polystyrene in the shape of either a ribbon, which is then corrugated [255], or a knotless net [256]. Another approach involves the molding of 1.5-lb/ft^3 foam from 8-mil-diameter filaments of polystyrene (1 in.

long) that have been steeped in petroleum ether [257]. In each of these processes the expandable polymer is provided in a form sufficiently bulky to substantially fill the mold prior to heating.

Expandable beads can be molded to give hollow foams if the heated mold is rotated about two axes and steam is generated or injected into the mold. By the inclusion of powdered low-density polyethylene and sodium tetraborate hydrate (as a source of water) with expandable beads, a tough, durable skin of polyethylene is obtained [258].

2. By Internal Heating

The exothermic heat of polymerization can be employed as a source of heat for expanding polystyrene. The synthesized polymer intermixed with the expanded beads serves as a binder and gives a hard surface. Foams with densities of 3.3 to 4.5 lb/ft^3 have been made in this manner from mixtures of expandable-polystyrene beads with an alcoholic solution of amine-catalyzed epoxy resin [259, 260]. Similarly foams of 2.5-lb/ft^3 density were obtained by mixing expandable beads with persulfate-catalyzed aqueous solutions of acrylamide and methylenebismethacrylamide [261, 262]. The latter system is proposed for foam-in-place filling of vertical shafts in buildings and ships.

C. Postshaping

In addition to being cut by hot wires or saws into insulation board, polystyrene-foam billets are also transformed into ceiling tiles, packaging components, decorative items, and combustible patterns for metal castings [263]. From cylindrical bead board continuous sheets are cut with blades [264] by making a peripheral cut tangentially, as in the peeling of veneer from logs. These sheets can be postformed mechanically in the presence of steam [265]. Polystyrene foams can be shaped by internally heated gypsum molds [266].

D. Flexibilizing

Polystyrene foams are classified as rigid foams, but with decreasing density they feel more resilient, especially below about 0.70 lb/ft^3. However, foams of higher density can be rendered flexible by opening the mold while the hot foam still exerts pressure, thus allowing postexpansion [267], by keeping the molded foam well above room temperature before cutting [268], by poststeaming [269], by crushing [270, 271], or by a combination of heating and crushing [272-274]. These procedures are designed to convert some of the closed cells to open ones. Commercial activity in flexibilized foams has been mostly in peeled bead-form sheet for packaging, cushioning, and insulation applications.

E. Adhesion

Adhesives for polystyrene foam are limited to those that do not dissolve the foam. For foams that will not be immersed in water, a polyvinyl acetate emulsion adhesive is satisfactory. For construction purposes asphalt adhesives, as emulsion or hot melt, or hydraulic cement are applied. For optimum resistance to heat and moisture a reactive type of adhesive, such as an epoxy, is recommended.

F. Surface Coatings

1. Decoration

Latex- or alcohol-based paints are commonly used for polystyrene foams. Etched or roughened effects are obtained by applying emulsions of solvents [275]. Pigments, glitter, or flakes dispersed in a solvent give both roughened and decorative effects [276].

2. Protection

Thick protective, as well as decorative, coatings of reactive epoxy or polyurethane resins are applied to polystyrene foams. Styrene solutions of unsaturated polyester resins must be avoided because styrene is a strong solvent for the foam. However, an undercoat of polyurethane [277] or polyvinyl acetate permits a top coat of the styrene-polyester solution. Replacement of styrene by certain nonsolvent monomers [278, 279] permits the direct application of unsaturated-polyester-resin coatings.

G. Panels and Laminates

1. Structural Panels by Steam-Molding Expanded Beads Between Facings

Polystyrene foam has gained considerable acceptance as the core of structural sandwich panels. The commercial development of polystyrene-bead-foam panels was reviewed in 1965 by Sarchet [280]. These panels consist of foam intimately bonded to relatively thin faces and are designed to replace conventional exterior finish, structural framing, insulation, vapor barrier, and interior finish. Design criteria for structural panels were developed in a cooperative program between the Koppers Co. and the Rensselaer Polytechnic Institute in 1955 to 1960. In 1959 the Federal Housing Administration granted approval for the use of plywood- and asbestos-cement-faced polystyrene-foam-core panel walls as load-bearing members in residential construction.

The panels were produced by molding preexpanded polystyrene beads of 1.1-lb/ft^3 density between adhesive-coated facing sheets and reinforcing rails. Steam was introduced through perforated metal tubes, that is, by the probe-molding process [231] described in Section VI. A. 1. b. In addition to expanding and fusing the beads, the steam also caused the heat-activatable adhesive to set and join the facings to the rails and the foam. Aluminum foil or sheet served as vapor barrier. Special locks and caulking methods were developed for joining. Such panels have been marketed for the construction of homes, cold-storage warehouses, and dust-free "clean rooms."

2. Continuous Production of Laminates

Laminates are produced commercially by continuously adhering paper, plastic film, aluminum foil, or polystyrene-foam sheet to polystyrene bead foam as it emerges from a continuous board-molding machine. These laminates are used in insulation, packaging, and decorative applications where a more durable or attractive surface than that of polystyrene foam is required. Damp expanded beads can be continuously molded between and bonded to paper by the application of dielectric heating [281]. A packaging material can be made by sandwiching expandable beads between two films, at least one of which is metallic, so that subsequent induction heating in the wrapping causes the beads to expand [282].

VII. EXPANDED BEADS AS FUNCTIONAL COMPONENTS

A. Foamed Coatings

A number of patents describe the use of expanded beads or expandable polystyrene for coatings. Nagel [283] coated a heated substrate, such as cloth, paper, wire, metal foil, or plastic film, by passing it through a fluidized bed of expanded beads. Hackett and Santelli [284] coated with expanded beads by dropping them through a heating zone onto a rotating heated substrate, such as a bottle. The simultaneous spraying and foaming of a polystyrene melt through a heated spray gun was described in another patent [285].

Bracht [286] sprayed ammonia gas and a polystyrene solution (in ethyl chloride) onto a subtrate, giving a film of ammonia-impregnated polystyrene when the ethyl chloride evaporated. On steaming, the coating expanded. Another patent [287] described the foaming of a coating of expandable polystyrene applied from solution onto yarns. Ingram [288] described an apparatus for spraying expanded beads in the presence of mild solvents. The softened beads stuck to the substrate and to each other. A solvent-softened mixture of expanded beads and sand has been used for caulking or finishing masonry surfaces [289].

Expandable beads have been made to adhere to various substrates and subsequently heated to give a coating of foam [290, 291]. Expanded beads have been deposited onto adhesive-coated paper, foil, or plastic film to make a laminate useful for packaging [292]. Nielsen [293] devised a spray gun for applying an insulative wall coating from a mixture of expanded beads, hydraulic cement, and latex. Hamilton [294] used chopped, large expanded beads in wall and ceiling coatings for decorative and acoustic purposes.

B. Bonded Expanded Beads

Many different binder systems for expanded beads have been broadly claimed: organic, inorganic, thermoplastic, or thermosetting [295, 296]. In structures made from bonded expanded beads the foam particles are cemented, rather than welded together as in molded-bead foams.

Expanded beads bonded with a flexible acrylate latex were calendered to give a sheet with good sound-absorbing properties [297]. This sheet could also be laminated to paper or plastic film. Expanded beads coated with a rubber or polyvinyl acetate latex could be bonded with slight pressure [298]. Wehr and Guziak [299] used expanded beads bonded with a styrene-butadiene latex as an underlayer in a synthetic ski slope. A similar material [300] was bonded with bitumen, latex, and the like for sports fields and playgrounds. A process [301] designed for in-situ packaging and on-site insulating involves spraying a mixture of a latex- and hydraulic-cement-coated expanded beads. The cement absorbs the water from the latex so that both materials set.

Shaped articles have been made from expanded beads bonded with bitumen, wax, or resins that melted at a lower temperature than the expanded beads [302]. Expandable beads and a low-softening-point binder have been worked into a sheet that could subsequently be expanded and molded with heat [303]. Vogel [304] developed foamed insulating building components by combining wood chips, expandable beads, and an adhesive, then steam-molding to give the final material.

Several reactive-type thermosetting binders have been disclosed for binding expanded polystyrene-foam particles; that is, by molding a mixture of expanded beads and a phenol-formaldehyde resin [305], polyurethane precursors [306], or a filled epoxy resin [307]. Rose [308] described a lightweight splint involving a vinyl bag containing expanded beads and a thermosetting resin, such as an epoxy or a polyurethane. The injured limb was wrapped after the bag was kneaded to mix in a hardener. Also, a mixture of expandable beads and an unsaturated polyester system was heated rapidly to foam the beads and harden the resin [309]. A porous agglomerate container for growing and transporting plants has been developed from expanded beads and a urea-formaldehyde resin [310].

C. Composite Foams

In composite foams expanded beads are distributed in the foam of a different material. Several Badische Anilin patents [311-315] teach the on-site preparation of insulation from mixtures of expanded beads and expandable reactive liquid resins, particularly urea-formaldehyde. Companion foams of the phenolic and urethane type are also mentioned. Expanded mineral fillers can also be included with the expanded polystyrene [316]. Rubber foams with improved load-bearing characteristics have been prepared by mixing expanded beads with a frothed rubber latex [317].

A foam sheet was formed by impregnating paper with fine polystyrene particles, impregnating the intermediate material with a mixture of pentane and acetone, and then heating [318]. Similarly a paper that could be steam-expanded to board was prepared from a mixture of pulp, latex, and expandable beads [319].

D. Lightweight Filler for Hydraulic Cements

A number of patents have described the use of expanded polystyrene particles as a lightweight filler in hydraulic cements. Such compositions have contained portland cement [320-332], concrete [333-338], silicate cements [330, 339-341] or plaster of paris [322, 327-330, 342-344]. Although polystyrene is usually added as exapanded beads, it can also be incorporated as expandable beads [340-341] to be heated later, as ground scrap [333, 343] or as molded-bead foam in the form of disks [327]. Such additives as carboxymethyl cellulose, a suspending agent [329, 342], or expanded perlite [332] have been used. Added binders included glue [323], synthetic adhesives [323], organosiloxanols [320], epoxy resins [322, 325, 326], and natural latex [331]. The foam-cement compositions, besides being lightweight, were claimed to resist weather cracking because of their lower thermal shrinkage and expansion [335] and to have good heat- and sound-insulating properties [320, 326-328, 331, 336, 337, 344].

Foam-filled concretes and mortars can be used as the unfilled materials are used. In addition, a number of applications are specifically mentioned in patents. Such structural objects as blocks [323, 332, 333, 336, 337], building elements [321, 322], and roadbeds [322, 335] have been cast. Insulation [324, 327, 340, 343], construction panels [326, 328, 330, 336, 339, 340, 343, 344], and tiles [344] have been molded, often with pressure. Glass fibers were used to strengthen [330] or face [343] these panels. In a unique application [341] a container was made suitable for holding acetylene by filling with a mixture of expandable beads and calcium silicate cement and then heating to expand the beads, set the cement, and drive off excess water, leaving a porous mass in the container.

Polystyrene-foam particles have also been used to lower the density and improve the insulation value of frothed cements. Sefton [345-347] disclosed various froth-stable cement slurries for mixing with expanded beads to produce building panels of 15- to 35-lb/ft^3 density. Klein [348] used slow-speed mixing to incorporate expanded beads with frothed plaster, cement, or chalk. Thiessen [349] prepared an aerated concrete containing expandable beads. After the concrete set, it was heated, expanding the polystyrene to fill the air spaces.

E. Other Applications as Lightweight Fillers

Fernhof [350] and others [351-354] described the preparation of bricks containing tiny expanded beads. On baking, the polystyrene burned out to leave lightweight refractory bricks. Dimensionally accurate refractory castings and riser sleeves used in metal molding were prepared by the same technique [355]. Correcta Werke [356] claimed that foundry sand made more uniform moldings with a smoother surface when expanded beads were added.

Stastny et al. [357] used expanded sulfonated polystyrene beads as a soil conditioner. Similarly foamed polystyrene was used to loosen soil [358].

An X-ray contrast medium containing expanded beads for use in the gastrointestinal tract has been described [359].

Cratsa [360] and others [361] described floating soap containing expanded beads.

Expanded beads were used to cover an oil surface to protect the oil from oxidation [362].

VIII. EXTRUDED FILM AND SHEET

In the late 1960s polystyrene-foam sheet emerged as a major commercial product with a rapid rate of growth. Sales increased from 15 million lb in 1965 to 70 million lb in 1969 [1]. The products most responsible for this growth were trays for meat, fruit, and vegetables.

The technical development of foam egg cartons did not progress as rapidly as that of meat trays. However, consumer reactions have been very favorable, and the optimistic forecasts for polystyrene sheet in this market appeared to be justified [363].

Polystyrene-foam trays are usually produced in the following steps:

1. Simultaneously extruding and expanding through a circular die a mixture of polystyrene, blowing agent, and cell-nucleating agent.

2. Pulling the sheet over a sizing mandrel.

3. Slitting and winding the sheet.

4. Postheating and vacuum-forming pieces of sheet, then trimming.

The early development of extruded polystyrene sheet was done by extruding mixtures of cell-nucleating agents and expandable polystyrene in the form of beads or pellets. Pellets were preferred at that time, because extruders were not designed to handle beads conveniently. In order to produce sheet at the lowest possible cost, extruders were later developed for introduction of the blowing agent into the barrel, thereby eliminating the more expensive expandable pellets as a raw material.

A. Feed to the Extruder

1. Polymer

Unplasticized, high-molecular-weight, crystal-grade polystyrene pellets are commonly used as feedstock for sheet. Beads cost less than pellets, but they require a forced-feeding device or compactor to ensure uniform and consistent delivery to single-screw extruders. Shell Research Ltd. reported [364] that a twin-screw extruder (Windsor RC 100) can be adapted to feeding expandable beads by the installation of slightly modified screws. Polystyrene has been made in suspension to form disks rather than beads [365]. These disks, with or without encapsulated blowing agent, fed well to conventional extruders.

2. Cell-Nucleating Agent

Extrusion of a mixture of polystyrene and pentane in the absence of a cell-nucleating agent produces a coarse, brittle, blistered extrudate of very high density (40 lb/ft^3). A practical method for converting expandable polystyrene beads to a fine-celled (i. e., cells with diameters of less than 10 mils) extruded foam was demonstrated by Houston and Tress [366] in 1954. Mixtures of powdered sodium bicarbonate and a stoichiometric amount of powdered organic acids with beads containing 6. 1% of petroleum ether (pentane fraction) were extruded into rods of foam with a density of about 4 lb/ft^3. The "in situ" generation of water and carbon dioxide was believed to form sites for pentane to collect in and then expand. Particularly suitable combinations were 0. 2 to 0. 3% each of sodium bicarbonate and citric acid hydrate. A refinement [367] of this system involves encapsulation of 0. 3% citric acid in expandable pellets (made by the quenched-pellet process described in Section II. E) and a later addition of 0. 1 to 0. 3% of sodium bicarbonate prior to extrusion. A disadvantage of the above system is the need to add fresh nucleating agent when scrap sheet is re-extruded. This difficulty was overcome by using 1 to 3% of a 2:1 styrene-maleic anhydride copolymer as cell-nucleating agent [368]. This material presumably performs as a nucleating agent by crystallizing exothermally as the melt leaves the extruder. Although the external cells of sheet made by this method are very fine, the internal cells are relatively large.

In addition to several modifications of the mixture of sodium bicarbonate and citric acid, other nucleating systems reported for extruding foam sheet are water with wetting agents [369], nitrogen gas [370], aluminum silicate [371], water from hydrated salts [372], amino acids [373], ammonium citrate [374], acid phosphate baking powders [375], potassium metasilicate [376], calcium silicate [377], mixtures of nitrogen-releasing blowing agents and calcium oxide [378], mixtures of magnesium silicate and silica [379], fluorocarbon polymers [380], and carbon dioxide [381].

3. Blowing Agent

Most polystyrene film and sheet has been produced with pentanes (predominantly n-pentane) as blowing agents, but since 1967 the use of fluorocarbon 11 (trichlorofluoromethane) and fluorocarbon 12 (dichlorodifluoromethane) has become more common [382]. The advantages of pentane (i.e., as n-pentane, isopentane, or petroleum ether or mixtures thereof) are their low cost (about 3¢ per lb versus 19-21 ¢ per lb for fluorocarbons) and proved performance. The advantages of the fluorocarbons are their lack of flammability and the capability (particularly of fluorocarbon 12) to expand polystyrene to foam of lower density. Mixtures of fluorocarbon 12 and pentanes are reported to give good-quality sheet [383].

As already mentioned, pentane is introduced either as a component of the polymer charge or under pressure into the molten polymer in the barrel of the extruder. (Fluorocarbon 12 is pressured into the barrel.)

Inorganic pigments of high surface area have been proposed as carriers for volatile hydrocarbons into the extruder: for example, hexane on silica [384] and pentane on diatomaceous earth [385]. Pentane or hexane can also be added in dry capsules (8-35 mesh) with breakable walls of a film-forming material like polyvinyl alcohol [386].

Polystyrene-foam sheet has been produced commercially with a 4:1 mixture of propane and tetrachloroethylene as blowing agent and citric acid and sodium bicarbonate as nucleating agent [387]. With titanium dioxide as nucleating agent, nitrogen gas in addition to pentane (3%) has been employed for making foam film [388].

B. Production of Film or Sheet

The ensuing discussion of the conversion of expandable-polystyrene pellets to sheet is based in part on a lecture by Martens [389, 390] of the Sinclair-Koppers Co. at Wayne State University in 1967.

1. Extruder

a. For Expandable-Polystyrene Pellets. Pentane is a plasticizer for polystyrene and thus permits the use of a relatively low extrusion temperature with moderate working of the material. Whereas unplasticized-polystyrene melt temperatures in an extruder are 175 to 190° C, those of expandable ploystyrene are as low as 115°C.

The earliest attempts to produce foam sheet were made with polyethylene blown-film equipment. However, to produce sheet of quality sufficient for thermoforming, many refinements were required in the extruder, dies, and take-off systems.

Extruders for foam range in size from 2.5 to 6.0 in. in diameter, with 4.5 in. being the most common. These have a minimum length-to-diameter ratio of 20:1, with 24:1 or 28:1 providing even better control. The extruder barrel should be electrically heated and water-cooled in at least five zones on a 24:1, and the minimum temperature control provided should be through proportioning-type instruments. Since very small temperature changes seriously affect gauge uniformity, stepless-type instruments on the die provide even better control. The extruder barrel should also be equipped with at least four points for pressure measurement. These are not normally included in a standard extruder; but, with foam, pressure indication provides an important day-to-day guide to extrusion uniformity and consistency.

The single factor that contributes most to successful extrusion is extruder-screw design. During Sinclair-Koppers' early attempts to design a screw that would be capable of producing quality foam sheet at optimum rates on a 2.5-in. extruder several serious problems were encountered. Probably the most significant was the occurrence of areas of unfoamed polystyrene sheet. At low extrusion rates a quality sheet could be produced, but at higher rates the sheet contained unfoamed blotches, and in all cases the unfoamed areas were accompanied by a loss of pressure in the rear extruder zones. Another serious problem was a sine-wave type of flow pattern in the sheet, corresponding to the revolution of the screw. This pattern was so serious in some instances that it affected the gauge uniformity. The screw pattern could be minimized somewhat through the use of a screen pack, but this caused excessive overheating of the melt. After numerous design changes were made, an extruder screw was developed that eliminated all of the early problems and was capable of producing quality foam-polystyrene sheet at optimum extrusion rates. It had an 0.5-in.-deep feed section of 4 flights, a short transition section of 3 flights, and a long deep metering section of 17 flights with a depth of 0.210 in. One significant change in this design is that of the nose or screw end. The conventional breaker plate and blunt screw end were eliminated and replaced with a tapered tip that extended into the die adapter. Best results were obtained by matching the angle of the screw tip to that of the die adapter by maintaining the clearance between the tip and the adapter equal to the metering depth of the screw. This new screw design eliminated the sine-wave pattern in the sheet and allowed the extrusion rate to be increased by about 25%. The major factor that now limits the extrusion rate attainable with this screw design is that of cooling capacity of the extruder.

Typical processing conditions for foam-polystyrene extrusion are shown in Table 5. These were established on an extruder of 2.5-in. diameter using the screw design described here.

TABLE 5

Typical Extruder Conditions for Expanded-Polystyrene Sheet

Extruder:	
Diameter, in.	2.5
Length-to-diameter ratio	24:1
Screw	17-flight metering
Die, bottom feed, in.	5
Die gap, in.	0.020
Sizing mandrel, in.	13
Feed	Expandable-polystyrene pellets
Barrel temperature, °F:	
Zone 1	220
Zone 2	265
Zone 3	210
Zone 4	200
Zone 5	200
Adapter temperature, °F	210
Die temperature, °F:	
Zone 1	220
Zone 2	225
Stock temperature, °F	225
Barrel pressure, psi:	
Zone 1	100
Zone 2	500
Zone 3	1000
Zone 4	1300
Zone 5	2200
Zone 6	3200
Zone 7	2000

TABLE 5 (continued)

Extruder drive:	
Screw, rpm	40
Current, A	26
Output, lb/hr	120
Line speed, ft/min	14.8
Sheet:	
Thickness, mils	55-62
Density, lb/ft^3	6.2
Width, in.	20.25

b. For Injection of Blowing Agent into Molten Polystyrene in the Barrel. In order to realize lower raw-material costs from crystal polystyrene in place of expandable-polystyrene beads or pellets, new extruders have been designed specifically for the injection of hydrocarbons or fluorocarbons into molten polystyrene. Over 95% of sheet is produced by injecting pentanes and/or fluorocarbons into the extruder barrel. Two functions are performed in such extruders: (a) intimate mixing of blowing agent and nucleating agent in the polymer melt and (b) cooling the melt to the foaming temperature [391]. These functions are usually carried out in two tandem single-screw extruders [392]: the first, for intensive high-speed mixing; and the second, with a wider screw, for cooling the melt at a relatively slow speed. Very long single screws with length-to-diameter ratios of 28:1 to 32:1 have been operated by controlling the pressure and feed rate of the polymer before adding pentane [393], but these units have not performed as well as two single-screw extruders in tandem. Mixing of pentane with polystyrene in a single-screw extruder is facilitated by dividing the molten polymer into separate small streams and injecting the pentane into each stream and the recombining the streams [394]. In another mixing device the molten resin is extruded into the top of a chamber through a perforated plate, forming a large number of narrow streams around which the foaming agent is vaporized [395]. Excellent control of mixing is claimed for a single-screw extruder if pressure zones are so arranged that the mass can be recycled to one or more positions [396].

The latest and most promising development involves the injection of blowing agent into the polymer during passage between a pair of intermeshing screws rotating in the same direction [397]. A shear rate about 80% less than that for single-screw machines is realized, using corotating 16:1 4.5-in. screws at only 15 to 20 rpm [398]. Consequently processing temperatures and energy requirements are significantly lower. Very little cooling is required before expanding. Furthermore a twin-screw machine (LMP) can be fed with beads as well as pellets.

2. Dies and Takeoff

The extrudate expands in all directions on emerging from a die. Thus a narrow-slot die gives a corrugated sheet of foam. Although these corrugations can be smoothed out by immediately passing the sheet through cold rolls [399], this technique raises the density of the foam prohibitively.

Foam sheet is ordinarily made by extrusion through a circular die, consisting of a block with a conical cavity in which is nested a conical mandrel. In the mandrel is a helical passage for cooling water arranged so that cooling is concentrated at the center rather than the edge[400]. An essential requirement is streamline flow, with no rapid changes in cross-sectional area within the flow channel. In addition, the die should have a gradually decreasing cross-sectional area throughout the flow channel to prevent premature expansion of the melt.

For foam film less than 20 mils thick the extruded tube is continuously blown by entrapped compressed air (in a manner similar to the production of polyethylene blown-film) while being simultaneously surface-cooled near the die by a surrounding stream of cool air [401]. The bubble of film is collapsed between rolls, and the edges are slit to obtain two flat sheets.

Most polystyrene sheet is produced at a thickness greater than 20 mils. For this purpose the sizing-plug technique is preferred to the blown-film method. The sheet is pulled over a mandrel (the so-called sizing plug), usually constructed of aluminum and provided with a means of cooling to remove frictional heat generated by the moving foam. Coming from the sizing plug, the film is slit, usually in two places, as illustrated in Fig. 28. Each of the sheets is pulled off individually by rotating rubber-covered rolls so that the sheet travels in an S-pattern with minimum compression.

Sheet characteristics can be regulated by the extent of cooling of the emerging hot extrudate by blowing cold air on it [402-404]. The effect of air cooling is to decrease the expandability of the foam at the surface, thus controlling density, stiffness, gloss, and hardness.

C. Posttreatment of Sheet

1. Requirements of Sheet for Thermoforming

Table 6 lists the typical physical properties of 50-mil polystyrene-foam sheet. However, sheet with these properties may still not meet the critical requirements of thermoforming. Disorientation of the sheet during takeoff can result in objectionable distortion on thermoforming. A study of cell structure versus thermoformability revealed that the cells should have a "free-blown" structure or preferably a slight biaxial stretching. If the cells are oriented more toward one axis than the other, the heated sheet will shrink excessively in that direction and expand in the other. Thermoformability improves with increasing cell size, but as cell size increases, the sheet becomes brittle. Therefore concentration of cell-nucleating

FIG. 28. Extrusion of polystyrene-foam sheet over a sizing mandrel. Reprinted from Ref. [389] by courtesy of Wayne State University.

agent must be adjusted to maintain a compromise in cell diameters at 4 to 8 mils so that the sheet can be thermoformed without being glassy and brittle. Finally, in order to be thermoformable without the tendency to collapse, the sheet must have aged sufficiently to permit air to diffuse into the cells. As seen in Fig. 29, sheet does not attain maximum expandability, in this case by radiant heating, for at least 2 days of aging at ambient conditions.

2. The Thermoforming Process

Most sheet is fed continuously in rolls to thermoforming machines. Generally the techniques for thermoforming nonfoamed plastics apply to foam sheet, but the operation is more critical because of the low density, heat sensitivity, rapid cooling, and low elongation of the foam. The thermoforming operation consists of first heating sheet simultaneously on both sides [405] by radiant-type heaters, preferably 12 in. away, and then transferring to a vacuum former for shaping to a male or female mold, usually with the assistance of a plug.

TABLE 6

Typical Properties of Expanded-Polystyrene Sheet

Thickness, mils	50
Density, lb/ft^3	6.5
Basis weight, $lb/1000\ ft^2$	27
Tensile strength, lb/in^2	400-500
Ultimate elongation, %	1.8-2.0
Tensile elastic modulus, lb/in^2	4000-5000
Burst strength, lb/in^2	26
Grease resistance, transudation time, sec	>1800
Water absorption, lb/ft^2	0.003
Water-vapor transmission, $g/24\ h\text{-}100\ in^2$:	
At 23° C and 50% R.H.	0.9
At 35° C and 90% R.H.	2.9
Cell size, mils	4-8

The heating portion of the thermoforming process is probably the most critical phase. The forming-temperature range for foam sheet is very narrow. There are only a few tenths of a second between the time that the sheet is pliable enough to be formed and the time that the sheet is overheated and the cell structure is completely broken down. In almost all cases the optimum combination of forming temperature and time results in a sheet that has achieved the maximum attainable expansion. Further heating of the sheet results in a slight breakdown or glossy appearance in the sheet surface and a resultant loss of thickness and increase in density. On the other hand, when the maximum-expansion point has not been

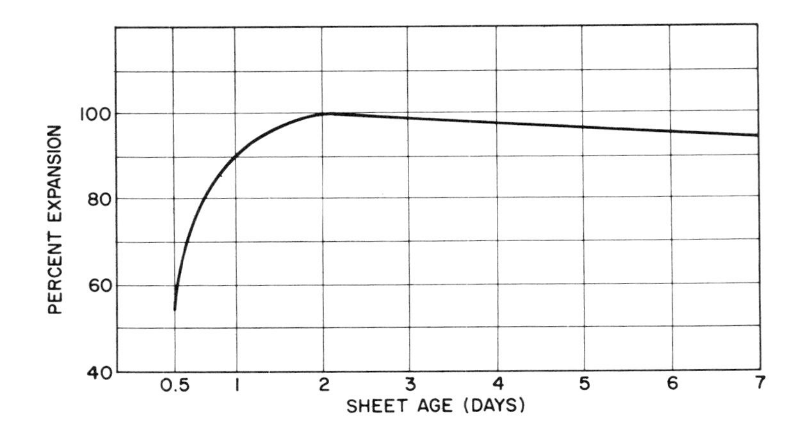

FIG. 29. Expandability of sheet versus age. Reprinted from Ref. 389 by courtesy of Wayne State University.

reached, the sheet is normally too cold to form, and tearing will result. Therefore, the sheet expansion is used as a guide for determing the proper combination of cycle and temperature.

In selecting a proper combination of foam-sheet properties, particularly thickness and density, cost is obviously the most important consideration. But the forming limitations and other sheet properties, such as expandability, must also be kept in mind. Foam sheet expands to approximately twice its original thickness. The expanded thickness, particularly in shallow-draw parts, should be considered in specifying the sheet characteristics. In some cases, when relatively deep draws of 0.8:1 or greater are anticipated, it may be necessary to use a slightly higher density sheet, even though this will increase the unit cost. Generally speaking, for reasons of both economy and structural strength in the formed part, it is best to select the lowest possible density and let the sheet thickness be the determining factor.

After the sheet has been selected, the next step is to determine the optimum processing conditions. Essentially this means determining the shortest time cycle at which a quality part can be made. Because of the insulating properties of the foam, high temperatures and short cycle times often result in a sheet that has been overheated and broken down on the outside surfaces, yet is too cold to form on the inside. Some compromise must therefore be achieved by lowering the heater temperatures and lengthening the cycle to obtain a sheet that is heated as uniformly as possible throughout. Normally 1200° F is considered the maximum heater temperature and 500° F the minimum. The part design and sheet thickness determine what combination of temperature and cycle time can be used. For shallow draws and relatively thin sheet, high temperatures and short cycle times may be used. For more critical draws and thicker sheet, however, a longer cycle and lower temperatures are required. Typical forming conditions for 60-mil sheet and a relatively shallow draw are listed in Table 7.

TABLE 7

Typical Thermoforming Conditions for Expanded-Polystyrene Sheet

Mold configuration:	
Size, in.	18 x 18
Number of cavities	4
Sheet characteristics (unformed):	
Thickness, mils	60
Density, lb/ft^3	6.5
Width, in.	20
Forming conditions:	
Top heater, ° F	700
Bottom heater, ° F	650
Plug temperature, ° F	70
Mold temperature, ° F	70
Total cycle, sec	3.5
Formed parts:	
Thickness, mils	110-115
Density, lb/ft^3	3.2

Although foam sheet cools very rapidly compared with other thermoplastic sheet materials, the cooling portion of the thermoforming cycle is still extremely critical. Insufficient cooling time will result in warped or distorted foam parts just as it will with parts formed from polyethylene.

In most cases the mold temperatures should be kept as cold as possible, and particular attention should be given to the mold design to ensure good cooling around the perimeter of each of the mold cavities. In some cases, however, the mold is heated slightly to aid in sheet distribution. Generally, with more efficient and faster continuous thermoforming machines, the cooling cycle represents by far the greatest percentage of the overall forming cycle. Even though complete and efficient cooling of the thermoformed part is achieved, however, some shrinkage will still take place during the time that the formed part and web pass from the mold to the trimming station. Because shrinkage is dependent not only on cooling efficiency but also on the design and configuration of the formed part, it is impossible to define an exact value for the coefficient of shrinkage. Best estimates are about 0.008 in./in.

3. Steam Expansion of Sheet

Extruded sheet can be continuously expanded to low densities (2 lb/ft^3) by exposure to steam. Distortion is minimized by passing the sheet over rotating drums. In one design [406] the steam passes out of a porous drum in contact with the sheet. In another method [407] the drums are internally heated and steam is sprayed from nozzles onto the sheet. These soft, low-density sheets, particularly when embossed with a dimpled pattern [408], have found use as shock-dampening plates for packaging.

4. Lamination

The impact and abrasion resistance of foam sheet is greatly improved by lamination to a thin, nonfoamed film, particularly crystal or rubber-reinforced polystyrene. These laminates are used mostly for thermoforming into lightweight, disposable dishware. Laminated sheet is produced continuously by extruding the hot foam layer against one [409] or two [410] layers of polystyrene film (e.g., 10 mils thick) to which the foam is bonded by heat and light pressure. The production of laminates from parallel-extruded streams of nonexpandable and expandable polystyrene has also been disclosed [411].

Foam sheet can be laminated continuously to foil or paper with the assistance of adhesives [412, 413].

5. Blow Molding

Foam-walled containers of up to 1 gal capacity have been blow-molded from expandable polystyrene nucleated with mixed citric acid and sodium bicarbonate [414]. Favorable conditions of temperature, pressure, screw design, and blow-molding cycle were defined, using a polyethylene blow-molding machine adapted to a 2.5-in. foam extruder. Blowing times were short (6-8 sec), and foam structures were attractive (6-7 lb/ft^3 with 2- to 4-mil cells), but no significant commercial applications were found for this process.

D. Generation and Recovery of Scrap

Approximately 20% of scrap is generated in commercial sheet-thermoforming operations. Although some scrap arises from startups, product changes, and mechanical failures, most scrap comes from trimming thermoformed parts. The sale or reuse of this scrap is essential to the economic production of meat trays, produce trays, and egg cartons because trimming losses are inevitable. Generally the scrap is ground into pieces, densified by heat and pressure, cooled, crushed, and refed to the extruder with fresh feed [415]. Zinc stearate may be added to inhibit fusion of the pieces during densification [416]. Part of the extruder charge (i.e., 10 parts of polystyrene with or without blowing agent) may be added to the scrap (90 parts) before grinding [417]. Pieces of scrap can be mixed with preexpanded beads and steam-molded, for example, in a block molding to be cut up for ceiling tiles [418]. Pieces stored in a closed space with pentane can be subsequently steam-molded to form a well-fused foam with a density of 2.5 lb/ft^3 [419].

E. Open-Celled Sheet

The extrusion process has been employed to make porous sheets of relatively high density. A mixture of 90 parts of polystyrene, 10 parts of sodium bicarbonate, and 0.2 part of Vaseline was extruded to form a film of 12- to 32-mil thickness with a fibrous structure and a density of 20 to 40 lb/ft^3 [420]. The film was permeable to gases and was proposed for packaging or filtering applications. In another system a mixture of 100 parts of polystyrene, 5 parts of polybutadiene, and 20 parts of polyoxyethylene (Polyox WSR 205) was blow-molded to an opaque film of 2-mil thickness [421] After extraction with water, the opaque sheet (50 lb/ft^3) resembled paper in appearance, feel, and physical properties, including water permeability.

IX. EXTRUDED BOARD

A. Solvents as Blowing Agents

The first commercial polystyrene foam was extruded board as originated by Munters and Tandberg [422] and developed by The Dow Chemical Co. [423-431]. Molten polystyrene containing methyl chloride was extruded into large foam logs, which were then cut into the desired boards, planks, or other shapes. This product, Styrofoam, was introduced in the United States for military applications during World War II. Commercial acceptance proceeded at a good rate after 1946, primarily for insulation. Significant quantities were also used in construction and in the floral trade.

In comparison with bead board, the disadvantages of extruded board are its limited formability (i. e., limited to the extrusion profile and post-cutting) and expandability (about 1.5-lb/ft^3 minimum density). The advantages of extruded board are its brittleness (making it suitable as a floral foam capable of being pierced by flower stems), resistance to water penetration, and consistent quality. For general purposes extruded board is offered at about 1.7 lb/ft^3, but foams of higher densities are available for structural uses requiring greater strength.

Because of its extensive use in construction, extruded foam is supplied in a self-exinguishing grade [429]. A promising, unique method, called "spiral generation," for the rapid on-site construction of domed buildings by continuously heat-sealing precut slabs of foam was described by Ziegler [431]. Spiral-generated domes have been used for industrial offices, warehouses, tank covers, trickling-filter covers, and recreation enclosures. For critical insulation purposes a "low-k-factor" grade was developed. It contains fluorocarbon gases, from the blowing agent, trapped in the foam cells.

Equipment for producing extruded board with methyl chloride or other normally gaseous solvent-type blowing agents is described in patents disclosing a holding vessel with a bottom port [423, 424] and a conveyor [425] with internal mixing to prevent channeling, discharging into a tunnel through which the foam expands and is shaped. Friction of the foam is avoided by a flow of hot water [426] or by air-cooled rollers [427] on the wall of the tunnel. Cell size is controlled by the inclusion of a stearate (Zn, Ca, etc.) [428] or porous powders [432] as nucleating agents. In foams containing added organic bromine compounds (e.g., acetylene tetrabromide) blue pigments (indigotin or copper phthalocyanine) serve both to nucleate cell formation and to identify the self-extinguishing grade made thereby [429]. The problem of warpage of freshly extruded board was overcome by mixing the appropriate amount of dichlorodifluoromethane with methyl chloride, so that diffusion of the blowing agent out of the fresh foam was within 0.75 to 6 times the diffusion of air into the foam [430].

The use of normally liquid blowing agents (i.e., methylene chloride optionally with or without benzene and/or acetone) was claimed in a process whereby several extruders feed one holding vessel [433].

B. Nonsolvents as Blowing Agents

Section VIII discussed the extrusion of foam sheet in tubular form from mixtures of polystyrene, nucleating agents, and hydrocarbons and/or fluorocarbons. Attempts have been made to produce board from such sheet-forming compositions, but this procedure has not attained the commercial success of the Styrofoam process.

Several die designs have been disclosed for the extrusion of board with hydrocarbons as blowing agents. Wiley, DeBell, and DeBell [434] extruded a mixture of polystyrene and propane through multiple small orifices to form a bundle of foamed stands that fused together. In a modification [435] of this system an additional die was used for shaping. Placing transverse obstructions inside a circular die gave a foam log with less warpage and wrinkling [436]. Expansion through slits in a gratelike pattern was used to form ribbons of butane-expandable polystyrene which foamed and coalesced [437]. Other die designs were a Greek cross [438], peripherally grooved orifices [439], and a fish tail [440]. Extrusion through a flared die with a chilled surface gave foam with a fibrous skin [441]. Corrugations were claimed to be removed by shaping the hot foam with cold concave metal surfaces [442].

The following materials have been claimed as nucleating agents for board extrusion with pentane as blowing agent: 10% kaolin [443], 20 to 80% polyvinyl chloride powder [440], and 2.5% powdered perlite [444]. With butane as blowing agent, a mixture of 1% silica and 1% zinc stearate was used as nucleating agent [440].

A 50:50 mixture of mineral fertilizer and polystyrene was extruded in the presence of pentane through a die to give a foamed soil conditioner that slowly released its nutrients [445].

X. STRUCTURAL FOAMS

Structural foams are molded by injection and expansion casting methods. These materials have a strong continuous skin and a foam core. Their densities are quite high, generally 20 to 40 lb/ft^3, in contrast to the 1- to 5-lb/ft^3 range of the steam-molded and extruded polystyrene foams discussed in the preceding sections. For equivalent stiffness a structural foam part can replace a substantially heavier nonfoam part. Structural foams are made by similar methods from polystyrene, rubber-modified "high-impact" polystyrene, and acrylonitrile-butadiene-styrene (ABS) polymers, as well as from polyethylene and polypropylene. Several reviews [446-451] of methods for molding these foams have been published. Uses for styrene polymers in this relatively new commercial development are in furniture parts (e.g., for chairs, tables, drawers, cabinets, beds, and lamps), shoe trees, shoe lasts and heels, poultry crates, marine parts, automotive parts, tool handles, and toys. Although they can be molded to resemble intricate wood carvings, structural foams do not absorb water or change in dimension as does wood. Structural foam molding permits the production in relatively inexpensive molds of large, heavy pieces that could not otherwise be made by conventional injection molding of unfoamed thermoplastics.

Both low- and high-pressure methods are employed for the injection molding of structural foams. The low-pressure method relies on the expansion of an expandable plastic melt after it has been extruded or shot into the mold. The pressure to fill the mold is supplied almost entirely by the expanding resin, so that mold pressures rarely exceed a few atmospheres. The advantage of this process is the use of inexpensive, rapidly produced molds (e.g., aluminum, epoxy). A disadvantage arising from the low pressure is that a smooth surface texture is hard to obtain.

The Union Carbide low-pressure process [452] uses an accumulator to receive a nitrogen-containing extrudate and store it at high enough pressure to prevent expansion. After the accumulator gets enough material for one shot, its contents are quickly discharged into the mold. The nitrogen blowing agent may be injected into the molten polymer in the barrel of the extruder, or it may be generated from a powdered nitrogen-releasing compound added with the thermoplastic pellets.

A low-pressure process of the Phillips Petroleum Co. employs a retracting-screw extruder to push the expandable plastic into the mold. The feed is a mixture of about 10 parts of regular pellets and 1 part of a "foam concentrate" (i.e., pellets containing a blowing agent).

In the Engelit process [453] the polymer is melted on a heated revolving disk and then scraped off with a doctor blade into the barrel of an extruder through an opening at the bottom of the barrel. Before the melt enters the die, a mixture of solid, powdered blowing agent and any pigment is added under pressure directly to the melt.

Dow's Frostwood [454, 455] was the first expandable-styrene polymer developed specifically for injection molding. Expandable particles containing a volatile liquid blowing agent were molded into a retractable mold form with a movable cavity. Pigmented polymers produced an attractive frostlike surface. Screw injection molders had to be modified to prevent loss of blowing agent through the feed zone.

Injection molding of mixtures of polystyrene and azodicarbonamide (a solid nitrogen-releasing blowing agent) to form lightweight, nailable heels for ladies shoes was disclosed by Baxter [456] and by Boutillier and Gourlet [457].

High-pressure methods of injection molding involve fully injecting an expandable mass of polystyrene into the mold. Then part of the polystyrene is withdrawn, permitting the remainder to expand. The Allied Chemical Corp. process [458, 459] permits the extruder screw or ram to retract, allowing some resin fo flow back into the extruder. Egress ports may also be used to permit part of the expanding resin to escape. The feed can be a mixture of resin and azodicarbonamide. Another method (USM Corp.) [448] has movable mold surfaces, so that expansion occurs as they are retracted. All of these methods require the high pressures normal to injection molding, and some require other expensive equipment. They have the advantages that smooth-surfaced moldings can be obtained and thin parts can be molded.

Acrylonitrile-butadiene-styrene plastic foams are molded by either injection molding or by the expansion-casting method developed for Marbon Cycolac resins [460]. In the expansion-casting method a mold is filled with expandable pellets containing a blowing agent. On heating, the pellets expand and fuse to give foams with densities of 25 to 35 lb/ft^3. This process shares with low-pressure injection molding the advantage of low-cost molds and the disadvantage of pebbly surfaces. Furthermore, the output per mold is low. Expansion casting can be carried out in an oven, in a heated press, or in modified rotational molding equipment.

XI. PROPERTIES OF POLYSTYRENE FOAMS

The data presented in this section were obtained from bulletins of the Sinclair-Koppers Co. [461] and The Dow Chemical Co. [462].

A. Mechanical Strength

Typical values for the tensile, flexural, shear, and compressive strengths of molded-beam foams are shown as a function of foam density in Figs. 30, 31, 32, and 33. For each property the relationship of strength to density is linear.

The compression set in foamed expandable polystyrene under increasing degree of compression is shown in Table 8.

Figure 34 depicts the effect of static loads of 4 to 10 lb/in^2 at 0 to 1000h on the deformation of foamed expandable polystyrene of 1.5-lb/ft^3 density.

Table 9 lists typical average mechanical properties for extruded polystyrene-foam boards of various densities and cell sizes.

Typical compressive stress-strain curves of extruded 1.8-lb/ft^3 foams of greatly different cell sizes are shown in Fig. 35. Compression was applied in the vertical direction, that is, normal to the major plane of the boards.

The mechanical properties of polystyrene foam are not adversely affected at subzero temperatures. For example, the impact strength of unnotched bead foams remains the same when cooled to the temperature of Dry Ice (-78.5° C), and the compressive strength of extruded foam at liquid-nitrogen temperature (-184° C) is approximately 25% higher than it is at room temperature.

Since polystyrene is thermoplastic, the mechanical properties of polystyrene foams decrease in an accelerated manner as the temperature is raised to near the heat distortion point. The zero-strength temperature of bead foams was determined to be approximately 88° C.

TABLE 8

Compression Set of Foamed Expandable Polystyrene[a, b]

Nominal density (lb/ft^3)	Compression set (%) after deformation of			
	10%	20%	30%	40%
1	2.3	5.4	10.3	14.3
2.5	3.1	9.6	16.8	24.2
3	3.8	9.8	17.7	24.3
4.5	4.0	11.7	18.6	27.7

[a]Reprinted from Ref. [461], p. 3, by courtesy of the Sinclair-Koppers Co.

[b]Method of test: Specimens, each 4 x 4 x 1 in, were compressed on the 1-in dimension and were allowed to recover for 90 minutes before measuring the compression set. Testing speed 0.05 in/min at 23°C and 50% relative humidity.

B. Thermal Properties

1. Heat Distortion

The maximum temperature for serviceability of polystyrene foams depends on residual expanding agent and unequal orientation stresses. This temperature, therefore, increases with increasing age and decreasing density. For unloaded extruded foams the approximate maximum temperature at which distortion does not occur is about 74° C.

The surface of molded-bead foams may blister at elevated temperatures because of the retention of blowing agent by the relatively dense surface. Foam specimens of 4 x 4 x 1 in. were molded at 1-, 2-, 4-, and 6-lb/ft^3 density and exposed for 48 h at 51.7, 65.6, and 79.4° C in an air-circulating oven. Although the foams of 1- and 2-lb/ft^3 density did not blister at 79.4° C, the 4- and 6-lb/ft^3 foams blistered slightly at 65.6° C and considerably at at 79.4° C.

TABLE 9

Average Mechanical Properties of Extruded Polystyrene Foam at 23.9° C[a]

Foam Type	Average density (lb/ft³)	Average cell size (mils)	Strength (psi)				Modulus (psi)		
			Compressive[c]	Tensile[d]	Shear[e]	Flexural[f]	Compressive[c]	Bending[f]	Shear[e]
FR	1.8	28	30[g]	70	35	60	1000	2200	1100
FB	1.8	40	25[g]	60	30	45	800	1500	900
HD 300	3.3	6	120[g]	225	65	85	6000	2000	875
SB	2.2	14	30[h]	60	35	140	1200	7500	1200
SM	2.2	14	30[h]	60	35	140	1200	7500	1200
RM	2.3	14	30[h]	60	35	140	1200	7500	1200
DB	1.8	60	18[h]	55	30	45	800	1500	900
BB	1.8	72	18[h]	55	30	40	800	1300	900

[a]Reprinted from Ref. [462], p. 4., by courtesy of The Dow Chemical Co.
[b]Styrofoam brand designation. Key: FR and HD 300, insulation board; FB, insulation billets; SB, perimeter insulation; SM, insulation sheathing; RM, roof insulation; DB, decorative billets and board; BB, buoyancy billets.
[c]According to ASTM Designation D1621-64.
[d]According to ASTM Designation D1623-64.
[e]According to ASTM Designation C273-53.
[f]According to ASTM Designation C203-58.
[g]Value at yield of 5% deformation, whichever occurs first.
[h]Value at 0.1 in deformation.

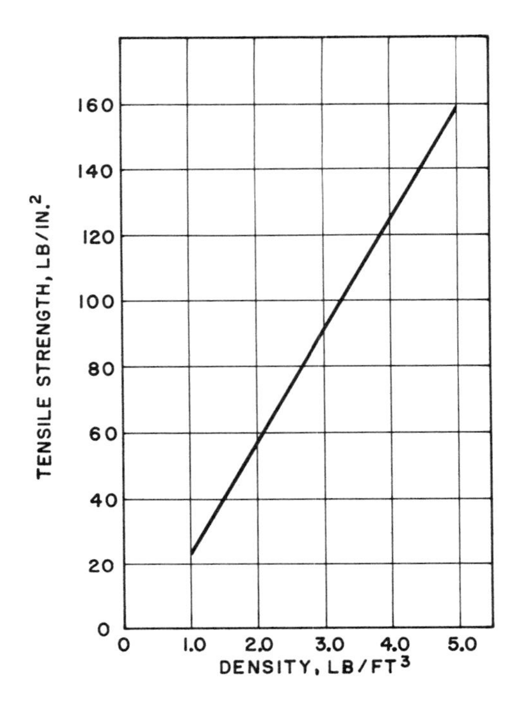

FIG. 30. Tensile strength of foamed expandable polystyrene. ASTM Designation D1623-64, type A specimen. Testing speed 0.05 in./min at 23° C and 50% relative humidity. Reprinted from Ref. [461], p. 2, by courtesy of the Sinclair-Koppers Co.

2. Coefficient of Expansion

The coefficient of linear thermal expansion at 1- to 4-lb/ft^3 density is 3 to 4 x 10^{-5} in./in.-° F in the temperature range -30° C to +30° C by ASTM Designation D696-44.

3. Thermal Insulation

The thermal conductivity of a homogeneous material is the rate of heat flow, under steady conditions, through a unit area of a unit thickness per unit temperature gradient in the direction perpendicular to the area. Because of their relatively low thermal conductivity, polystyrene foams have been applied successfully as low-temperature insulation.

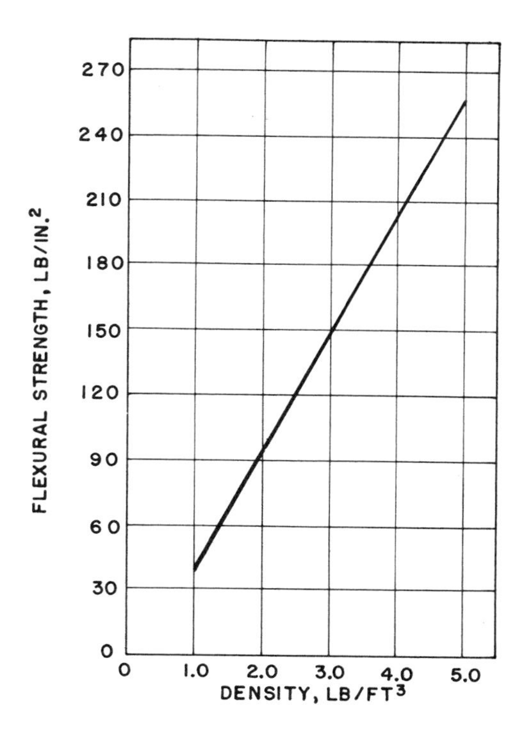

FIG. 31. Flexural strength of foamed expandable polystyrene. Military Specification MIL-P-19644A. Testing speed 0.05 in./min at 23° C and 50% relative humidity. Reprinted from Ref. [461], p. 2, by courtesy of the Sinclair-Koppers Co.

a. Molded Beads. The thermal conductivity, or k factor of bead foams was determined by the guarded-hot-plate method according to ASTM Designation C177-63. Figure 36 shows the change in value of k with density at a mean temperature of 23.9° C.

Comparison of the k factor of polystyrene foams with that of other thermal insulation materials should be made only at the intended operating temperature. Although the k factor of polystyrene foams decreases with lower mean temperatures, this is not always the case with other thermal insulation materials. At a mean temperature of -168°C the k factor of molded-bead foams of 1- to 2-lb/ft^3 density is only 0.09. Typical k values in the 1- to 2-lb/ft^3-density range at various mean temperatures are shown in Fig. 37.

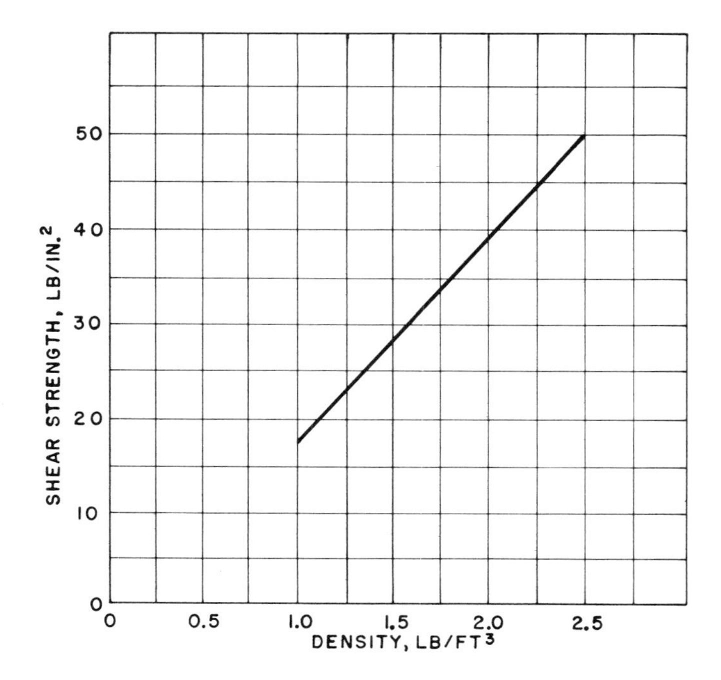

FIG. 32. Shear strength of foamed expandable polystyrene. ASTM Designation C273-61, 12 x 4 x 1-in. specimen. Testing speed 0.06 in./min at 23° C and 50% relative humidity. Reprinted from Ref. [461], p. 3, by courtesy of the Sinclair-Koppers Co.

Foamed-polystyrene particles can be used as loose fill for thermal insulation. The k factor of such loose fill at a density of 1 lb/ft^3 decreases from 0.30 to 0.09 as the mean temperature decreases from 23.9° C to -157° C.

The k factor of molded insulation board of 1-lb/ft^3 density meets the Federal Housing Administration requirement for perimeter insulation (±15% change from original value) after repeated wetting-and-drying cycles.

b. Extruded Boards. The thermal conductivity of extruded polystyrene foam within the densities produced is a function of cell size and mean temperature. The values given in this section are for cells ranging from approximately 4 to 72 mils in diameter at mean temperatures of 4 to 24° C.

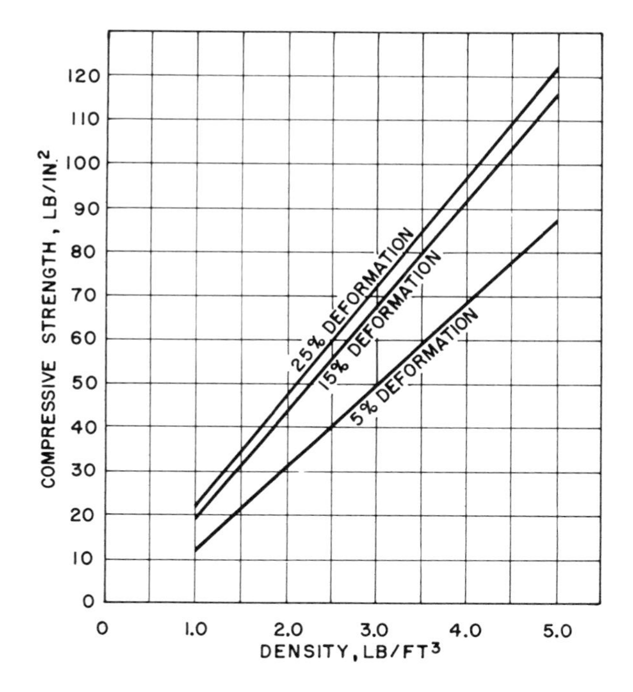

FIG. 33. Compressive strength of foamed expandable polystyrene. ASTM Designation D1621-64, procedure A. Testing speed 0.1 in./min-in. thickness at 23°C and 50% relative humidity. Reprinted from Ref. [461], p. 3, by courtesy of the Sinclair-Koppers Co.

The initial k factors of extruded foams are lower than those published because of residual blowing agent in the cells. The k factors listed in Table 10 were obtained on aged samples on a heat-flow-meter test apparatus, which was calibrated to ASTM Designation C177-63 (guarded-hot-plate method).

The specific heat of board of 1.8-lb/ft^3 density was determined to be 0.27 Btu/lb-°F at 4°C.

C. Water Absorption

Since polystyrene itself absorbs negligible amounts of water and since the cells of polystyrene foam are noninterconnecting, the resistance of polystyrene foams to penetration by liquid water is excellent.

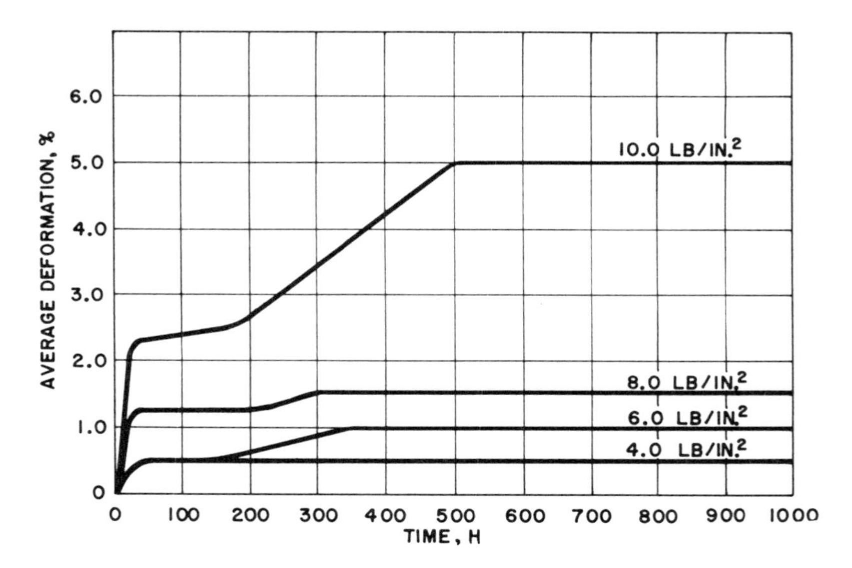

FIG. 34. Static compressive creep of foamed expandable polystyrene at a density of 1.5 lb/ft^3. Reprinted from Ref. [461], p. 4, by courtesy of the Sinclair-Koppers Co.

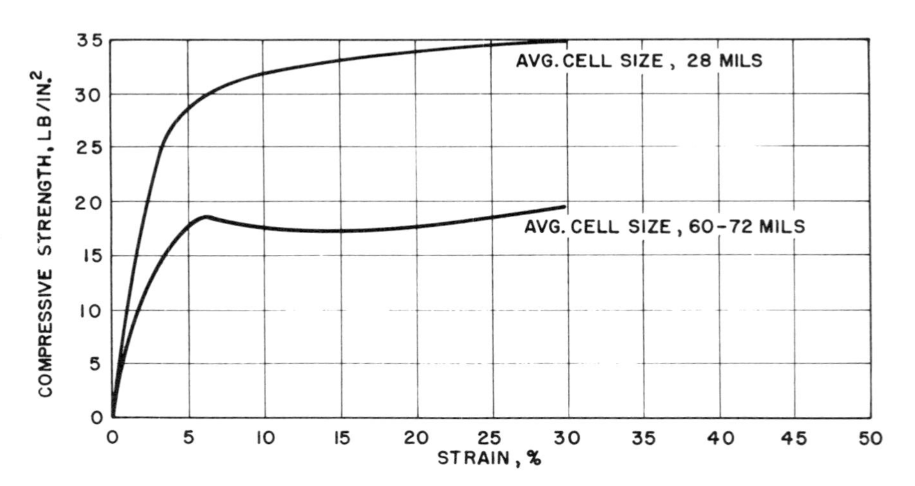

FIG. 35. Typical compressive stress-strain curve for extruded polystyrene foams of 1.8-lb/ft^3 density. Reprinted from Ref. [462], p. 5, by courtesy of The Dow Chemical Co.

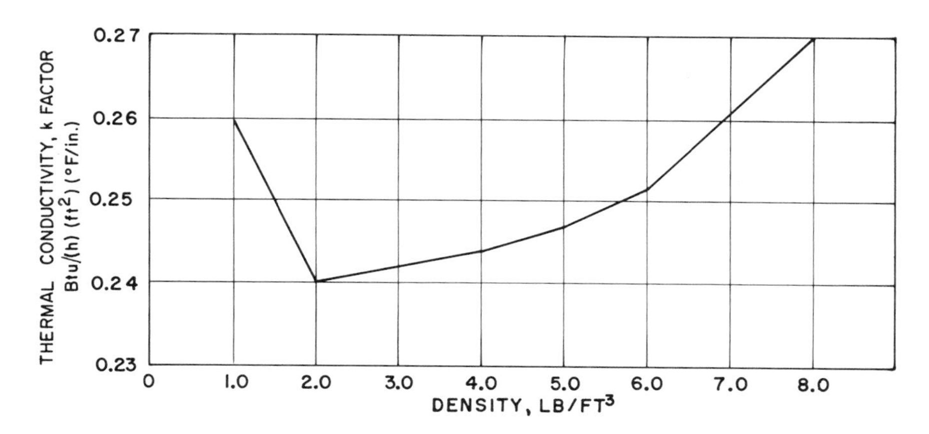

FIG. 36. Thermal conductivity of foamed expandable polystyrene as a function of density at a mean temperature of 29.3° C. ASTM Designation C177-63. Reprinted from Ref. [461], p. 5, by courtesy of the Sinclair-Koppers Co.

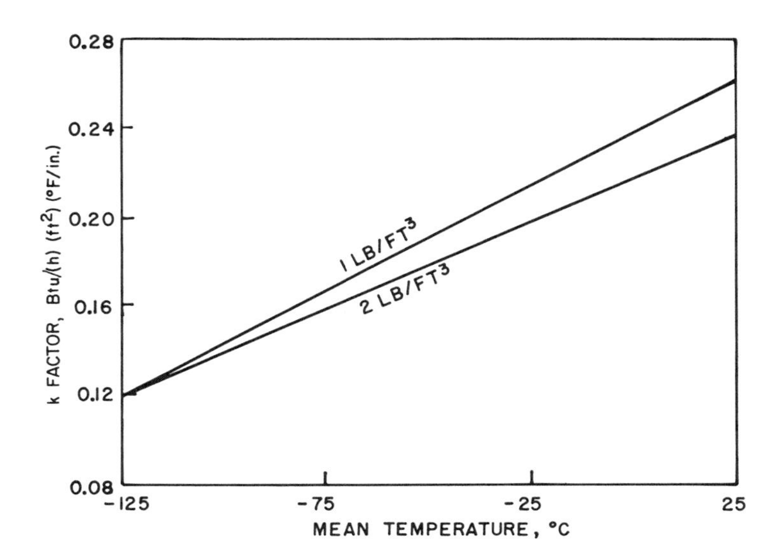

FIG. 37. Thermal conductivity of foamed expandable polystyrene. ASTM Designation C177-63. Density range 1 to 2 lb/ft^3. Reprinted from Ref. [461], p. 5, by courtesy of the Sinclair-Koppers Co.

TABLE 10

Thermal Conductivity (k Factor) of Extruded Polystyrene Foam[a]

Styrofoam brand	Average cell size (mils)	Average k factor (Btu/(h) (ft^2) (°F/in.)) At 4° C[b]	At 24° C[b]
FR (Insulation board)	28	0.24	0.26
FB (Insulation billets)	40	0.26	0.28
BB (Buoyancy billets)	72	--	0.35
DB (Decorative billets and boards)	60	0.28	0.30
HD 300 (Insulation board)	6	0.19	0.21
SB (Perimeter insulation)	14	0.18	0.20
SM (Insulation sheathing)	14	0.18	0.20
RM (Roof insulation)	14	0.18	0.20

[a]Reprinted from Ref. [462] p. 8, by courtesy of The Dow Chemical Co.
[b]Mean temperature.

1. Molded Beads

Molded specimens, tested by the method of Military Specifications MIL-P-19644A (submerged at room temperature for 48 h at a hydrostatic pressure equivalent to 10 ft of water), absorbed less than 0.06 lb of water per square foot of specimen surface. Further information on water absorption is given in Table 11.

Water absorption has an unfavorable effect on thermal conductivity; but since the water absorption of foamed polystyrene is quite low, the effect on the k factor is minor. A thermal conductivity test was performed according to ASTM Designation C177-63 at a mean temperature of 23.9° C on a 1-lb/ft^3 foam specimen before and after submersion under a 10-ft head of water for 48 h. The k factor increased only from 0.26 to 0.28 as a result of submersion. No water drainage was observed from the specimen during the thermal conductivity determination.

2. Extruded Board

The water picked up by submerged extruded foams is almost entirely in the open cells at the cut surface. Water absorption after submersion for 24 h just under the surface of water according to a modification of ASTM Designation C272-53 was less than 0.10% for foams with an average cell size of 40 to 72 mils and less than 0.50% for foams with an average cell size of 6 to 28 mils.

TABLE 11

Water Absorption of Molded Polystyrene Foam[a, b]

Nominal density (lb/ft^3)	Water absorbed (lb/ft^2) of specimen surface[c]	
	Actual	Specification[d]
1	0.05	-
1.5	0.04	-
2	0.04	0.12
2.5	0.04	-
3	0.04	0.12
5	0.03	0.10

[a]Reprinted from Ref. [461], p. 7, by courtesy of the Sinclair-Koppers Co.
[b]Specimen size: 4 x 4 x 1 in.; surface character: skin or smooth surface removed; water temperature: 18.3 to 32.2° C; type of submersion: hydrostatic pressure equivalent to 10 ft of water (19.05 $lb/in.^2$); submersion time: 48 h.
[c]Tested by the method of Military Specification MIL-P-19644A.
[d]Maximum allowed in specification.

D. Water-Vapor Transmission

The suitability of a material to function as a vapor barrier between environments with different humidities is measured by its water-vapor transmission (WVT). The water-vapor transmission for foamed polystyrene is quite low, although it is not a water-vapor barrier. Water-vapor transmission decreases with increasing density and decreasing cell size, as shown in Table 12.

E. Sound Absorption

Polystyrene foam, because of its small, noninterconnecting cell structure and its rigid nature, is not an efficient sound-absorbing material. It is approximately equal to wood, plaster, brick, and other common building materials in sound absorption at higher frequencies, but is not as good as commercially available sound-absorbing materials at the commonly tested frequencies.

TABLE 12

Water-Vapor Transmission of Polystyrene Foams[a]

Type	Average cell size (mils)	Nominal density (lb/ft^3)	Average perms[b, c]
Molded	2-8	1.0	1.2-2.2
Molded	2-8	2.0	0.6-0.8
Molded	2-8	2.5-3.0	0.5-0.7
Extruded	14	2.2-2.3	0.6
Extruded	28-40	1.8	1.0
Extruded	60-72	1.8	1.6

[a]Data from Refs. [461] and [462].

[b]Tested according to ASTM Designation C355-64.

[c]Grains per hour through 1 f^2, 1 in. thick, under a partial pressure of 25 mm Hg.

The test results in Table 13 were obtained by the ASTM reverberation-room method on mountings as indicated. All specimens were 1 in. thick.

F. Energy Absorption

Because of its excellent energy-absorption properites and low cost, expanded polystyrene has become the workhorse material among cellular plastics for protective packaging. Common forms of polystyrene foam are (a) molded expanded beads; (b) fabricated board, both molded and extruded; (c) impenetrable loose fill; and (d) thermoformed sheet. The greatest acceptance of foam for packaging has been in the form of molded expanded beads. We shall consider here the properties that characterize molded-bead foams for cushioning against impact. Knowing the fragility of the item to be packaged and certain dynamic properties, an appropriate package can be designed.

TABLE 13

Sound Absorption of Extruded Polystyrene Foam[a]

Frequency (Hz)	Sound absorption (%)	
	Mounting No. 4	Mounting No. 7
125	1	38
250	3	25
500	5	17
1000	14	15
2000	49	9
4000	19	15

[a]Reprinted from Ref. [462], p. 10, by courtesy of The Dow Chemical Co.

The function of a cushion is to decelerate packaged items gradually by absorbing part of the impact by means of deflection of the cushion. An efficient cushion will lengthen the time required for a packaged item to decelerate to zero velocity, as indicated by a low G value. The G value of the cushion represents that portion of the impact energy unabsorbed by the cushion and transmitted to the packaged item and is expressed as a ratio:

$$G = \frac{\text{deceleration at impact}}{\text{acceleration of gravity}} .$$

The G value of the cushion is measured by dropping a weighted platen on the cushion surface. Deceleration at impact is observed on a recording oscilloscope connected to an accelerometer on the dropping head. By varying the weight of the platen, G values over a broad static stress range can be obtained.

By the use of equipment designed in accordance with ASTM Designation D1596-59T, "Dynamic Properties of Package Cushioning Materials," specimens of polystyrene foam were tested at densities of 1.2, 2.0, and 3.0 lb/ft^3 and in thicknesses of 1, 2, and 4 in. Test specimens were placed on the base plate and impacted with varying loads (static stresses) from a height of 30 in. Each specimen was subjected to five drops at each load. For each load a new specimen was used.

To study the effect of foam density on cushioning properties, graphs of G versus static stress for three densities (thickness constant) were plotted. Data for the fifth drop of varying loads on 2-in.-thick foams are plotted in Fig. 38. Foam of 1.2-lb/ft^3 density has the lowest G values over the broadest range. Foam of 2-lb/ft^3 density exhibits lower values for a limited portion of the curve, but the improvement in G is not great enough to justify the increased material cost.

A similar analysis was made of dynamic deflection and permanent set (Fig. 39). Although 1.2-lb/ft^3 foam has the greatest dynamic deflection, it also has the least permanent set. Therefore there is little incentive for using foams of higher densities since the best cushioning properties and the least permanent set are both obtained at 1.2-lb/ft^3 density.

Further studies were made of 1-, 2-, and 4-in.-thick foams of 1.2-lb/ft^3 density to define the change in G with the number of drops at several static stress values. The values of G were found to increase with each successive drop (Fig. 40). Therefore the use of the first-drop data is not realistic for package design. Fifth-drop data, although conservative, are more meaningful up to a certain static stress limit. As shown in Fig. 41, for 1-in. specimens this limit is 0.66 lb/in.^{2}; for 2 in. it is 1.40 lb/in.^{2}. Beyond these levels the change in G with repeated drops is too great to permit the use of fifth-drop data in package design.

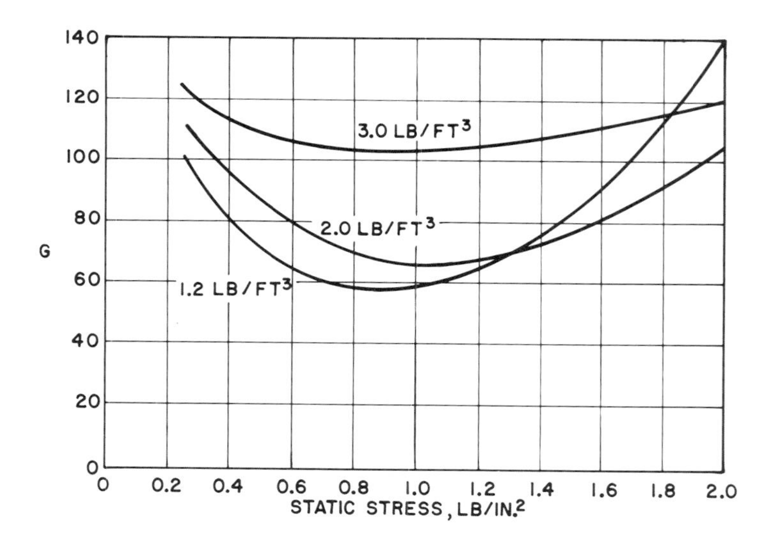

FIG. 38. Fifth-drop G values as a function of static stress. Thickness 2 in.; density 1.2, 2.0, and 3.0 lb/ft^3; drop height 30 in. Reprinted from Ref. [463], p. 5, by courtesy of the Sinclair-Koppers Co.

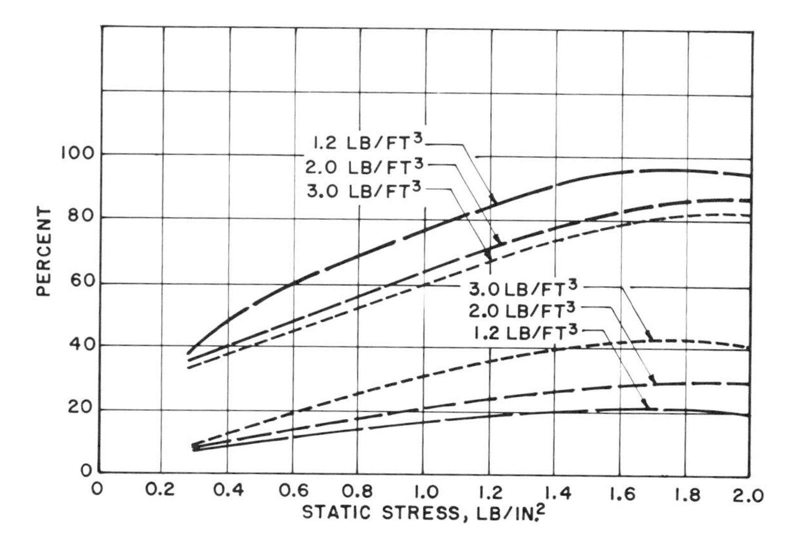

FIG. 39. Fifth-drop dynamic deflection and permanent set versus static stress. Thickness 2 in.; density 1.2, 2.0, and 3.0 lb/ft^3. Reprinted from Ref. [463], p. 6, by courtesy of the Sinclair-Koppers Co.

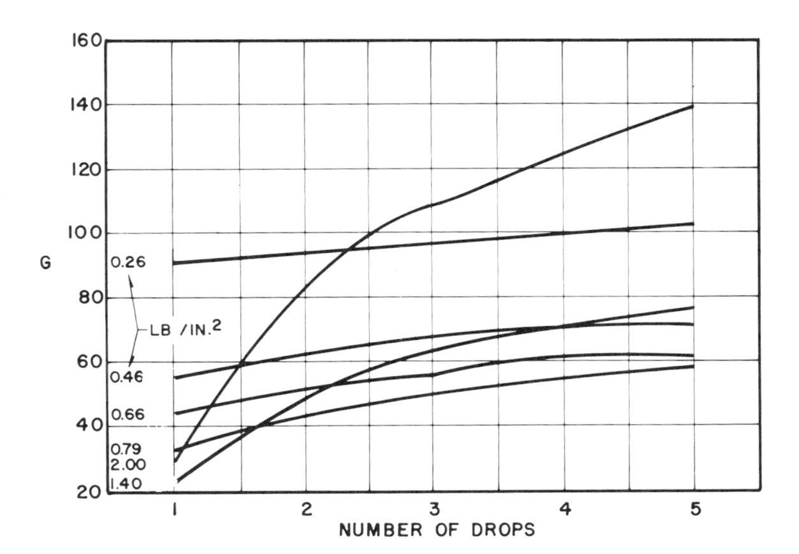

FIG. 40. Values of G versus number of drops. Thickness 2 in.; density 1.2 lb/ft^3; drop height 30 in. Reprinted from Ref. [463],p. 6, by courtesy of the Sinclair-Koppers Co.

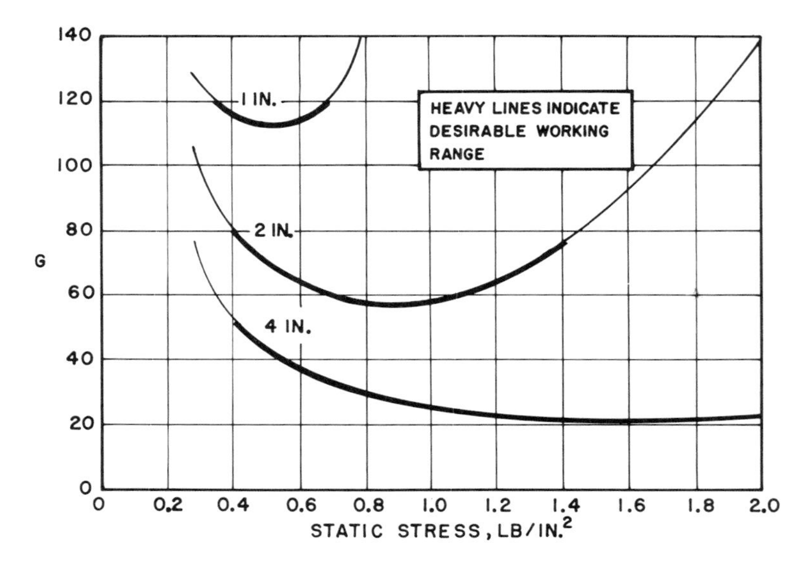

FIG. 41. Fifth-drop G values as a function of static stress. Thickness 1, 2, and 4 in.; density 1.2 lb/ft^3. Reprinted from Ref. [463], p. 8. by courtesy of the Sinclair-Koppers Co.

A designer, knowing or estimating the G factor of the item to be packaged, selects the minimum thickness of foam that provides a lower G value of the package. Since the foam is most efficient in specific static stress ranges, he must then adjust either the bearing area between the item and the cushion or the contact area on the package exterior to be within these efficient limits. Such adjustment is most easily accomplished by a series of ribs.

Seven different rib designs (e.g., semicircular, triangular, rectangular, and trapezoidal) were investigated for cushioning efficiency and durability at one and five drops. The most effective designs were trapezoids C and D (Figs. 42 and 43). By using such ribs substantial savings in packaging costs and shipping damage can be realized.

To summarize, optimum foamed-polystyrene packages can be designed with the assistance of the following:

1. Fifth-drop data (G versus static stress) on foams of 1.2-lb/ft^3 density.
2. Rib heights equal to or greater than the anticipated dynamic deflection.
3. Rib shapes preferably of trapezoidal form C or D (Fig. 42).

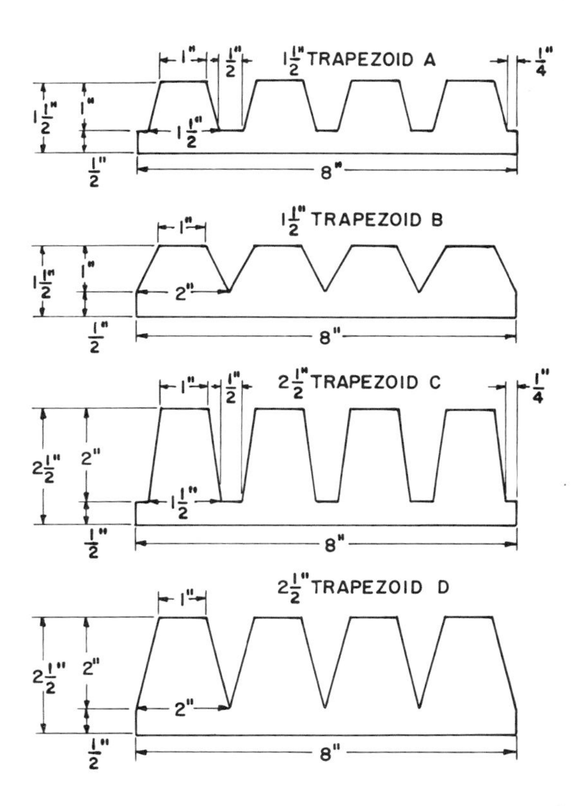

FIG. 42. Trapezoidal rib design. Reprinted from Ref. [463], p. 9, by courtesy of the Sinclair-Koppers Co.

G. Inertness

Inertness is of prime importance in packaging and other applications where foamed polystyrene is in contact with metal. Polystyrene is completely inert to metals, and hence any corrosion of metals in contact with foamed polystyrene must be caused by other factors.

To minimize corrosion due to water or humidity, the foam piece should be dried in an oven or in a heated, ventilated room until the residual moisture from molding is reduced to an acceptable level.

To demonstrate inertness, molded-foam pieces were submerged in boiling water for 30 min. The water extract was found to have a pH of 7.2, compared with a pH of 7.1 for a water "blank."

TABLE 14

Electrical Properties of Foamed Expandable Polystyrene[a,b]

Dielectric constant:	
At 60 Hz	1.19
At 1 kHz	1.07
At 1 MHz	1.02
Dissipation factor:	
At 60 Hz	0.0005
At 1 kHz	0.0005
At 1 MHz	0.0005
Loss factor:	
At 60 Hz	0.0006
At 1 kHz	0.0006
At 1 MHz	0.0006
Resistivity:	
Volume, ohm-cm	3.8×10^{13}
Surface, megohms	0.9×10^{6}
Dielectric strength, V/mil:	
Short-time	48-50
Step-by-step	49

[a]Data from Ref. [461].

[b]Nominal density 2 lb/ft^3. Specimens conditioned at 23° C and 50% R. H.

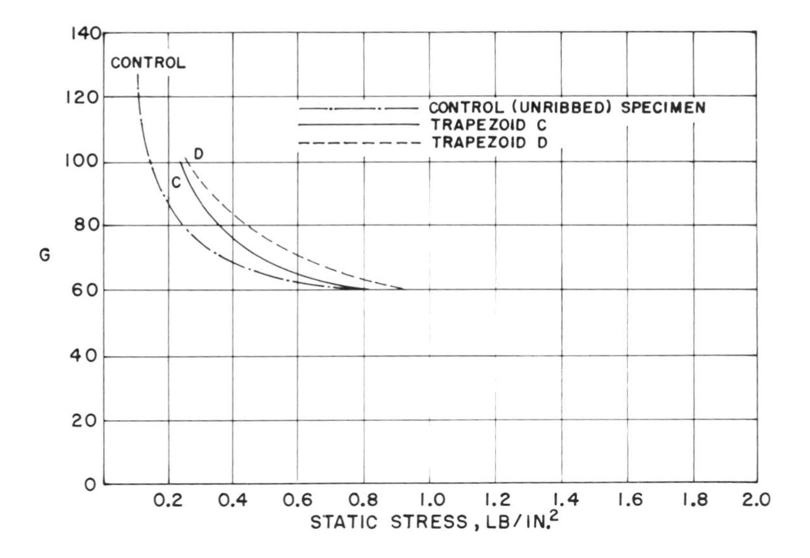

FIG. 43. Fifth-drop G values as a function of static stress. Trapezoidal ribs, 2.5 in., density 1.2 lb/ft^3. Reprinted from Ref. [463], p. 13, by courtesy of the Sinclair-Koppers Co.

H. Electrical Properties

The electrical characteristics of polystyrene foams are remarkably similar to the characteristics of air, since the foam is predominantly air. This applies to arc resistance as well as to other electrical properties. The foam melts about the path of an arc as soon as the arc penetrates the foam.

Other electrical properties of polystyrene foam of 2-lb/ft^3 density are listed in Table 14.

The dielectric loss of foamed polystyrene is quite low and is relatively unaffected after 5 months' storage (see Table 15).

TABLE 15

Dielectric Loss of Foamed Polystyrene

Specimen conditioning	Dielectric loss
As received	0.00007
After 5 months' storage, followed by 11 days at 25° C and 50% R. H.	0.00005

I. Resistance to Attack

1. Chemical Reagents

The resistance of foamed polystyrene to chemical reagents is the same as that of unfoamed polystyrene. Most acids and their water solutions do not attack the foam, although strong oxidizing acids, such as nitric or perchloric acid, will decompose polystyrene. Fuming sulfuric acid (20-40% oleum) sulfonates polystyrene, and chlorine and bromine also react detrimentally with it. Alkalies and salts in water solution do not chemically affect foam regardless of concentration, temperature, and duration of exposure.

2. Solvents

The thin cell walls and large exposed surface of foamed polystyrene make it especially sensitive to attack by solvents. Solvent exposure that affects unfoamed polystyrene to a minor degree can cause collapse of the foam cells.

Chlorinated and aromatic hydrocarbons, esters, and ketones are excellent solvents for polystyrene. The aromatic-hydrocarbon portion of gasoline, solvent naphtha, fuel oil, and mineral oil exerts a solvent action on polystyrene. Foamed polystyrene does not soften or swell after 70-h exposure to lubricating oil, conforming to Military Specification MIL-L-6082. Essential oils of high terpene content, like oil of lemon or orange, as well as turpentine, are excellent solvents for polystyrene. The lower aliphatic alcohols and glycols exert little or no solvent action on polystyrene, but higher alcohols cause relatively slow softening or swelling, as do acetone, aldehydes, ether alcohols, glacial acetic acid, some unsaturated hydrocarbons, and essential oils of low terpene content. Whenever the presence of solvents is suspected, tests of adequate duration at the anticipated use temperature should be performed to determine the effect, if any, of the solvent on the foamed polystyrene.

3. Fungi and Bacteria

Fungus attack has not been observed on foamed polystyrene, and the foam does not support bacterial growth. Fungal or bacterial growth on foam is evidence that the surface of the foam has been soiled. Only the soilage, and not the foam, supplies nutrients for fungal or bacterial growth.

A mixed-spore suspension of fungi was sprayed on specimens of foam that were then incubated for 14 days at 30° C and 95% relative humidity. The suspension contained the spores most commonly specified in military specifications for cushioning materials: *Chaetomium globosum*, USDA 1042.4; *Aspergillus niger*, USDA TC-215-4247; *Aspergillus terreus*, PQMD 82-J; *Penicillium citrinum*, ATCC 9849; *Fusarium moniliforme*,

USDA 1004.1. At the end of the incubation period the specimens were inspected microscopically for growth. There was no evidence of fungal growth or attack on the foamed polystyrene.

Foamed polystyrene has been tested for fungus resistance in accordance with Federal Housing Administration (FHA) "Test Procedures to Determine the Acceptability of Perimeter Insulation for Concrete Floor on Ground," June 11, 1956, paragraph V. In this method culture dishes with garden soil and feeder blocks (pondersoa pine sapwood) were steam-sterilized for 30 min at 15 psig (121° C). The specimens (unsterilized), culture dishes, garden soil, and feeder blocks were sprayed with a single-spore suspension of each of three fungi and incubated at 26.7° C and 70% relative humidity for 90 days, by which time the feeder blocks were overgrown. The spores were Lenzites trabea (Madison), No. 617; Polyporous versicolor, No. 6917; Coniophora puteana, No. 515. At the end of the incubation period the foams were inspected microscopically, and there was no evidence of fungal growth or attack.

Foam was inoculated with a culture of Sporocytophaga myxococcoides and incubated for 7 days at room temperature. At the end of the incubation period the specimen did not display any bacteriological deterioration.

4. Ants, Termites, Rodents, and Marine Borers

Foamed polystyrene is not attractive to ants, termites, and rodents since it has no food value. However, it is not a barrier to them. Experiments have shown that ants, termites, and rodents will chew through foamed polystyrene to reach food. Marine borers can attack polystyrene foam, as they do wood, and the foam should be protected by an antifouling paint over a suitable primer.

J. Flammability

The heat of combustion of polystyrene is 16,000 Btu/lb. The combustion prodcuts are carbon dioxide, carbon monoxide, water, and soot (carbon).

Regular grades of foam are characterized as Class 1 by ASTM Designation D2125-62T, "Tentative Specification for Cellular Polystyrene." Class 1 is defined as "natural or colored and not nonburning or self-extinguishing when tested in accordance with the Method of Test for Flammability of Plastic Foams and Sheeting (ASTM Designation D1692)." Self-extinguishing grades of foam are characterized as Class 2 by ASTM Designation D2125-62T. Class 2 is defined as "natural or colored and nonburning or self-extinguishing when tested in accordance with ASTM Designation D1692." Such foams may be classed as flame retardant. The term "flame-retardant" denotes "combustible materials which have been subjected to treatment to prevent or retard ignition or the spread of fire under conditions of use for which they were designed."

K. Light Stability

Exposure of polystyrene foams to direct sunlight will cause deterioration of the surface cellular structure. Such degradation is usually evidenced by discoloration. Degradation of the self-extinguishing and some colored grades is more rapid than that of regular, noncolored grades. The degraded surface should be removed by brushing or air blasting before applying any material requiring adhesion to the foam.

L. Toxicological Properties

1. Molded Beads

a. Personnel Contact. The components of expandable polystyrene are generally regarded as safe by competent toxicologists. Years of experience under varied exposure conditions in numerous foam-producing plants have demonstrated the harmlessness of expandable-polystyrene beads and foam.

Foam made from flame-retardant grades of expandable polystyrene cannot be given the same toxicological clearance as foam made from regular grades of expandable polystyrene.

b. Food Contact. The foam made from regular grades of expandable polystyrene "under conditions of good manufacturing practice" enjoys an exempt status under the Food Additives Amendment of 1958 to the Food, Drug and Cosmetic Act of 1938. The exempt status is an official acknowledgement of safety for such uses. This exemption does not extend to foam made from flame-retardant expandable polystyrene. Likewise, it does not extend to foam that contains colorants or other additives if such added substances may migrate into food and are neither "generally recognized as safe," nor "prior sanctioned," nor used in minimum quantity or of appropriate grade of purity. Official recognition by regulation as to the safety of use in contact with food does not relieve the user of the risk that the food may be adulterated, as shown by change in taste, odor, or other characteristics as the result of contact with the foam. To avoid this liability food containers should be tested before quantity use to be sure that the food does not become tainted or adulterated.

2. Extruded Board

Extruded polystyrene-foam board itself presents no toxicity problem, although certain fabricating methods may. These can be adequately dealt with through the use of proper ventilating equipment. Fabricated pieces are nonirritating and present no health hazard in handling.

ACKNOWLEDGMENTS

The authors wish to express their gratitude to the following people for their assistance and advice in the preparation of this chapter: P. B. Nelson, J. P. Sipe, T. P. Martens, R. H. Immel, R. R. Cobbs, R. J. Hochschild, G. C. Kiessling, and J. H. Ashcom, all of the Sinclair-Koppers Co.; and E. M. Fettes, H. A. Wright, F. M. Angeloni, E. W. Kifer, E. P. Meckly, J. M. Bialosky, L. C. Couchot, W. Kurvach, and M. A. Deem, of the Research Department of the Koppers Co.

REFERENCES

1. Mod. Plastics, 47, No. 1, 76 (1970).
2. Mod. Plastics, 31, No. 9, 103 (1954).
3. A. R. Ingram, paper presented at the 155th Meeting, American Chemical Society, San Francisco, April 1968; published in Addition and Condensation Polymerization Processes (R. F. Gould, ed.), Advances in Chemistry Series 91, American Chemical Society, Washington, D.C., 1969, Chapter 33.
4. J. J. P. Staudinger, B. K. Kelly, and A. Cooper (one-half to Distillers Co. and one-half to Expanded Rubber Co.), British Pat. 605,863, (1948).
5. J. Sunde, Norwegian Pat. 95,882 (1960).
6. R. Colwell (to Monsanto), U.S. Pats. 2,857,339, 2,857,340, and 2,857,341 (1958); N. Platzer (to Monsanto), U.S. Pat. 2,857,342 (1958).
7. N. Platzer (to Monsanto), U.S. Pat. 3,072,581 (1963).
8. Stamicarbon N. V., Belgian Pat. 661,777 (1965).
9. A. A. Eng (to Union Carbide), U.S. Pat. 3,304,274 (1967).
10. F. Stastny and K. Buchholz (to BASF), German Pat. 951,299 (1956).
11. F. Stastny and R. Gaeth (to BASF), German Pat. 1,009,391 (1957).
12. S. T. Day (to Crown Machine and Tool), U.S. Pat. 3,126,354 (1964).
13. J. M. Harrison (to Crown Machine and Tool) U.S. Pat. 3,127,360 (1964).
14. D. A. Russell and E. E. Hardy (to Monsanto), U.S. Pat. 3,262,625 (1966).
15. A. A. Aykanian (to Monsanto), U.S. Pat. 3,324,210 (1967).
16. R. K. Shelby (to Monsanto), U.S. Pats. 3,300,551, 3,300,552, and 3,300,553 (1967).
17. A. A. Aykanian and R. K. Shelby (to Monsanto) U.S. Pat. 3,424,826 (1969).
18. H. E. Knobloch, F. Meyer, and F. Stastny (to BASF), German Pat. 1,214,395 (1966).

19. C. E. Bushnell, H. F. Long, and R. C. Stadden (to Armstrong Cork), U. S. Pat. 3,389,199 (1968).
20. Badische Anilin- und Soda-Fabrik AG, Belgian Pat. 604,776 (1961).
21. J. D. Griffin (to Dow Chemical), Belgian Pat. 651,577 (1965).
22. G. F. D'Alelio (to Koppers), U. S. Pat. 2,983,692 (1961).
23. K. Buchholz, F. Stastny, and R. Gaeth (to BASF), U. S. Pat. 2,950,261 (1960).
24. W. J. Cleland, R. G. Thomas, and E. Seijo (to Styrene Products Ltd.), U. S. Pat. 2,893,963 (1959).
25. J. Lintner, H. Petrovicki, and F. Schaffernack (to K. Holl, J. Lintner, H. Petrovicki, and F. Schaffernack), U. S. Pat. 3,085,073 (1963).
26. W. E. Hall and A. C. Poshkus (to Armstrong Cork) U. S. Pat. 3,088,925 (1963).
27. Societa Edison, Canadian Pat. 738,490 (1966).
28. Rexall Drug and Chemical Co., British Pat. 1,155,904 (1969).
29. E. B. Ulmanis (to Dow Chemical), British Pats. 988,289 and 988,290 (1965).
30. A. K. Jahn (to Dow Chemical), British Pat. 1,009,130 (1965).
31. D. J. Rode and G. Greenawald (to Koppers), U. S. Pat. 3,342,760 (1967).
32. F. Rosenthal (to Radio Corp. of America), U. S. Pat. 2,533,629 (1950).
33. F. Stastny and R. Gaeth (to BASF), U. S. Pat. 2,681,321 (1954).
34. L. C. Rubens and W. B. Walsh (to Dow Chemical), French Pat. 1,476,888 (1967).
35. F. Stastny and K. Buchholz (to BASF), U. S. Pat. 2,744,291 (1956).
36. H. Mueller-Tamm and J. Grohmann (to BASF), German Pat. 1,163,025 (1964).
37. K. Buchholz, J. Grohmann, and E. Stahnecker (to BASF), Belgian Pat. 668,325 (1965).
38. L. Zuern, H. Mueller-Tamm, and K. Buchholz (to BASF), U. S. Pat. 3,324,052 (1967).
39. P. V. Burt (to Monsanto Chemicals Ltd.), British Pat. 1,104,926 (1968).
40. E. Stahnecker, H. Mueller-Tamm, L. Zuern, J. Grohmann, and K. Buchholz (to BASF), German Pat. 1,152,261 (1963).
41. K. W. Doak (to Koppers), U. S. Pat. 3,192,169 (1965).
42. Foster Grant Co., Netherlands Pat. Appl. 6,606,609 (1966).
43. I. Hatano, K. Senuma, T. Kasanatsu, and M. Nishino (to Kanegafuchi Chemical Industry Co.), U. S. Pat. 3,265,643 (1966).
44. Societa Edison, British Pat. 1,062,011 (1967).
45. Produits Chimiques Pechiney-Saint-Gobain, British Pat. 1,063,333 (1967).

46. O. Marek, J. Mrazek, M. Tomka, and M. Cerny, British Pat. 978,631 (1964).
47. P. Wolff and S. H. Kaaber, British Pat. 1,021,250 (1966).
48. Compagnie Internationale pour le Commerce et l'Industrie, Belgian Pat. 566,994 (1958).
49. J. M. Grim (to Koppers), U.S. Pat. 2,594,913 (1952).
50. The Dow Chemical Co., Belgian Pat. 677,173 (1966).
51. The Dow Chemical Co., Belgian Pats. 673,061 and 674,357 (1966).
52. E. G. Barber and A. N. Roper (to Shell International Research Maatschappij N. V.), British Pat. 1,056,470 (1967).
53. K. Buchholz (to BASF), U.S. Pat. 2,888,410 (1959).
54. K. Wilkinson (to Monsanto Chemicals Ltd.), British Pat. 1,095,410 (1967).
55. M. H. Roth (to Monsanto), U.S. Pat. 2,816,827 (1957).
56. L. C. Rubens (to Dow Chemical), U.S. Pat. 2,878,194 (1959).
57. A. R. Ingram (to Koppers), Canadian Pat. 718,088 (1965).
58. K. Wilkinson (to Monsanto Chemicals Ltd.), British Pat. 1,101,884 (1968).
59. A. K. Jahn and E. J. Gillard (to Dow Chemical), U.S. Pat. 3,043,817 (1962).
60. T. O. Harmon (to Dow Chemical), British Pat. 1,093,804 (1967).
61. A. V. Yanishevskii and K. V. Voevodin, J. Appl. Chem. USSR (Eng. Transl.), 40 (3), 533 (1967).
62. F. R. Spencer (to American Cyanamid), U.S. Pat. 3,018,257 (1962).
63. H. E. Knobloch, F. Meyer, and F. Stastny (to BASF), U.S. Pat. 3,117,941 (1964).
64. Chemische Werke Hüls AG, British Pat. 1,016,714 (1966).
65. F. Meyer, H. E. Knobloch, H. Freyschlag, and R. Dieter (to BASF), U.S. Pat. 3,154,605 (1964).
66. L. C. Rubens and E. F. Engels (to Dow Chemical), U.S. Pat. 3,026,272; E. F. Engels (to Dow Chemical), U.S. Pat. 3,026,273; W. J. McMillan and G. L. Nicholls (to Dow Chemical), U.S. Pat. 3,026,274 (1962).
67. Badische Anilin- und Soda-Fabrik AG, British Pat. 948,300 (1962).
68. The Dow Chemical Co., Belgian Pat. 622,808 (1963).
69. C. W. Schroeder and M. E. Fuller (to Shell Oil), Canadian Pat. 682,464 (1964).
70. F. H. Collins (to Sun Chemical) U.S. Pat. 3,250,834 (1966).
71. S. J. DelBene (to Koppers), U.S. Pat. 3,121,132 (1964).
72. J. Karpovich (to Dow Chemical), U.S. Pat. 3,072,584 (1963).
73. Badische Anilin- und Soda-Fabrik AG, German Pat. 1,479,980 (1969).
74. J. Lintner, H. Petrovicki, and F. Schaffernack (to K. Holl, J. Lintner, H. Petrovicki, and F. Schaffernack), Canadian Pat. 663,596 (1963).

75. N. Sassa (to Kanegafuchi Chemical), British Pat. 1,106,581 (1968).
76. S. M. Kline and G. F. Lafferty (to Koppers), Canadian Pat. 683,415 (1964).
77. H. Petrovicki, F. Schaffernack, K. Holl, and J. Lintner, German Pat. 1,248,934 (1967).
78. A. R. Ingram and H. W. Jurgeleit (to Koppers), U.S. Pat. 3,222,343 (1965).
79. H. Urbach and H. Mueller-Tamm (to BASF), Belgian Pat. 623,387 (1963).
80. H. Ohlinger, W. Guenther, W. Single, K. Buchholz, and H. Wild (to BASF), U.S. Pat. 3,287,286 (1966).
81. V. Bonin and H. Bartl (to Farbenfabriken Bayer), German Pat. 1,160,616 (1964).
82. G. Will, U.S. Pat. 3,256,219 (1966).
83. Chemische Fabrik Kalk G.m.b.H., German Pat. 1,495,227 (1968).
84. A. R. Ingram and H. A. Wright, Mod. Plastics, 41, No. 3, 152 (1963).
85. H. A. Wright (to Koppers), U.S. Pats. 3,027,334, 3,027,335, and 3,060,138 (1962).
86. I. Hatano and K. Senuma (to Kanegafuchi Chemical), U.S. Pat. 3,265,642 (1966); I. Hatano, K. Senuma, T. Kasamatsu, and M. Nishino (to Kanegafuchi Chemical), U.S. Pat. 3,265,643 (1966).
87. R. A. Reed, R. C. H. Spencer, and W. G. Barb (to Whiffen & Sons Ltd.), British Pat. 790,312 (1958).
88. E. A. Hedman and S. R. Mather (to the Richardson Co.), U.S. Pat. 3,138,478 (1964).
89. E. Pilz et al., East German Pat. 51,424 (1966).
90. G. F. D'Alelio (to Scott Paper), U.S. Pat. 3,224,983 (1965).
91. W. C. Overhults (to W. R. Grace and Co.), British Pat. 938,639 (1963).
92. Badische Anilin- und Soda-Fabrik AG, Belgian Pat. 661,324 (1965).
93. F. L. Saunders and J. H. Oswald (to Dow Chemical), U.S. Pat. 3,449,270 (1969).
94. N. E. Scheffler (to Dow Chemical), U.S. Pat. 3,449,268 (1969).
95. Borg-Warner Corp., Netherlands Pat. Appl. 6,413,848 (1965).
96. L. C. Rubens (to Dow Chemical), U.S. Pat. 2,952,594 (1960).
97. S. J. DelBene and A. J. Fox (to Koppers), U.S. Pat. 3,285,865 (1966).
98. A. R. Ingram and E. H. Gleason (to Koppers), U.S. Pat. 3,288,731 (1966).
99. Y. Motoishi (to Kurashiki Rayon), Japanese Pat. 16,437 (1967).
100. O. Marek, M. Tomka, J. Mrazek, and M. Cerny, British Pat. 1,013,215 (1965).
101. The Dow Chemical Co., Netherlands Pat. Appl. 6,514,797 (1966).
102. L. C. Rubens, J. Cell. Plastics, 1, 311 (1965); L. C. Rubens (to Dow Chemical), U.S. Pat. 2,848,427 (1958).

103. L. C. Rubens (to Dow Chemical), U. S. Pat. 2,848,428 (1958).
104. L. C. Rubens and J. S. Warren (to Dow Chemical), U. S. Pat. 3,033,805 (1962).
105. F. Scalari and G. Sabatini (to Montecatini Societa Generale per l'Industria Mineraria e Chimica), Italian Pat. 664,755 (1964).
106. G. F. D'Alelio (to Koppers), U. S. Pat. 2,994,670 (1961).
107. H. A. Wright (to Koppers), U. S. Pats. 3,259,594 and 3,259,595 (1966).
108. Shell International Research Maatschappij N. V., Netherlands Pat. Appl. 6,506,473 (1965).
109. E. Stahnecker, H. Mueller-Tamm, and H. Friederich (to BASF), British Pat. 1,015,709 (1966).
110. Asahi Electro-Chemical Company Ltd., Japanese Pat. 2,148 (1967).
111. R. F. Lindemann, Ind. Eng. Chem., 61 (5), 70 (1969).
112. J. Eichhorn (to Dow Chemical), U. S. Pat. 3,058,926 (1962).
113. J. Eichhorn, J. Appl. Polymer Sci., 8, 2497 (1964).
114. A. K. Jahn and J. W. Vanderhoff, J. Appl. Polymer Sci., 8, 2525 (1964).
115. K. Buchholz and F. Stastny (to BASF), U. S. Pat. 3,001,954 (1961).
116. C. F. Raley, Jr., and W. R. Nummy (to Dow Chemical), U. S. Pat. 3,004,935 (1961).
117. H. Mueller-Tamm, K. Buchholz, and F. Stastny (to BASF), U. S. Pat. 3,093,599 (1963).
118. Badische Anilin- und Soda Fabrik AG, Belgian Pat. 691,252 (1967).
119. H. A. Wright (to Koppers), U. S. Pat. 3,359,220 (1967).
120. J. Eichhorn (to Dow Chemical), U. S. Pat. 3,284,544 (1966).
121. J. Eichhorn (to Dow Chemical), U. S. Pat. 3,296,340 (1967).
122. A. R. Ingram (to Koppers), U. S. Pat. 3,274,133 (1966).
123. E. Priebe, H. Weber, and H. Willersinn (to BASF), German Pat. 1,245,593 (1967).
124. H. Laib and H. Burger (to BASF), German Pat. 1,244,396 (1967).
125. J. Eichhorn (to Dow Chemical), U. S. Pat. 3,271,333 (1966).
126. H. Weber and H. Burger (to BASF), U. S. Pat. 3,420,786 (1969).
127. H. Burger, G. Daumiller, J. Grohmann, E. Kastning, H. Weber, and H. Willersinn (to BASF), U. S. Pat. 3,457,204 (1969).
128. R. Ilgemann and R. D. Rauschenbach (to BASF), French Pat. 1,411,363 (1965).
129. H. Willersinn, R. D. Rauschenbach, and R. Ilgemann (to BASF), French Pat. 1,425,972 (1966).
130. Badische Anilin- und Soda-Fabrik Ag, Belgian Pat. 642,915 (1964).
131. M. L. Zweigle (to Dow Chemical), U. S. Pat. 3,056,752 (1962).
132. A. K. Jahn (to Dow Chemical), U. S. Pat. 3,086,885 (1963).

133. Shell International Research Maatschappij N. V., Belgian Pat. 626,772 (1963).
134. P. D. Marsden (to Shell Oil), U. S. Pat. 3,429,737 (1969).
135. T. H. Ferrigno (to Minerals and Chemicals Phillip Corp.), U. S. Pats. 3,300,437 and 3,301,812 (1967).
136. W. H. Voris (to Koppers), U. S. Pat. 3,369,927 (1968).
137. A. R. Ingram and A. J. Zupanc (to Koppers), U. S. Pat. 3,370,022 (1968).
138. A. R. Ingram and H. A. Wright (to Koppers) U. S. Pat. 3,389,097 (1968).
139. R. H. Immel and P. B. Nelson (to Sinclair-Koppers), Canadian Pat. 784,270 (1968).
140. B. M. Sergot, L. A. Muirhead, and B. H. Williams (to Shell International Research Maatschappij N. V.), British Pat. 1,132,147 (1968).
141. Shell International Research Maatschappij N. V., South African Pat. 68/4206 (1969).
142. H. Mueller-Tamm and E. Stahnecker (to BASF), German Pat. 1,298,274 (1969).
143. J. Bianco (to Koppers), U. S. Pat. 3,020,247 (1962).
144. A. Yahata and H. Kitada (to Sanyo Kako Co.), Japanese Pat. 6,806,547 (1968).
145. H. L. Nicholson (to Koppers), U. S. Pat. 3,399,025 (1968).
146. Geigy AG, Belgian Pat. 669,399 (1965).
147. E. Hein, German Pat. 1,221,008 (1966).
148. Farbwerke Hoechst AG, Belgian Pat. 710,630 (1968).
149. E. Stahnecker and H. Mueller-Tamm (to BASF), Belgian Pat. 627,053 (1963).
150. Badische Anilin- und Soda-Fabrik AG, French Pat. 1,463,625 (1966).
151. Badische Anilin- und Soda-Fabrik AG, Netherlands Pat. Appl. 6,516,742 (1966).
152. A. N. Roper and E. G. Barber (to Shell Oil), U.S. Pat. 3,224,984 (1965).
153. S. Hosokawa, Y. Horikawa, K. Yoshiwara, and S. Miyazawa (to Sekisui Chemical Co. and Sekisui Sponge Industry Co.) Japanese Pat. 68,06522 (1968).
154. M. H. Roth (to Monsanto), U. S. Pat. 2,911,381 (1959).
155. M. H. Roth (to Monsanto), U. S. Pat. 2,910,446 (1959).
156. E. G. Pollard and L. C. Rubens (to Dow Chemical), U. S. Pat. 3,013,996 (1961); E. D. Andrews, D. V. Francis, and D. J. Rode (to Koppers), U. S. Pat. 3,207,712 (1965).
157. A. R. Ingram, paper presented at the Wayne State University Polymer Conference on Cellular Materials, Detroit, May 1967.

158. S. J. Skinner, S. Baxter, and P. J. Grey, Plastics Inst. (London) Trans. J., 32 (97), 180 (1964).
159. S. J. Skinner, S. Baxter, S. D. Eagleton, and P. J. Grey, Mod. Plastics, 42 (5), 171 (1965).
160. A. R. Ingram, E. Lyle, and E. M. Fettes, in Macromolecular Synthesis (N. G. Gaylord, ed.), Vol. 3, Wiley, New York, 1969, pp. 135-139.
161. G. A. Pogany, Brit. Plastics, 37, 506 (1964).
162. R. N. Haward, H. Breuer, and G. Rehage, J. Polymer Sci., Part B, 4, 375 (1966).
163. W. Biltz, Raumchemie der festen Stoffe, Voss, Leipzig, 1934.
164. J. D. Ferry, Viscoelastic Properties of Polymers, Wiley, New York, 1961, pp. 219 and 225.
165. A. R. Ingram, J. Cell. Plastics, 1, 69 (1965).
166. K. Yamamoto and Y. Wada, J. Phys. Soc. Japan, 12, 374 (1957).
167. K. H. Illers and E. Jenckel, J. Polymer Sci., 41, 528 (1959).
168. A. R. Ingram, R. R. Cobbs, and L. C. Couchot, in Resinography of Cellular Plastics (R. E. Wright, ed.), ASTM STP 414, American Society for Testing and Materials, Philadelphia, 1967, pp. 53-67.
169. F. Angeloni and E. W. Kifer (Koppers Co.), unpublished work, 1969.
170. H. Rodman, Jr. (to Koppers), U. S. Pat. 3,023,175 (1962).
171. T. Couchman (to D. Bloom), U. S. Pat. 3,139,272 (1964).
172. F. B. Brockhues and W. Muhm, U. S. Pat. 3,257,103 (1966).
173. B. Oxel (to Isolerings Aktiebolaget WMB), Canadian Pat. 706,305 (1965).
174. Compagnie de Saint-Gobain, French Pat. 1,561,173 (1969).
175. P. E. Fischer, W. A. Heatley, Jr., and F. H. Lambert (to Champlain-Zapata Plastics Machinery), U. S. Pat. 3,155,379 (1964).
176. Poron Insulation Ltd., Australian Pat. 64,003/65 (1967).
177. Koppers Co., British Pat. 849,117 (1960).
178. K. E. Anderson, J. E. Welford, and G. J. Edney (to B. X. Plastics Ltd.), British Pat. 1,032,036 (1966).
179. Rohpappen-Fabrik Worms, Belgian Pat. 666,397 (1965).
180. K. R. Denslow (to Dow Chemical), U. S. Pat. 3,428,720 (1969).
181. G. L. Attanasio and F. H. Lambert (to Champlain-Zapata Plastics Machinery), U. S. Pat. 3,162,704 (1964).
182. A. Columba, Italian Pat. 727,577 (1967).
183. E. A. Edberg and J. J. Tress (to Koppers), U. S. Pat. 3,015,479 (1962).
184. I. J. Stanchel, U. S. Pat. 3,273,873 (1966).
185. R. A. V. Raff and M. K. Adams, SPE Journal, 24 (11) 56 (1968).
186. R. A. Kraus and E. J. Kraus, U. S. Pat. 3,262,686 (1966).

187. J. J. Boot and W. de F. Anderson (to Formica International Ltd.), British Pat. 922,547 (1963).
188. W. J. McMillan and K. R. Denslow (to Dow Chemical), U. S. Pat. 2,884,386 (1959).
189. G. Schuur (to Shell Oil), U. S. Pat. 3,126,432 (1964).
190. Compagnie de Saint-Gobain, French Pat. 1,440,076 (1966).
191. Compagnie de Saint-Gobain, French Pat. 1,552,124 (1969).
192. Badische Anilin- und Soda-Fabrik AG, French Pats. 1,497,194 and 1,497,195 (1967).
193. The Dow Chemical Co., Belgain Pats. 672,335,672,336, and 672,349 (1965).
194. M. R. Oswald and R. L. Kinsey (to Dow Chemical), U. S. Pat. 3,147,321 (1964).
195. D. C. Paulson (to Dow Chemical), U. S. Pat. 3,165,303 (1965).
196. R. P. Lowry (to Dow Chemical), U. S. Pat. 3,263,981 (1966).
197. C. Kienzle (to Foster Grant Co.), French Pat. 1,459,599 (1966).
198. J. R. Sare, R. L. Ropiquet, and D. G. Bergeron (to Alta Industries), U. S. Pat. 3,400,037 (1968).
199. J. W. C. Miller, Rubber and Plastics Weekly, 277 (1963).
200. G. Schuur, Brit. Plastics, 38, 219 (1965); Plastica, 18 (2), 74, (1965).
201. S. Baxter, F. Heath, and W. B. Brown (to Monsanto Chemicals Ltd), U. S. Pat. 3,328,497 (1967).
202. E. J. Knapp (to Corning Glass Works), U. S. Pat. 3,417,170 (1968).
203. Compagnie de Saint-Gobain, French Pat. 1,555,781 (1969).
204. F. L. Landon, U. S. Pat. 3,446,882 (1969).
205. G. Kracht (to Swedish Crucible Steel), U. S. Pat. 3,309,440 (1967).
206. D. L. Russell (to Dow Chemical), U. S. Pat. 3,347,961 (1967).
207. S. J. Skinner (to Monsanto), British Pat. 1,005,137 (1965).
208. G. S. Scarvelis and G. R. Zaloudek (to Owens-Illinois Glass), U. S. Pat. 3,207,820 (1965).
209. A. Bracht, French Pat. 1,336,261 (1963).
210. M. F. Fronko (to Koppers), U. S. Pat. 3,056,753 (1962).
211. P. C. Chaumeton (to Styrene Products Ltd.), U. S. Pat. 2,986,537 (1961).
212. P. B. Nelson and R. H. Immel, paper presented at the Wayne State University Polymer Conference on Cellular Materials, Detroit, May 1966.
213. F. Stastny (to BASF), U. S. Pat. 2,787,809 (1957).
214. S. J. Skinner and S. D. Eagleton, Plastics Inst. (London) Trans. J., 32, 231 (1964).
215. S. D. Eagleton and S. J. Skinner (to Monsanto Chemicals Ltd.), British Pat. 1,012,277 (1965).
216. Badische Anilin- und Soda-Fabrik AG, Netherlands Pat. 6,609,560 (1967).

217. L. Zuern, E. Stahnecker, J. Grohmann, and K. Buchholz (to BASF), German Pat. 1,256,888 (1967).
218. S. D. Eagleton and S. J. Skinner (to Monsanto Chemicals Ltd.), British Pat. 1,083,040 (1967).
219. B. Roberts and R. J. Stephenson (to Monsanto Chemicals Ltd.), British Pat. 1,093,899 (1967).
220. P. V. Burt (to Monsanto Chemicals Ltd.), British Pat. 1,082,966 (1967).
221. Badische Anilin- und Soda-Fabrik AG, Belgian Pat. 691,326 (1967).
222. J. E. Becker, paper presented at the 7th Meeting of the Society of the Plastics Industry, New York, April 1963.
223. L. S. Norrhede and E. T. Linde (to Isolerings Aktienbolaget WMB), Swedish Pat. 171,095 (1960).
224. B. Oxel (to Dyfoam), U.S. Pat. 3,262,151 (1966).
225. B. Oxel, Plastics Des. Proc., 6, No. 2, 36 (1966).
226. R. S. Schultz, E. F. Polka, and J. A. Cherney (to American Can), U.S. Pat. 3,170,010 (1965).
227. W. A. Dart (to Dart Manufacturing), U.S. Pat. 3,178,491 (1965).
228. C. R. Wiles and K. R. Hilton (to Dow Chemical), U.S. Pat. 3,196,484 (1965).
229. R. S. Robbins, W. T. Palmer, and J. F. Newell, Jr. (to Crown Machine and Tool), U.S. Pat. 3,224,037 (1965).
230. R. G. Bridges, R. N. Aleson, D. B. Hutchings, R. E. Whited, and M. L. Dyrness (to Tempo Plastic), U.S. Pats. 3,225,126 and 3,224,040 (1965).
231. E. A. Edberg (to Koppers), U.S. Pat. 3,042,967 (1962).
232. Societe Commerciale de Distribution et de Representation, French Pat. 1,551,168 (1968).
233. Foley Packaging and Insulation Ltd., French Pat. 1,555,352 (1969).
234. I.O. Salyer, J. L. Schwendeman, C. J. North, and L. E. Erbagh, J. Cell. Plastics, 5, 99 (1969).
235. J. M. Harrison and R. E. Smucker (to Crown Machine and Tool), U.S. Pat. 2,951,260 (1960).
236. C. E. Plymale (to Owens-Illinois Glass), U.S. Pat. 3,111,710 (1963).
237. R. E. Smucker and J. M. Harrison (to Crown Machine and Tool), U.S. Pat. 3,162,705 (1964).
238. R. A. Kraus and E. J. Kraus, U.S. Pat. 3,167,811 (1965).
239. E. A. Edberg and R. H. Immel (to Koppers), U.S. Pat. 2,998,501 (1961).
240. F. Stastny and B. Ikert (to BASF), British Pat. 924,767 (1963).
241. E. A. Edberg and R. H. Immel (to Koppers), U.S. Pat. 3,242,238 (1966).

242. H. Mandel (to Darplastex AG), U.S. Pat. 3,341,638 (1967).
243. B. Ikert, Kuntstoffe, 55, 181 (1965).
244. Plastics Week, 26 (48), 4 (1966).
245. C. R. Wiles, E. F. Builford, and L. R. Stanford (to Dow Chemical), U. S. Pat. 3,173,975 (1965).
246. Tokyo Keiki Seizosho K. K., British Pat. 1,117,885 (1968).
247. F. Stastny (to BASF), French Pat. 1,428,153 (1966).
248. Badische Anilin- und Soda-Fabrik Ag, Netherlands Pat. Appl. 6,609,560 (1967).
249. Compagnie de Saint-Gobain, French Pat. 1,440,106 (1966).
250. Societa Edison, British Pat. 1,049,936 (1966).
251. P. Oddi (to Montecatini S. p. A.) U. S. Pat. 3,368,009 (1968).
252. Compagnie de Saint-Gobain, French Pat. 91,244/1,440,106 (1968).
253. R. H. Immel (to Koppers), U. S. Pat. 3,278,658 (1966).
254. Badische Anilin- und Soda-Fabrik AG, French Pat. 1,500,623 (1967).
255. R. C. Medhurst (to Standard Oil Co. of Indiana), U. S. Pat. 3,435,103 (1969).
256. F. B. Mercer (to Plastic Textile Accessories Ltd.), British Pat. 1,058,637 (1967).
257. H. Suzaki, M. Okuda, and E. Hattori (to Asahi Chemical Industry Co.), Japanese Pat. 67,11513 (1967).
258. R. A. Nonweiler (to Lakeside Plastics), U. S. Pat. 3,457,205 (1969).
259. R. F. Newberg and R. O. Newman (to Dow Chemical), U. S. Pat. 2,958,905 (1960).
260. D. L. Graham, R. N. Kennedy, and E. L. Kropscott (to Dow Chemical), U. S. Pat. 2,959,508 (1960).
261. L. Zuern and H. Mueller-Tamm (to BASF), Belgian Pat. 641,661 (1964).
262. Badische Anilin- und Soda-Fabrik AG, German Pat. 1,287,794 (1969).
263. H. F. Shroyer, U. S. Pat. 2,830,343 (1958).
264. E. & L. Crabbe, British Pats. 963,005 and 963,006 (1964).
265. Badische Anilin-und Soda-Fabrik AG, Belgian Pat. 675,893 (1966).
266. Rheinstahl Hutten-Werke AG, German Pat. 1,221,401 (1966).
267. T. J. Grabowski (to Armstrong Cork), U. S. Pat. 3,058,162 (1962).
268. Swedish Crucible Steel Co., British Pat. 1,110,350 (1968).
269. Swedish Crucible Steel Co., British Pat. 1,073,813 (1967).
270. The Dow Chemical Co., Australian Pat. 48,917/64 (1966).
271. Continental Can Co., British Pat. 1,021,902 (1966).
272. Poron Insulation Ltd., Australian Pat. 63,263/65 (1967).
273. Compagnie de Saint-Gobain, South African Pat. 67/6596 (1968).
274. Olsonite Co., Canadian Pat. 780,265 (1968).
275. F. Stastny (to BASF), French Pat. 1,372,348 (1964).

276. A. M. Donofrio (to International Assemblix) U. S. Pat. 3,192,063 (1965).
277. F. M. Wright and E. E. Parker (to Pittsburgh Plate Glass), U. S. Pat. 3,256,133 (1966).
278. Badische Anilin- und Soda-Fabrik Ag, Belgian Pat. 681,382 (1966).
279. M. E. Baum (to Koppers), U. S. Pat. 3,431,319 (1969).
280. B. R. Sarchet, J. Cell. Plastics, 1, 478 (1965).
281. Koppers Co., German Pat. 1,236,174 (1967).
282. J. Durand, French Pat. 1,528,760 (1969).
283. F. J. Nagel (to Polymer Corp.), U. S. Pat. 2,981,631 (1961).
284. J. W. Hackett and T. R. Santelli (to Owens-Illinois Glass), U. S. Pat. 3,212,915 (1965).
285. Forschungszentrum der Luftafahrtind, British Pat. 970,271 (1964).
286. A. Bracht, German Pat. 1,183,675 (1964).
287. Sekisui Adoheya Kogyo K. K., British Pat. 971,655 (1964).
288. A. R. Ingram (to Koppers), U. S. Pat. 3,307,788 (1967).
289. R. J. Raymond (to Dow Chemical), U. S. Pat. 3,291,762 (1966).
290. M. J. Hiler (to Midland Chemical), U. S. Pat. 2,862,834 (1958).
291. J. Durand, French Pat. 1,528,760 (1969).
292. Jiffy Manufacturing Co., British Pat. 1,035,030 (1966).
293. P. O. Nielsen (to Polymer Engineering), U. S. Pat. 3,257,229 (1966).
294. W. D. Hamilton, U. S. Pat. 3,338,848 (1967).
295. J. L. M. Newnham, P. Merriman, and D. J. Simcox (to Dunlop Rubber), U. S. Pat. 3,251,916 (1966).
296. H. Schmidt and E. G. Haller, Belgian Pat. 671,009 (1965).
297. H. Reinhard, H. Thurn, and W. F. Beckerle (to BASF), French Pat. 1,388,565 (1965).
298. K. Holl, J. Lintner, H. Petrovicki, and F. Schaffernak (to J. Lintner, H. Petrovicki, and F. Schaffernak), German Pat. 1,227,649 (1966).
299. H. W. Wehr and N. J. Guziak (to Dow Chemical), U. S. Pat. 3,091,998 (1963).
300. Badische Anilin- und Soda-Fabrik AG, Netherlands Pat. 6,614,046 (1967).
301. Polymer Engineering Corp., French Pat. 1,505,196 (1967).
302. Badische Anilin- und Soda-Fabrik AG, German Pat. 1,219,221 (1966).
303. Aero-Commerce G. m. b. H., German Pat. 1,479,929 (1969).
304. R. E. Vogel, German Pat. 1,281,679 (1968).
305. R. D. Beaulieu and B. Freedman (to Monsanto), U. S. Pat. 3,245,829 (1966).
306. E. W. Oak (to Baxenden Chemical), British Pat. 1,033,702 (1966).
307. Contraves AG, South African Pat. 67/1632 (1967).
308. F. L. Rose (to Avco), U. S. Pat. 3,375,822 (1968).

309. Badische Anilin- und Soda-Fabrik Ag, German Pat. 1,494,973 (1969).
310. Societe Produits Chemiques Pechiney-Saint-Gobain, French Pat. 1,521,403 (1968).
311. F. Stastny, L. Unterstenhofer, F. Graf, and H. J. Loeffler (to BASF), U. S. Pat. 3,336,184 (1967).
312. L. Raichel, L. Unterstenhofer, and W. Krieger (to BASF), Belgian Pat. 639,111 (1964).
313. Badische Anilin- und Soda-Fabrik AG, Belgian Pat. 668,322 (1965).
314. Badische Anilin- und Soda-Fabrik AG, French Pat. 88,683/1,383,882 (1967).
315. Badische Anilin- und Soda-Fabrik AG, Belgian Pat. 635,484 (1964).
316. La Vermiculite et la Perlite S. A., French Pat. 1,474,167 (1967).
317. J. L. M. Newnham and D. J. Simcox, (to Dunlop Rubber) U. S. Pat. 3,277,026 (1966).
318. G. E. Lightner and N. R. Nickolls (to Monsanto) U. S. Pat. 3,144,492 (1964).
319. Monsanto Chemicals Ltd., British Pat. 1,118,221 (1968).
320. Badische Anilin- und Soda-Fabrik AG, Belgian Pat. 651,869 (1964).
321. S. Bastian, British Pat. 1,041,209 (1966).
322. Badische Anilin- und Soda-Fabrik AG, Belgian Pat. 700,081 (1967).
323. A. J. Crous, South African Pat. 66/6767 (1967).
324. W. Knabe, East German Pat. 54,624 (1967).
325. Badische Anilin- und Soda-Fabrik AG, German Pat. 1,266,680 (1968).
326. Badische Anilin- und Soda-Fabrik AG, Belgian Pat. 718,081 (1968).
327. Manufacture de Machines du Haut-Rhin S. A., Belgian Pat. 705,217 (1968).
328. E. Trombetta, French Pat. 1,520,699 (1968).
329. M. Jeffery, British Pat. 1,118,621 (1968).
330. C. Thevenot and R. A. A. Jeannin, French Pat. 1,534,689 (1968).
331. B. Paille, French Pat. 1,534,941 (1968).
332. A. Stingl, German Pat. 1,471,472 (1969).
333. S. F. Fronda, British Pat. 1,050,892 (1966).
334. Synfibrit G. m. b. H., South African Pat. 65/4843 (1966).
335. G. Kropfhammer, German Pat. 1,471,292 (1968).
336. F. A. Thomas, French Pat. 1,537,078 (1968).
337. L. Stingl, German Pat. 1,471,468 (1969).
338. Lanco and Enterprise Castagnetti Freres & Cie, Belgian Pat. 721,105 (1969).
339. Badische Anilin- und Soda-Fabrik AG, Netherlands Pat. 6,607,684 (1966).
340. Badische Anilin- und Soda-Fabrik AG, Belgian Pat. 657,782 (1964).

341. S. Morgensen and B. G. Nilsson (to Svenska Aktiebolaget Gasaccumulator), U. S. Pat. 3, 254, 144 (1966).
342. Intervam N. V., Netherlands Pat. 6, 607, 033 (1967).
343. R. Trautvetter et al., East German Pat. 63, 319 (1968).
344. BPB Industries Ltd., German Pat. 1, 282, 534 (1968).
345. R. C. Sefton (to Koppers), U. S. Pat. 3, 214, 393 (1965).
346. R. C. Sefton (to Koppers), U. S. Pat. 3, 257, 338 (1966).
347. R. C. Sefton (to Koppers), U. S. Pat. 3, 272, 765 (1966).
348. G. Klein, French Pat. 1, 451, 555 (1966).
349. G. Thiessen (to Koppers), U. S. Pat. 3, 021, 291 (1962).
350. S. Fernhof, U. S. Pat. 2, 996, 389 (1961).
351. E. Einstein and D. O. McCreight (to Harbison-Walker Refractories), U. S. Pat. 3, 176, 054 (1965).
352. Zirconal Ltd., British Pat. 1, 067, 625 (1967).
353. B. Becker & Ziegelwerg Rottenburg Gebruder Meier & Kunze, German Pat. 1, 471, 030 (1968).
354. Didier-Werke AG, German Pat. 1, 471, 086 (1968).
355. Sekisui Suponji Kogyo Kabushiki Kaisha & Atsumi Ohno, Australian Pat. 1643/66 (1967).
356. Correcta Werke G. m. b. H., Swedish Pat. 465, 777 (1969).
357. F. Stastny, W. Schwenke, W. Kunzer, and J. Jung (to BASF), German Pat. 1, 284, 683 (1968).
358. Badische Anilin- und Soda-Fabrik AG, German Pat. 1, 245, 205 (1967).
359. S. J. G. Sandmark and E. H. Hagstam, U. S. Pat. 3, 368, 944 (1968).
360. S. Cratsa (to Koppers), U. S. Pat. 3, 359, 206 (1967).
361. Colgate-Palmolive Co., Belgian Pat. 692, 881 (1967).
362. O. Eckerle, German Pat. 1, 218, 358 (1966).
363. W. O. Weber, SPE Journal, 25, 70 (1969).
364. Appl. Plastics, 12 (9), 19 (1969).
365. A. R. Ingram and H. W. Jurgeleit (to Koppers), U. S. Pat. 3, 222, 343 (1965).
366. J. C. Houston and J. J. Tress (to Koppers), U. S. Pat. 2, 941, 964 (1960).
367. C. H. Pottenger (to Koppers), U. S. Pat. 3, 089, 857 (1963).
368. D. W. Simpson (to Dow Chemical), U. S. Pat. 3, 231, 524 (1966).
369. F. A. Carlson, Jr. (to Monsanto), U. S. Pat. 2, 797, 443 (1957).
370. K. R. Nickolls (to Monsanto), U. S. Pat. 2, 998, 396 (1958).
271. N. Platzer (to Monsanto), U. S. Pat. 2, 861, 898 (1958).
372. R. A. Barkhuff and N. Platzer (to Monsanto), U. S. Pat. 2, 911, 382 (1959).
373. S. P. Nemphos (to Monsanto), U. S. Pat. 2, 956, 960 (1960).
374. F. Stastny et al. (to BASF), German Pat. 1, 144, 911 (1963).
375. R. P. Beaulieu, S. P. Nemphos, and D. A. Popielski (to Monsanto), U. S. Pat. 3, 069, 367 (1962).

376. F. A. Carlson and N. Platzer (to Monsanto), U. S. Pat. 3,084,126 (1963).
377. The Dow Chemical Co., British Pat. 994,610 (1965).
378. Shell International Research Maatschappij N. V., British Pat. 1,005,119 (1965).
379. U. S. Sato, South African Pat. 67/04377 (1968).
380. Shell International Research Maatschappij N. V., Belgian Pat. 712,767 (1968).
381. Chemische Werke Hüls AG, French Pat. 1,541,938 (1968).
382. L. I. Naturman, Plastics Technol., 15 (10), 41 (1969).
383. The Dow Chemical Co., Belgian Pat. 721,152 (1969).
384. F. A. Carlson, Jr. (to Monsanto), U. S. Pat. 2,962,456 (1960).
385. E. Erdman (to Haveg Industries), British Pat. 1,005,477 (1964).
386. Continental Can Co., Belgian Pat. 706,398 (1968).
387. K. Nakamori (to Sekisui Sponge Industry), U. S. Pat. 3,293,196 (1966).
388. K. Azuma, U. S. Pat. 3,368,008 (1968).
389. T. P. Martens, paper presented at the Wayne State University Polymer Conference on Cellular Materials, Detroit, May 1967.
390. T. P. Martens and A. J. Fox, in Thermoplastic Foams for Rigid Packaging, SPE Technical Papers, RETEC, Hartford, Conn., 1965, pp. 6-15.
391. A. A. Aykanian, E. E. Hardy, and G. A. Latinen (to Monsanto), U. S. Pat. 3,160,688 (1964).
392. W. A. Jacobs and F. H. Collins (to Sun Chemical), U. S. Pat. 3,151,192 (1964).
393. G. C. DeWitz and J. Dekker (to Shell Oil), U. S. Pat. 3,344,215 (1967).
394. P. E. Vesilind (to Koppers), U. S. Pat. 3,287,477 (1966).
395. K. Azuma, Canadian Pat. 744,304 (1966).
396. WMB International Aktienbolaget, German Pat. 1,269,334 (1968).
397. Haveg Industries, Inc., Belgian Pat. 723,827 (1969).
398. F. Martelli, Mod. Plastics, 46 (9), 146 (1969).
399. A. A. Aykanian and F. A. Carlson, Jr. (to Monsanto), U. S. Pat. 2,905,972 (1959).
400. A. J. Fox (to Koppers), U. S. Pat. 3,331,103 (1967).
401. A. R. Kudlack (to Koppers), German Pat. 1,200,534 (1965).
402. J. H. Lux (to Haveg Industries), U. S. Pat. 3,189,243 (1965).
403. C. A. Richie (to Owens-Illinois Glass), U.S. Pat. 3,194,864 (1965).
404. American Can Co., British Pat. 1,009,842 (1965).
405. S. M. Kline (to Koppers), U. S. Pat. 3,137,747 (1964).
406. C. A. Richie (to Owens-Illinois), U. S. Pat. 3,335,207 (1967).
407. R. D. Pitsch (to American Excelsior), U. S. Pat. 3,364,519 (1968).

408. American Excelsior Corp., Netherlands Pat. 6,612,913 (1967).
409. N. P. Suh, D. F. Stewart, and R. M. Lamade (to Sweetheart Plastics), U. S. Pat. 3,159,698 (1964).
410. W. A. Stewart (to Koppers), U. S. Pat. 3,217,070 (1965).
411. The Dow Chemical Co., Canadian Pat. 814,321 (1969).
412. J. B. Sisson (to St. Regis Paper), U. S. Pat. 2,917,217 (1959).
413. Shell International Research Maatschappij N. V., Belgian Pat. 678,584 (1966).
414. H. H. Goldsberry and A. J. Fox, SPE Journal, 18, 448 (1962).
415. H. E. N. Collard (to Koppers), Canadian Pat. 623,804 (1961).
416. D. V. Francis (to Koppers), U. S. Pat. 3,344,212 (1967).
417. E. Erdman and F. J. Halligan (to Haveg Industries), British Pat. 1,082,875 (1967).
418. Shell International Research Maatschappij N. V., Belgian Pat. 639,257 (1964).
419. Shell International Research Maatschappij N. V., Netherlands Pat. 6,517,250 (1967).
420. Etablissements Kuhlmann, French Pat. 1,483,880 (1967)
421. Sekisui Chemical Co., French Pat. 1,501,127 (1967).
422. C. G. Munters and J. G. Tandberg (to C. G. Munters), U. S. Pat. 2,023,204 (1935).
423. O. R. McIntire (to Dow Chemical), U. S. Pat. 2,450,436 (1948).
424. O. R. McIntire (to Dow Chemical), U. S. Pat. 2,515,250 (1950).
425. J. L. McCurdy and C. E. DeLong (to Dow Chemical), U. S. Pat. 2,669,751 (1954).
426. F. E. Dulmage (to Dow Chemical), U. S. Pat. 2,537,977 (1951).
427. J. L. McCurdy and C. E. DeLong (to Dow Chemical), U. S. Pat. 2,774,991 (1956).
428. C. E. DeLong (to Dow Chemical), U. S. Pat. 2,577,743 (1951).
429. D. E. Ballast and J. D. Griffin (to Dow Chemical), U. S. Pat. 3,188,295 (1965).
430. M. Nakamura (to Dow Chemical), Belgian Pat. 642,533 (1964).
431. E. E. Ziegler, J. Cell. Plastics, 1, 494 (1965).
432. G. Durand, J. Mace, and C. Nicolaides (to Societe Nationale des Petroles d'Acquitaine), French Pat. 1,461,370 (1966).
433. Dynamit-Nobel AG, Belgian Pat. 654,348 (1963).
434. F. E. Wiley, F. DeBell, and J. DeBell (to Foster Grant Co.), U. S. Pat. 3,121,130 (1964).
435. Badische Anilin- und Soda-Fabrik AG, French Pat. 1,454,465 (1966).
436. S. J. Skinner (to Monsanto Chemicals Ltd.), U. S. Pat. 3,427,371 (1969).
437. S. Baxter and J. H. Gilbert (to Monsanto Chemicals Ltd.), British Pat. 1,034,120 (1966).
438. British Petroleum Company Ltd., Belgian Pat. 725,038 (1969).

439. Monsanto Chemicals Ltd., Belgian Pat. 714,564 (1968).
440. Grunzweig & Hartmann AG, German Pat. 1,288,308 (1969).
441. Monsanto Chemicals Ltd., French Pat. 1,436,966 (1966).
442. Monsanto Chemicals Ltd., Belgian Pat. 715,259 (1968).
443. K. Holl, J. Lintner, H. Petrovicki, F. Schaffernack, and E. Zemb, British Pat. 1,048,865 (1966).
444. K. Azuma, U.S. Pat. 3,374,300 (1968).
445. H. Eberle and G. Wuttke (to Grunzweig & Hartmann AG), U.S. Pat. 3,417,171 (1968).
446. R. G. Angell, Jr., J. Cell. Plastics, 3, 490 (1967).
447. R. V. Jones, P. B. M. Milam, Jr., and N. J. Edmunds, J. Cell. Plastics, 3, 445 (1967).
448. L. L. Scheiner, Plastics Technol., 14 (12), 35 (1968).
449. C. L. Weir, Mod. Plastics, 46 (3), 68 (1969).
450. R. G. Hochschild, in "Polymeric Foams" Symposium at the Program Design Institute, Stevens Institute, Hoboken, N.J., September 1969.
451. L. D. Cochran and C. W. Osborn, SPE Journal, 25 (9) 20 (1969).
452. R. G. Angell, Jr. (to Union Carbide), U.S. Pat. 3,268,636 (1966).
453. T. P. Engel (to Phillips Petroleum), U.S. Pat. 3,342,913 (1967).
454. C. E. Beyer and R. B. Dahl (to Dow Chemical), U.S. Pat. 3,058,161 (1962).
455. L. W. Meyer, SPE Journal, 18, 1399 (1962).
456. S. Baxter (to Monsanto Chemicals Ltd.), U.S. Pat. 3,200,176 (1965).
457. P. E. Boutillier and J. L. Gourlet (to M. P. C. N. E. Kuhlmann), U.S. Pat. 3,219,597 (1965).
458. A. Spaak and C. W. Weir (to W. R. Grace Co.), U.S. Pat. 3,211,604 (1965).
459. S. N. Weissman and C. L. Weir (to Allied Chemical), U.S. Pat. 3,384,691 (1968).
460. D. C. Woollard, J. Cell. Plastics, 4, 16 (1968).
461. Properties of Foamed Dylite Polystyrene, Sinclair-Koppers Co. Bull. 9-273, Pittsburgh, Pa. 1969, Chapter 2.
462. Styrofoam Brand Plastic Foam, Dow Chemical Co. Bull. Form No. 172-162-20M-11-67, Midland, Mich., 1967.
463. Principles of Packaging Design, Sinclair-Koppers Co. Bull. 9-273, Pittsburgh, Pa., 1969, Chapter 5a.

Chapter 11

PHENOLIC FOAMS

Anthony J. Papa
and
William R. Proops

Union Carbide Corporation
South Charleston, West Virginia

I. INTRODUCTION

The history of phenolic foams is a part of the development of the phenolic plastics industry itself. The polymerization of phenol-formaldehyde resins was first investigated by Baekeland [1, 2], who observed, as did others (see Ref. [1]), that the evolution of gaseous products during processing precluded the preparation of moldings of phenol-formaldehyde resins. Only undesirable products, characterized as "spongy, porous, and unfit for commercial use," were obtained. Baekeland found that the gaseous expansion of the resin could be overcome by the simultaneous application of heat and pressure [2] during molding; this discovery led to the commercial production of Bakelite.

One of the first recorded structural applications for phenolic foams was their reported use in Germany under the trade name Troporit P, as a replacement for balsa wood in the aircraft industry, during the early part of World War II [3-5]. At about the same time Thermazote phenolic foams were introduced in England [6]. A considerable amount of research on the development of low-density phenolic foaming resins and processing techniques was also initiated as early as 1945 by the Union Carbide Corp. in the United States.

Technical advancements [7, 8] leading to continuous production caused considerable activity in phenolic foam for sandwich construction panels in Germany [9] and France [10, 11]. However, in spite of their many desirable properties for structural applications, phenolic foams have been slow in gaining serious recognition as thermal insulation material in this country. Their brittleness and open-celled structure have put them at a disadvantage when compared with other more popular materials, such as the polyurethanes. As a result phenolic foams have been limited to specialty applications--for example, the decorative floral market [12] and fissile material containers [13].

Several other early reviews on phenolic foams are available [6, 8, 14-18]. This chapter is concerned mainly with recent activities in the field of phenolic foams, special attention being given to their processing and applications.

II. RAW MATERIALS USED IN PHENOLIC FOAMS

Phenolic foams are most commonly produced by the acid curing of resole resins initiated at room temperature. Novolac resins, when heated with a curing agent that liberates formaldehyde, also give phenolic foams. These two phenolic resins are generally produced by the condensation of phenol and formaldehyde. Foamable resin compositions also require a blowing agent and a surfactant. Frequently other modifying ingredients are included to control foam density, texture, and mechanical properties.

Only a few of the numerous early references to the synthesis of phenolic resins are cited in this section; up-to-date procedures can best be obtained by consulting the patents cited throughout this chapter.

A. Preparation and Properties of Foam Resins

1. Resole and Novolac Resins

Most phenolic foaming resins are composed of phenol-formaldehyde, prepolymers, which are of the condensation type and are classified into two major types: the resole (or one-step) and the novolac (or two-step) resins. The resole resins are conventionally prepared by carrying out the condensation in a ratio of 1 mole of phenol to about 1.5 to 3.0 moles of formaldehyde under alkaline conditions [19-21]. Oxides and hydroxides of barium [22], sodium and potassium; sodium carbonate and bicarbonate; ammonium hydroxide; and alkyl amines are typical alkaline catalysts. After the reaction has proceeded to the desired state, the alkaline catalyst is neutralized, and the neutralized resin is vacuum-distilled to remove water of condensation to the desired solids content. These liquid resole products are of prime concern for foaming and are potentially thermosetting; this reaction proceeds even at room temperature and can be accelerated by heat or acidic catalysts. Undoubtedly the resins are complex mixtures of mononuclear and polynuclear phenolic molecules containing reactive methylol groups. A typical resole structure is compound (1).

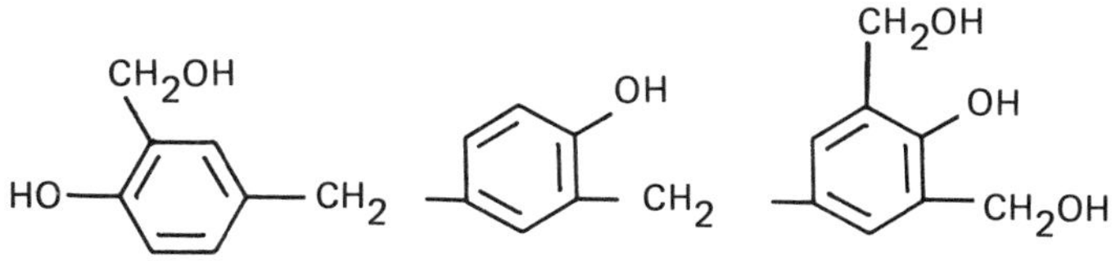

CH_2OH HO— —CH_2— OH —CH_2— CH_2OH OH —CH_2OH

(1)

In addition to phenol itself many other substituted phenols have been employed for the reaction: resorcinol [23-25] m-cresol [26, 27], 3,5-xylenol [28], and aminophenols.

The aldehydes employed are formaldehyde, furfural [29], acrolein [30], alkyl aldehydes, aryl aldehydes, and polyaldehydes.

Novolacs are prepared by employing a slight excess of phenol in the condensation reaction in the presence of such acidic catalysts as oxalic acid [19-21]. These resins are higher condensation products than resoles and are linear in structure. Because of the increased molecular weight and lack of methylol groups, they are no longer in the liquid form and soluble in water and alkali. Consequently in foaming compositions novolacs require the addition of a curing agent like hexamethylenetetramine, which provides formaldehyde for crosslinking by means of introducing additional methylene bridges to the final polymer stage. Final curing is effected by the application of heat to the novolac-hexamethyleneteramine mixture. Again the structure of a novolac is complex, but is typically illustrated by compound (2).

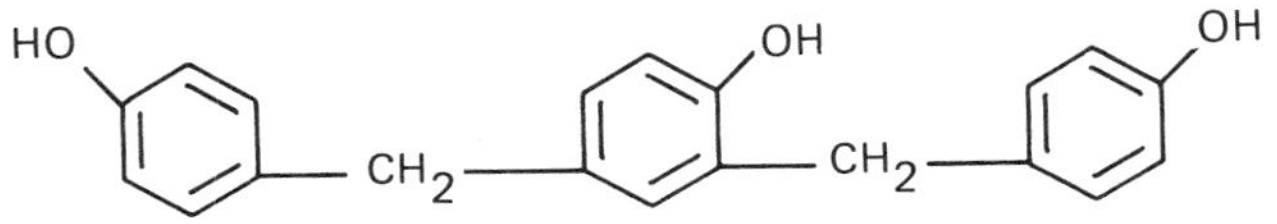

(2)

2. Polymerization

No information is available on the precise final polymer structure resulting from the polymerization of resole and novolac resins in foaming compositions. In fact little is known of the mechanism of curing and of the polymer structure of resole resins cured with an acid catalyst [31]. The extremely complex process in which numerous competing reactions occur simultaneously and the infusibility and insolubility of the products have hampered progress in understanding this resin chemistry. Model studies on resole hardening have been devoted mainly to uncatalyzed, heat-cured systems [19-21]. Research on the curing of catalyst-free [32, 33] and alkali-catalyzed [34] resoles has been recently conducted with infrared and ultraviolet spectrophotometry and differential thermal analysis.

It is generally concluded that the methylol groups of the resole initially condense with the elimination of water to give benzyl ethers, which in turn cleave under acidic curing conditions to form methylene bridges between the aromatic nuclei.

It is probable that the final polymer structures resulting from the curing of novolacs are quite similar to those from resoles. In any case, because of processing difficulties, novolacs hold a position of minor importance as foaming resins.

B. Commercial Resins

1. Composition and Properties

Phenol-formaldehyde foaming resole resins are supplied commercially in terms of the desired foam density. As already mentioned, the phenol-to-formaldehyde ratio may be varied. The method of resin preparation will influence the degree of condensation and polymerization, and this in turn will greatly affect the performance of the resin in foams. The viscosities of these resins (as measured at 25° C) are consistent with desired density and will range from about 500 to 6500 cP. For example, the lower viscosity range of 500 to 1000 cP is employed for foam with a density of less than 1 lb/ft^3 whereas a resin of 3000- to 6000-cP viscosity is used for foam desired in the density range 2 to 4 lb/ft^3. The Union Carbide Corp. [35] offers three resins designed for foam densities ranging from 0.2 to 65 lb/ft^3. This is accomplished by the appropriate blending of these resins for the described density. The Marblette Corp. [12] supplies resin for foams in the 10- to 15-lb/ft^3 density range.

Advertised sales specifications for a typical resole foaming resin like Union Carbide's BRLA-2760 include a viscosity of 2300 to 6500 cP at 25° C, a specific gravity of 1.275 to 1.300 at 25° C, and a pH of 6.0 to 6.3. Usually resins are supplied at a solids content of 60 to 80%, the major diluents being water and free phenol. The pH will range from 6.0 to 6.8. Because of their chemistry, resole resins tend to increase in viscosity and lose polymerizability on standing at room temperature. Ordinarily they are maintained at about 5° C or lower during storage to arrest their self-condensation. Addition of trioxane [36] to the resole resins has been reported to improve storage stability.

C. Catalysts

Because heat alone requires long curing cycles, strong acids are generally employed to cure resole resins for foaming. The catalyst functions basically as an instigator of the exothermic reaction. The heat generated by the condensation reaction is the basis of foaming. Weak acids do not provide sufficient heat to form other than dense foams. As with most ambient-temperature foaming systems, the quantity and choice of catalyst are critical in obtaining a stable foam with adequate properties. Because of the high functionality and inherent structural rigidity of phenolic resins, stable, rigid phenolic foam can result when only a low degree of condensation is achieved. High weight loss on aging and the necessity of postoven heating are prime evidence of insufficient cure. An excessive amount of catalyst will bring about an overly rapid cure and incomplete expansion, with resulting internal stresses. Cracking is evidence that internal stresses have been set up and is cause for rejection of the foam.

The most commonly used catalysts are hydrochloric, sulfuric, benzenesulfonic, toluenesulfonic, xylenesulfonic, phenolsulfonic, and phosphoric acids.

Mixtures of boric acid or boric anhydride with oxalic acid have been recently reported as effective catalysts [37]. Unlike the foregoing acid catalysts, this combination catalyst produced foam with substantially decreased corrosiveness and increased fire resistance, that is, they were nonpunking (see Section II.G). Similarly a number of metal halides, in particular aluminum chloride [38, 39] and ferric chloride [28], catalyze the cure of resole foaming resins. Incorporation of aluminum chloride reportedly imparts improved flame resistance and nonpunking character to the foams [28, 38, 39].

In another disclosure [40] certain inorganic compounds, such as phosphorus pentoxide, which are capable of reacting with the water present in the resin or evolved during the condensation, when used in combination with small amounts of hydrochloric acid, provide sufficient exotherm for foaming and curing. This catalyst mixture diminishes steam formation and produces a hard foam without the necessity for external heating.

Less commonly employed catalysts for resole resins include boron trifluoride etherates [41, 42], perchloric acid [42], peroxides [43], peroxides and heat [44], mixed organophosphoric acids with mineral acids [45], and in situ hydrolysis of chloroformates [46].

Although the initiation of the polymerization of resole resins by alkaline catalysts at room temperature has not been accomplished, curing can be induced with heating. For example, the heat of reaction of aluminum powder with sodium hydroxide has been used to expand and cure resole resins [47]. Also, alkaline catalysis with external heating has been described for foamable compositions containing isocyanates [48, 49]. Similarly isocyanates with the application of external heat to about 150° C [50] or in mixtures with various acids [37, 51, 52] are effective crosslinking reagents for the acceleration of cure. The isocyanates are incorporated either as the monomer or are used as isocyanate-terminated polyol prepolymers. Improvements in the friability resistance, compression strength, and water resistance of the foam are claimed.

Novolac resins are conventionally cured and crosslinked by heating a mixture containing an alkaline hardener like hexamethylenetetramine, a surfactant, and a blowing agent [53, 54]. This approach suffers from difficult processability and leads to foams of high density with limited application.

D. Surfactants

A variety of water-soluble, nonhydrolyzable, acid-stable surface-active agents can be used in phenolic foamable compositions. The nonionic types are the most commonly employed materials. Particularly useful are siloxane-oxyalkylene copolymers [48, 55, 56], such as the Union Carbide silicones L-530, L-5310, and L-5340. Polyoxyethylene sorbitan fatty acid esters [28], such as the Tween series (Atlas Chemical Industries), and the products of the condensation of ethylene oxide with castor oil and alkyl phenols [57] are also employed. The use of these agents has an appreciable effect on the cured foam's compressive strength, cell size, and texture, and the capability of a formulation to expand to produce low-density foam with uniform cells. Unlike polyurethane foaming compositions, the use of surfactants with phenolic resins does not appear to have a gross effect on the closed-cell content of the foams. The Tweens are reported to show pore-size control for a full range of viscosities of resole resins [28]. Other surfactants, such as various cationic [58] and anionic [59] types, are also effective, but these are of older vintage and have lost popularity to the silicones.

E. Blowing Agents

The exothermic resole condensation reaction produces water, in the form of steam, as a by-product that is capable of expanding the polymerizing phenolic resin. Water blowing alone causes nonuniform, coarse cellular texture, with large voids, or "blowholes," in phenolic foams, with resultant high density. Therefore volatile liquids and solid blowing agents are incorporated into phenolic foamable compositions to regulate cell size as well as to aid in resin expansion. Inasmuch as the final density of phenolic foams is not controlled predominantly by the amount of blowing agent, other contributing factors are the type of resin, the type and amount of catalyst, and the temperatures of the ingredients.

Vaporizable liquids like polyhalogenated saturated fluorocarbons [60] with a boiling point between 30 and 100° C are preferred. These blowing agents produce foams composed of a plurality of spherical, microscopic cells with improved flame retardancy and toughness. Due to the incompatibility of these blowing agents with the resole, emulsions are formed [55]. Other effective volatile liquids immiscible with the resole resin are such hydrocarbons as unleaded white gasoline, n-hexane, n-heptane, and cyclohexane [58]. Aliphatic ethers also produce finer and uniform cellular materials [61]. Foams in the density range of 0.2 to 3.0 lb/ft^3 have been produced with these soluble blowing agents. Methylene chloride in mixtures with ethers is also used to produce lightweight foams [39].

Use has also been made of a large number of blowing agents that are decomposed by the acid catalyst or heat. Effective materials include sodium carbonate, ammonium carbonate, sodium nitrite, sodium sulfite, and sodium bicarbonate. A variety of well-known types of organic compounds have been used as blowing agents: di-N-nitrosopentamethylenetetramine [62], N,N-disubstituted 5-amino-1,2,3,4-thiatriazoles [63], diesters of azodiformic acid [64], diazonium salts [65], sulfonhydrazides [66], N-alkyl N-nitrosodiacetoneamine [67 68], and ammonium nitrites [69].

Other blowing agents consist of maleic anhydride-calcium carbonate mixtures [43], carbon dioxide from the hydrolysis of chloroformates [46], oxygen from the decomposition of peroxides [70], and hydrogen gas liberated by the reaction of metal powders with acid catalysts [45, 71, 72].

F. Modifiers

Because of the notorious rigidity of phenolic polymers, considerable efforts have been expended on modifications to impart flexibility and toughness to phenolic foams. Some of the more frequently employed suitable approaches are discussed in this section. Although a commercial solution to this problem has not been attained, this area of activity holds much promise for future success.

Reactive modifiers have been employed that are incorporated during the resole-resin preparation and become an integral part of the polymer structure. Such ingredients include polyvinyl alcohol [22], polyvinyl alcohol-polyvinyl chloride [22], resorcinol [59, 73], m-cresol [59], furfuryl alcohol [74], and various polyols [45]. As already mentioned isocyanates are very effective modifiers.

Another example of this general approach is the utility of a second resin that is compatible with the resole and capable of being cured along with the resole under the influence of the acid catalyst. Typical resins are urea-aldehyde condensation products [72], epoxides [42, 72], polyepoxyorganosilicones [75], and novolacs [76]. Thermoplastic polymers like polyvinyl acetate, polyvinyl alcohol, and polyvinyl butyral have also been used as resin-viscosity modifiers to simplify the foaming process as well as to improve the properties of the final product [77, 78].

Phenolic foamable compositions can be modified by such physical additives as unreactive, high-molecular-weight organic polymers or certain inert fillers. Some polymers that are used are polyamides [79, 80], polyacrylonitrile [79], alkadiene-acrylamide copolymers [81], polyesters [79], polyvinyl chloride [82], nitrile rubbers [82], and styrene copolymers [82]. Finely divided fillers, such as talc, mica, asbestos, wood flour, cork flour, and carbon black [83], ordinarily improve the texture and homogeneity of the foams. However, higher density foams usually result.

Still another approach involves posttreatment of the foams with various plasticizers. Lower alkyl phthalates and phosphates [84] have been used as impregnants. The process is ordinarily carried out at 100° C and affords a product with improved elasticity. Treating the foam with steam results in a softening effect [85].

Foams can also be stabilized in the resilient form obtained immediately after preparation by treatment with ethylene oxide [86], siloxane-ethylene oxide copolymers [48], and 12- to 18-carbon alkyltrimethylammonium chloride [87].

G. Flameproofing Agents

Phenolic foams are highly flame resistant and nonignitable when exposed to an open flame source. However, under prolonged exposure to flame and heat these foams will smolder and char until consumption is nearly complete. This process is commonly referred to as "punking." In recent years efforts by various research groups have been directed at overcoming this deficiency by incorporating flame-retardant compounds. This property of phenolic foams will be mentioned further in Section IV.B.

As already mentioned, nonpunking in phenolic foams can be achieved by boric-oxalic acid and ferric-aluminum chloride as the foaming catalyst. The addition of flame-retardant boron and antimony compounds, such as boric acid and its esters [88] and antimony oxide [55], has been found to impart punking resistance to foams. Similar results were claimed when certain organic amides, such as urea, thiourea, dicyanodiamide, and melamine [89], were added to the foam composition.

Such reactive materials as polyhydroxylated and/or polyaminated derivatives of a polychlorinated diphenyl or polychlorinated polyhenylbenzene [90], when added at the phenol-aldehyde condensation state, also offer an approach to foam with high-temperature and punking resistance.

III. FOAMING PROCESSES

Because it can be prepared and cured at room temperature, phenolic foam can be fabricated by simple means. It has been molded in batchwise and continuous methods and sprayed in place.

Production of foam begins with the choice of a resole resin, which in turn is selected according to the desirable density. Generally speaking, the low-viscosity resins have rapid foaming rates and result in a very low density. Foaming is typically accomplished by adding the condensing or curing agent (catalyst) to a preblend of phenolic resin, surfactant, and blowing agent at room temperature. The liquid catalyst can be easily mixed into the resin-based material by a variety of methods. Optimum results are obtained by using a high-speed stirrer. Automatic metering and mixing equipment can be used for large-scale production. Mixing should be vigorous for 15 to 20 sec, and the catalyzed material should be poured into place within 30 to 60 sec. The composition is allowed to heat spontaneously to a temperature sufficient to cause foaming and curing of the cellular product.

In the case of novolac resins the composition must be heated to cause expansion. Depending on the formulation and quantity of material employed, the time of expansion and cure will vary. Usually the foam is strong enough to be handled within 5 min of the start of fabrication.

A. Batch Production

Large blocks of foams can be produced by a batchwise process in simple mold constructions in a short period of time. Disadvantages include the rather high manpower costs, poor uniformity of density throughout the article, and high raw-material losses, which run up to 40%, due to cutting, crusting, dust, scrap in mixing vessels, and poor surface characteristics.

A new process-equipment design [91] has been disclosed for the production of slab or panellike molded articles with improved uniformity in the physical characteristics of lightweight phenolic foams. The method utilizes vertical pour and high free rise in an open mold. The temperature of the foaming mixture is adjusted somewhat lower, for example, about 10° C, than the surface temperature of the molds to give a very thin but strong skin that is intimately bonded to the cellular material for surface protection. The high temperature differential increases the thickness of the skin to the point of being excessive, and the cohesive force between the skin and the underlying cellular material is weakened. Cooling the mold below the temperature of the mix has also been recommended for increasing skin thickness [16]. The molding operations are most commonly carried out with mixes at 30 to 35° C and the mold is maintained at 40 to 45° C. Experience has shown that it is desirable to coat the mold surfaces with Teflon or a silicone-based mold-release agent. This molding technique can also be employed to produce contour-shaped articles of lightweight phenolic foam.

Foaming compositions are normally introduced into the bottom of the forms through a tube. Another mold design makes use of a method of casting catalyzed resin from an internal mixing chamber in such a way as to avoid resin losses and stringing out of resin during transfer [92]. A floating-lid technique that is employed also conserves foam by providing squared blocks.

Molds can be lined with materials that become the surface layer of the molded article [93]. These laminated articles are particularly suited for building panels for roofs, partitions, walls, ceilings, and floors.

Several other related examples of batchwise open-molding operations and compositions employed are described in the patent literature [55, 62, 77, 78].

Phenolic novolac resin compositions modified with a nitrile rubber or polyvinyl chloride have been used to prepare molds for the manufacture of shoes [82]. The molding composition consists of a mixture of novolac resin, modifier, hexamethylenetetramine, and a solid blowing agent. The mold itself is a two-part closed mold. Once filled, the mold is heated at a temperature sufficient to cause the mixture to soften and expand under the action of the blowing agent. Typical mold conditions involve a temperature of 135° C for about 3 h. Rather dense, closed-celled, strong, tough-skinned cellular products result.

B. Continuous Slab Production

An apparatus suitable for the continuous production of large molded slabs of phenolic foam has been recently developed [94, 95] and is now being employed in France for a variety of applications [96]. A schematic view of the foam machine is seen in Fig. 1. This process, known as the "Paul Vidal process," is capable of foam-panel production rates as high as 6.5 ft/min. The equipment consists of a relatively narrow (up to 6 in.) continuous conveyor belt and two vertical paneled walls moving with the conveyor. After the mixture is dispensed onto the conveyor, and as the foamable composition expands vertically upward, it is carried by the conveyorized vertical walls and base surface through the apparatus. Panels can be made ranging in thicknesses from about 0.75 to 6 in., widths up to about 4 ft, and any practical length. Foams in the density range of 2 to 6 lb/ft^3 are most conveniently produced. This process offers foam with evenly distributed density, product reproducibility, low waste (to less than 5%), no objectionable odor, and a smooth, hard, flat surface.

Further automation is provided by the availability of a sawing machine [97].

C. Spraying

Because of the rapid thermosetting characteristics of resole resins, spraying of phenolic foam presents a special problem not associated with current commercial foam-spray systems. Technical developments in this area have been rather slow, and at present no commercial equipment is available. The chief obstacle is the lack of hydraulic, airless spraying apparatus for very reactive materials.

The composition and equipment suitable for the rapid application of a low-density foam in construction insulation have been described in a patent [98]. In this equipment the reactants are combined internally and

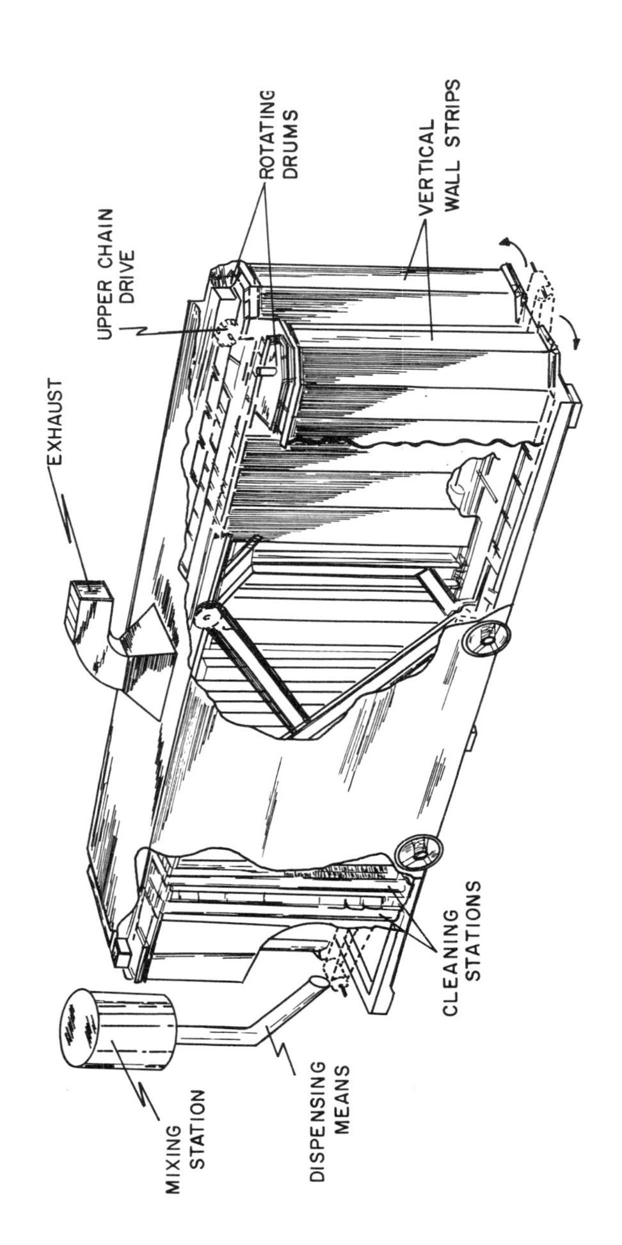

FIG. 1. Perspective view of the Paul Vidal processing machine for the continuous manufacture of slab-stock phenolic foam. Reprinted from Ref. [94].

hydraulically atomized. Essential features include minimal delay in expelling combined reactants, hydraulically operated valving, and such a flow pattern and sequence that only the resin stream wets the combining area walls during spraying, thus avoiding a buildup of reactive material inside the nozzle. The catalyst is injected into, and enveloped by, the resin stream inside the nozzle, and mixing is completed by the atomization of this two-component stream. Plugging during startup and shutdown is prevented by the automatic sequencing of reactant flow. No solvent is required for purging, even though nozzle-orifice lengths of six to eight times the diameter are used. Because the gun requires only two lines and is relatively light, operator fatigue is minimized.

Foam densities of 0.35 to 0.5 lb/ft^3 were normal for the formulation shown in Table 1 when so metered that the acid preblend weight was 20 parts per 100 parts of resin preblend. An open-celled, nonpunking, and nonburning foam results with a k-factor value of 0.28 to 0.29.

TABLE 1

Typical Phenolic Foam Spray Formulation[a]

Preblend	Parts by weight
Resin:	
BRLA-2759[b]	100
Silicone L-530[c], surfactant	1
Methylene chloride	3
Catalyst:	
Xylenesulfonic acid[d] (technical)	6
Methylene chloride	1

[a]Data from Ref. [98].

[b]Potentially very reactive, stable resole resin designed for foam with a density of 0.2 to 0.8 lb/ft^3, Union Carbide Corp.

[c]Union carbide Corp.

[d]The Richardson Co., Chemical Division.

IV. FOAM PROPERTIES

The general influence of phenolic foam composition and processing conditions on properties has been discussed to some extent in the preceding sections of this chapter. The several literature surveys cited in the introduction of this chapter present some data on the general properties of phenolic foams. Of these, the recent survey by Paul [16] is the most extensive.

In common with most cellular plastics, phenolic foam properties are strongly reflected in their structural variables, such as density, cell size and geometry, and overall polymer composition.

This section deals with the advantages and disadvantages relating to the use of phenolic foams in fabricated products. Particular emphasis is devoted to properties receiving recent research efforts.

A. Effects of Density and Cell Characteristics

Stable phenolic foam of widely varying density can be prepared by ambient-temperature-foaming techniques. Densities ranging from 0.2 to about 10 lb/ft^3 can be easily and conveniently prepared. Because of the polymerization chemistry and the cellular nature of the foams, the density of a phenolic foaming composition changes with the amount of blowing agent up to a threshold value and then remains essentially unchanged as the amount of blowing agent is increased further. Other influential factors on the final foam density include the type and amount of catalyst and surfactant employed.

Perhaps the most serious limitation of phenolic foam for thermal applications is its partly open-cell structure. This feature leads to foam with higher k-factor values and water absorption as compared with the polyurethanes. Low-density foams are essentially all open celled; higher density foams (2 lb/ft^3 and greater) can contain as much as 75% closed cells [55]. However, the closed-cell content of higher density foams is quite variable, and a reproducible empirical approach is lacking. The cell structure is homogeneous, extremely small, and spherical, and is unexcelled by that of any other cellular plastic. This property affords foam with certain beneficial end-use characteristics. These are discussed in the subsections that follow.

1. Mechanical Properties

The compressive and tensile strengths of phenolic foams catalyzed by hydrochloric and boric-oxalic acids as a function of foam density are presented in Table 2.

TABLE 2

Mechanical Properties of Phenolic Foams[a]

Property	Catalyst: Hydrochloric acid		Catalyst: Boric-oxalic acid	
Density, lb/ft^3	1.1	1.9	1.6	2.6
Closed cells, %	2	2	2	0
k-Factor at 23° C, parallel, Btu/(h)(ft^2)(° F/in.)	0.24	0.24	0.24	0.25
Compressive strength, psi:				
Parallel	12	22	14	32
Perpendicular	7	17	11	21
Tensile strength, psi:				
Parallel	15	--	10	16
Perpendicular	9	--	6	14
Friability (ASTM C367 and C421), weight loss, %:				
10 min	81	59	69	35
20 min	100	96	100	66
Aged compressive strength, psi:				
Cold, -30° C, 14 days	--	21	13	29
Dry, 70° C at 5% R. H., 14 days	--	21	13	30
Humid, 70° C at 100% R. H., 14 days	--	20	14	28

[a]Data from Ref. [99].

The values in Table 2, represent typical data obtained from core samples made by normal small-scale laboratory hand-batch foaming procedures. As expected, strength properties rise with increasing density for each catalyst. Compressive-strength values plotted along with the higher density values reported by Paul [16] are compared with those of typical rigid polyurethane foams in Fig. 2. From this comparison it is apparent that phenolics approach the strength properties of the polyurethanes at low densities. The differences have been attributable to the open-cell character of the phenolic foams.

Load-versus-deformation curves for 3.6- and 10.3-lb/ft^3 core samples from blocks of fiber-glass-reinforced phenolic foam catalyzed by boric-oxalic acid are shown in Fig. 3 [13]. Curves A and B show that a certain critical force level achieves a plateau that remains relatively constant with further increasing force. Raising the density of the foam increases the critical load that can be tolerated before yield, as shown in Fig. 4. This energy-absorption characteristic of phenolic foam is suited for certain applications, such as crash panels in automobiles [13].

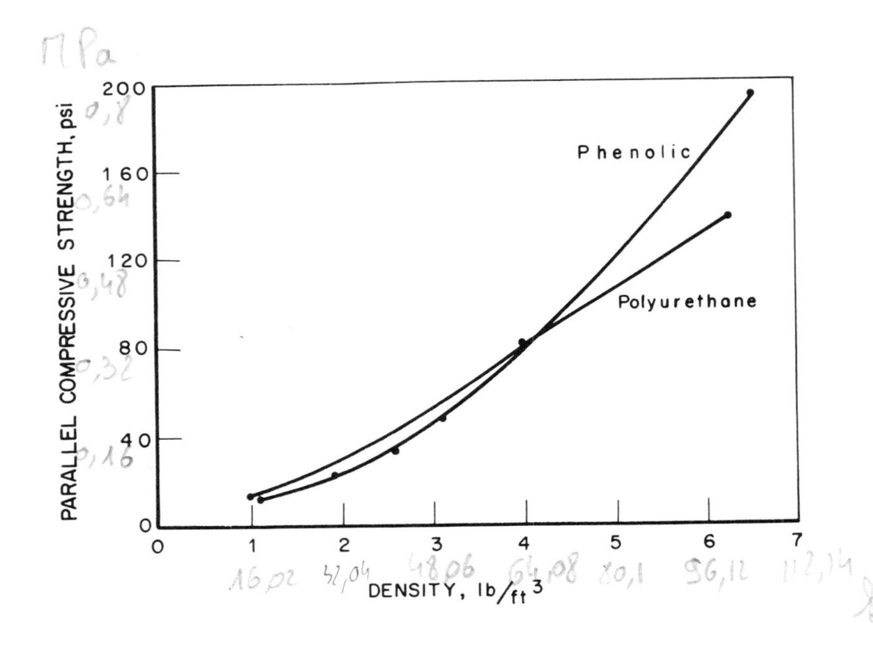

FIG. 2. Variation in parallel compressive strength with density of phenolic and polyurethane foams. Reprinted from Ref. [16] by courtesy of Lake Publishing.

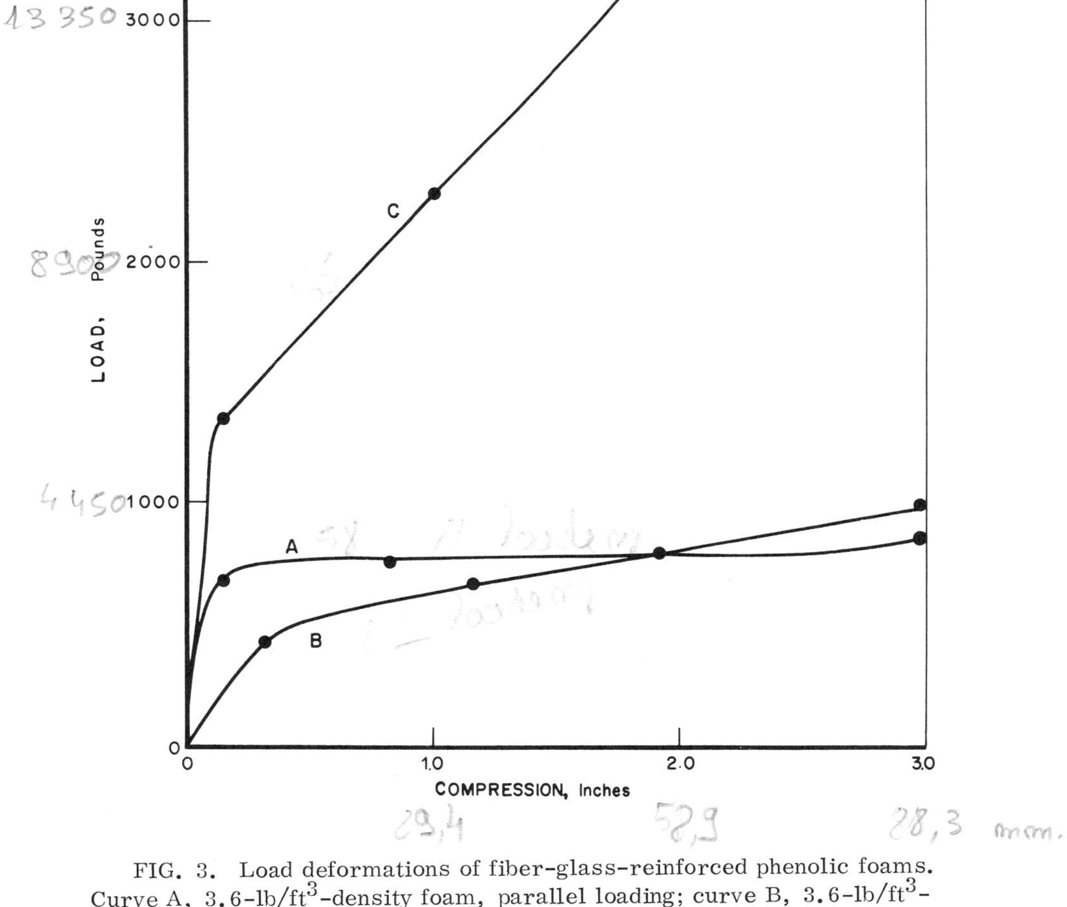

FIG. 3. Load deformations of fiber-glass-reinforced phenolic foams. Curve A, 3.6-lb/ft^3-density foam, parallel loading; curve B, 3.6-lb/ft^3-density foam, perpendicular loading; curve C, 10.3-lb/ft^3-density foam, parallel loading. Reprinted from Ref. [13], p. 8, by courtesy of The Society of the Plastics Industry, Inc. New York.

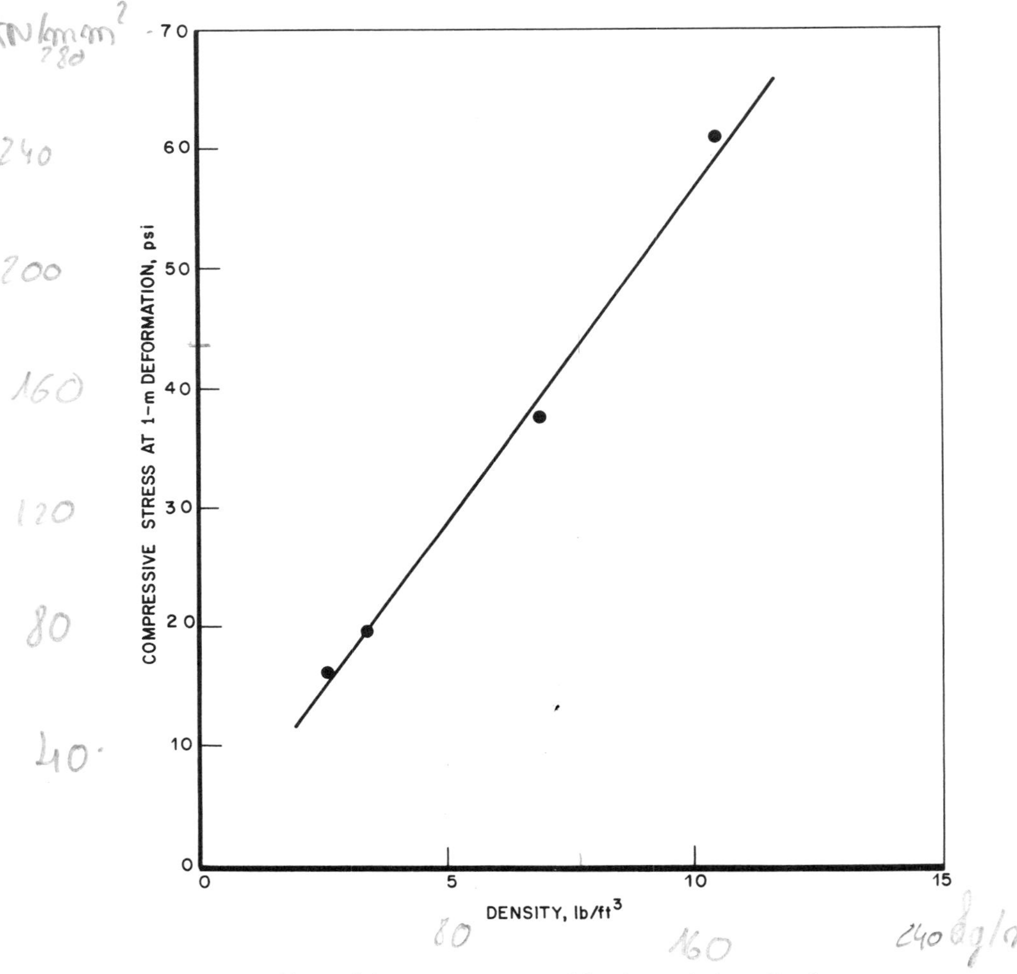

FIG 4. Effect of density on critical loading of phenolic foam. Reprinted from Ref. [13] , p. 8, by courtesy of The Society of Plastics Industry, Inc., New York.

Another impressive characteristic of phenolic foam is its good mechanical strength over a wide range of temperatures. Strengths remain essentially unchanged for years and can withstand temperatures ranging from -30 to +70° C. This property has been illustrated in Table 2.

Most phenolic foams prepared to date are much more friable than polyurethane foams. This is not unexpected, considering the notoriously brittle nature of phenolic resins in general. A comparison of friability (measured according to ASTM C367 and C421) as a function of density was given in Table 2. However, the laboratory testing by ASTM methods of the friability resistance of phenolic foams is frequency not an accurate measure of expected end-use performance. In many instances phenolic foams have lasted the lifetime of fabricated finished articles without showing such evidence of friability as settling or dusting (see trailer truck test in Section VI. A. 2). Perhaps factors relating to friability in phenolic foam depend on a more involved process than our current test methods measure.

2. Dimensional Stability

Among cellular plastics phenolic foam is outstanding in dimensional stability. The open-celled character contributes to this property. The effects of accelerated-aging tests on volume changes at varying conditions of humidity and temperature are shown in Table 3. Stability values ranging from below -70 [13] to above +175°C [16] approximate the usable-temperature range of the foam. The foam may be exposed to liquid-nitrogen temperature without shrinkage or cracking. High-temperature stability will be discussed more fully later in this section.

3. Moisture Sensitivity

The water-vapor transmission and absorption data for several density ranges of phenolic foam are presented in Tables 4 and 5. As

TABLE 3

Accelerated Aging of Phenolic Foams[a]

Density, lb/ft^3	1.6	2.0	2.6
Aging, volume change, %:			
Humid, 70° C at 100% R.H., 28 days	-4.4	0.1	-4.0
Dry, 70° C at 5% R.H., 14 days	-3.0	-2.5	-3.0
Cold, -30° C, 14 days	0.5	0	-3.0

[a]Data from Ref. [99].

already mentioned, the high values are a consequence of the partially open-celled nature of the foam. The magnitude of the values will depend on the density of the foam as well as environmental conditions. Weight loss at room temperature and low relative humidity is probably due to the volatilization of low-boiling-point materials in the foam. The phenolic polymer itself is insensitive to moisture. The rate of water absorption is reported to be significantly reduced by the presence of an outer skin on the foam [16]. Suitable surface protection is commonly employed to eliminate the effects of moisture.

TABLE 4

Water-Vapor Transmission of Phenolic Foam

Density (lb/ft^3)	Water-vapor transmission in 24 h (g/m^2)
2.0	300/51 mm[a]
3.5	300/51 mm[a]
2.2	12.27/4.9 mm[b]
3.4	4.46/4.9 mm[b]

[a]Data from Ref. [6].

[b]Data from Ref. [17].

TABLE 5

Water-Vapor Absorption of Phenolic Foam[a]

Density (lb/ft^3)	Weight change (%)
1.9	8, at 70° C, 100% R.H., 28 days
1.9	-5.6 at 70° C, 5% R.H., 14 days
1.9	-0.8 at -30° C, 14 days

[a]Data from Ref. [99].

4. Thermal Conductivity

The thermal insulating capability of phenolic foam is quite good when compared with other materials, except for those that contain retained fluorocarbon. For example, k-factors ranging from 0.19 to 0.22 Btu/ (h)(ft^2)(° F/in.) are obtained for foam with densities of 2.50 to 6.25 lb/ft^3, respectively, at 20° C [17]. This surprisingly low thermal conductivity is ascribed to the exceptionally fine cellular nature of phenolic foam. Even further lowering of k-factor values results with decreasing temperature, as expected for an air-filled foam. The effect of temperature on k-factor values for a foam of about 1.85-lb/ft^3 density is shown in Table 6 [100].

5. Acoustical Properties

The acoustical properties of phenolic foam are summarized in Table 7 [6]. The sound-absorption coefficients are very high in comparison with those of other rigid plastic foams. Measurements were made in a reverberation room at various frequencies for 2.3- and 4.0-lb/ft^3 density foam. This high sound-absorption property of phenolic foam is probably due to its open-cell structure. The values are only slightly less than those of commercial materials.

TABLE 6

Thermal Conductivity of Phenolic Foam[a]

Temperature (° C)	k Factor (Btu/(h)(ft^2)(° F/in.))
-101.1	0.14
-73.3	0.16
-45.6	0.18
0	0.20
+10	0.22
+37.8	0.25
+65.6	0.30

[a]Data from Ref. [100].

TABLE 7

Sound Absorption of Phenolic Foam[a]

Density (lb/ft^3)	Thickness (in.)	Reverberation-absorption coefficient frequency (Hz) 125	250	500	1000	2000	4000
2.3	1	0.05	0.25	0.65	1.00	0.80	0.80
4.0	1	0.15	0.20	0.70	0.90	0.80	0.85

[a]Data from Ref. [6]. Foam support: laid on floor.

B. Flammability

One of the most outstanding characteristics of phenolic foams is their nonflammability. Flammability results, obtained by various test methods, for a phenolic foam catalyzed by hydrochloric acid and a recently developed nonpunking boric-oxalic acid are presented in Table 8.

The flammability data of Table 8 are exceptionally good for an organic-based cellular substance. Also very significant is the low smoke-density rating for phenolic foam.

It is noteworthy that even at as low a density as 0.8 lb/ft^3, a sprayed phenolic foam (catalyzed by xylenesulfonic acid) exhibited a flame-spread rating of 25 by the Underwriters' Laboratories tunnel test, with a smoke-density value of 2 and a fuel contribution of 15 [101].

One longstanding problem associated with phenolic foams is their tendency to punk. Some foams will punk while others do not, and the ones with the greatest tendency to punk are those catalyzed by hydrochloric acid. Recent research efforts have shown that the choice of catalyst [37, 39] is the most important factor in ensuring nonpunking foams. Further resistance to punking has also been accomplished by employing certain flameproofing agents in foam formulations, as has been discussed. The mechanism of the punking of phenolic foams has not been elucidated.

At present there exists no standard test method to characterize the punking behavior of phenolic foam. The most commonly employed test is based on heating a 0.5-in.-thick sample of core foam with a 765°C flame until the temperature of the opposite side of the foam reaches 260°C [37]. A somewhat older procedure involves embedding a steel rivet heated at 900°C between two pieces of foam [6].

TABLE 8

Typical Flammability Properties of Phenolic Foam[a]

	Acid catalyst	
	Boric-oxalic	Hydrochloric
Foam density, lb/ft^3	2.0	1.9
ASTM D1692-67T rating	S. E.[b]	S. E.[b]
ASTM E84-60T, underwriter's tunnel test:		
Flame-spread rating	10	
Fuel contribution	23	
Smoke-density rating	0	
Bureau of Mines flame-penetration test:		
Penetration time, sec	595	53 (punked)
Weight loss, %	10.3	4.8
ASTM E162, radiant panel test,		
Flame-spread rating	No ignition	

[a]Data from Ref. [99].

[b]Self-extinguishing (no ignition).

C. Thermal Stability

Phenolic foams also possess good thermal stability over a broad temperature range. Typical high-temperature behavior of foam of 2-lb/ft^3 density is presented in Table 9. The maximum continuous-usage temperature has not been clearly defined. Values ranging from 130 [17] to about 175° C [16] have been cited for continuous service before evidence of foam shrinkage or embrittlement occurs. Foams have been reported to withstand discontinuous service-temperature exposures up to 180° C [96].

The high-temperature stability of phenolic foam can be markedly improved by incorporating inert fillers into the foam formulation. For example, a nonpunking foam containing as little as 10% glass-fiber rovings withstood a temperature of 816° C for 1 h without any evidence of shrinkage or cracking [13]. There is practically no low-temperature limit for useful operation of phenolic foam [100].

TABLE 9

Thermal Stability of Phenolic Foams[a]

	Heating conditions			
Sample	T (° C)	Duration (days)	Volume change (%)	Weight change (%)
1	100	1	-	-18.0
2	150	1	-5.0	-24.4
3	200	1	-15.1	-24.6
4	200	14	-30.4	-35.1
5	200	28	-47.1	-50.2

[a]Data from Ref. [99].

D. Corrosion Resistance

Phenolic foam is conventionally prepared with an acid catalyst and the acidic residue remaining after foaming can in some cases present a serious disadvantage in subsequent application where corrosion may be a problem. Phenolic foam catalyzed by boric-oxalic acid has been found to produce negligible rates of corrosion on copper, galvanized steel, 304 stainless steel, aluminum 3003, and mild steel after a 46-day exposure at 55° C and 100% relative humidity [99]. Foams prepared with xylenesulfonic or toluenesulfonic acid are also reported to be noncorrosive [60].

Serious corrosion is obtained when foam catalyzed by hydrochloric acid comes in contact with metallic parts. Foaming mixtures containing such basic materials as calcium oxide and sodium sulfite are reported to effectively neutralize the acid catalyst in the foam, thereby providing less corrosive foam [60]. Noncorrosive phenolic foam can also be produced by postneutralizing the residual acidity in the foam sample by gaseous ammonia or alkylamine [102]. Postdrying of the foam at 100 to 140° C about 1 to 3 h is recommended to drive off volatile residues, particularly acids, prior to neutralization in order to minimize the formation of corrosive ammonium salts [103]. The success of this procedure depends on the use of an easily volatilized acid catalyst, such as hydrochloric acid.

Another approach to eliminate corrosion due to acid uses an encapsulated neutralization material that is confined by a protective coating in the foam formulation and is inert to the acidic mixture [104]. The protective substance is removed at a temperature somewhat above 50° C during the foam-expansion step. An example of a substance that can be used is calcium oxide coated with calcium stearate.

V. RELATIONSHIP OF STRUCTURE TO PROPERTIES

Correlation of the physical properties of phenolic foam with structure depends basically on the availability of appropriate model polymers. Differences in molecular weight, crosslink density, and structure of constituents between crosslinks are, among the parameters that influence polymer properties. Unfortunately the technology of cured resole resins has not advanced to a stage at which well-defined products differing systematically in a specific structural feature are available for study.

In addition to polymer structure and composition, the mechanical properties of phenolic foam vary with cellular structure, density, cure temperature, and processing. Cellular factors that affect the performance of phenolic foams have been studied by Harding [105] and will be presented in Chapter 17.

Of interest are the extensive investigations of Conley and co-workers [106-109] on the mechanism of thermal degradation of cured resole-type resins. It was found that all the cured resinous products studied exhibited the same degradation pattern above 300° C. From this it was concluded that the thermal degradation processes of phenolic resins are primarily a function of the stability of the dihydroxydiphenylmethane (3) unit and are not dependent on the molecular species present in the resole resin before curing.

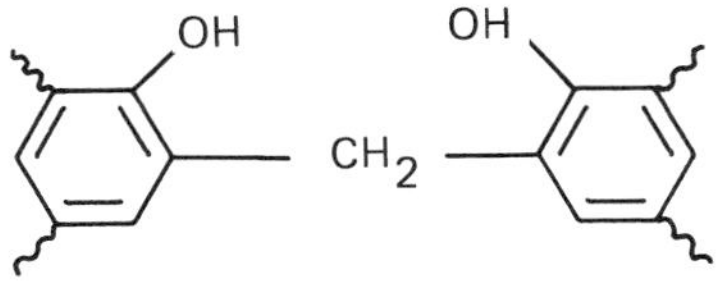

OH OH

CH_2

(3)

Moreover, it was found that the primary processes involved are oxidative, and not pyrolytic, in nature. Interestingly, the resin itself can act as a source of oxygen.

VI. APPLICATIONS

Phenolic foam is a versatile material with a wide field of utility. In addition to their principal use as thermal insulation, phenolic foams have been used in the floral trade, nuclear packaging, acoustical panels, carbonized structures, and filling voids. Other recommended applications for phenolic foams are for automotive and aircraft accessories [13, 77, 110, 111], shoe compositions [82, 112, 113], fibers and fabrics industry [114-119], adhesive foams [120, 121], decorative coatings [122, 123], bowling balls [124], electric battery separators [125], sealing openings in drilling wells [126], filling electric-light-bulb caps [127], briquettes [128], plant pots [129], and cellular carbon [130]. These latter uses are either not important commercially as yet or represent rather small-volume items.

A. Thermal Insulation

1. European Activity

Phenolic foams have found some acceptance as a thermal insulation material in Europe, particularly in Germany and England. Recent interest during the last 3 to 4 years is evident in France and Italy. Since World War II, when the German aircraft industry employed phenolic foam, a growing interest has been developed for various commercial applications [9, 17]. Although the low cost of phenolic foams was undoubtedly the prime incentive, the majority of these applications make use of the good insulating qualities of these foams as well as their particularly good fire retardance, low smoke generation, and good heat resistance. One common usage where these properties are especially beneficial is roof insulation. In this application the foam is applied as board stock to the roof. Sandwich panels for both indoor and outdoor use comprise another large European market for phenolic foam.

In addition to its use in the commercial and private building trade in France, phenolic foam is also employed to insulate freight cars, trucks, and the cabins and hulls of boats. As already mentioned, phenolic foam is produced commercially in a continuous operation as slab stock, and one French producer sells it under the Zol trade name [96]. Panels laminated as a sandwich of phenolic foam between layers of such materials as aluminum and various plastics are also available.

A variety of present applications pertaining to the construction industry in England have been outlined by Cooper [6]

In addition to the applications already cited, in Germany phenolic foams are also used as forms for concrete casting and for filling hollow structures [17].

Phenolic foam under the trade name Fenolite [100, 131], has recently been used in Italy for low-temperature industrial insulation in manufacturing plants, but it has not been widely adopted as yet. This interest had been instigated by the foam's outstanding ability to withstand dimensional change by temperature cycles (particularly between ambient and lower temperatures), outstanding chemical resistance, and good thermal insulation properties at lower temperatures.

As with other insulations for industrial plants, a sealer coating is necessary to prevent an undesirable penetration and condensation of moisture into the body of the insulation with resulting dimensional distortion and loss of thermal insulation. A number of protective coatings, such as neoprene and polyvinyl chloride, have good adhesive properties on phenolic substrates as weather barriers.

2. American Activity

In contrast to the Eurpoean market, the use of phenolic foam has not penetrated the United States marketplace to any extent as a thermal insulation material. However, because of its inherent nonflammability, this type of foam is gaining attention. For example, it was recently included along with polyurethane in a large-scale fire-test evaluation for industrial plant applications [132]. The phenolic foam exhibited no flame spread and provided significant fire protection with no fire hazard, but the cellular product was destroyed in the immediate fire area, where the internal temperature of the pipe reached 538° C after 35 min.

The use of phenolic foam for preformed-wall-panel construction for buildings has been studied by Baumann [133]. A building construction based on sandwich panels of phenolic foam cores and epoxy resin-glass cloth skins was erected and examined over a period of 10 years. Epoxy was used to glue panels to each other and to the concrete floor. The building is shown in Figs. 5 and 6. After 10 years the shear and compressive strengths of the foam cores were unaltered. Also, loading of the roof to about three times the design load did not cause structural failure. No warping or cracking of the foam panels developed during this time. The only problem with the hydrochloric acid-catalyzed foam was some corrosion of the baseboard electric heaters. This deficiency could easily be overcome by the recently developed noncorrosive systems we have discussed.

Phenolic foam has been found useful for insulating refrigerated trailer trucks [133]. Testing revealed that a trailer-truck bed that had been installed with a foamed-in-place phenolic foam system had withstood over 90,000 miles and 3 years of service with acceptable thermal insulation and no evidence of mechanical failure, such as settling. The aluminum floor and beams in contact with the sulfuric acid-catalyzed foam showed no signs of corrosion.

FIG. 5. Newly constructed plastic foam building [133]. Courtesy of J. A. Baumann, Union Carbide Corp.

B. Floral Foam

The major outlet for phenolic foam in the United States is in the floral field [134] , where stems of cut flowers are embeded into the water-soaked foam. In this case foam compositions that may lead to corrosion and punking are not problems. Formulated products are utilized where maximum open-cell content for water penetration and low friability resistance are prime requisites.

Another advantage of phenolic foam is its ability to prolong the life of cut flowers. This characteristic is particularly valuable for the preservation of cut flowers for shipment.

For floral utility the foam may be conveniently and quickly produced by the batch process in large blocks or slabs without the necessity of postcuring or neutralization treatments. Formulations are initially adjusted for maximum open-cell content without the formation of large voids or blowholes. The foam is easily cut and packaged with a minimum of loss due to waste.

FIG. 6. Testing structural strength of the plastic foam building of Fig. 5 after 10 years [133]. Courtesy of J. A. Baumann, Union Carbide Corp.

C. Packaging

The exceptionally low densities that can be achieved, the shock-absorption characteristics, and the low cost make phenolic foam well suited for a variety of general purpose packaging applications. These properties have led to suggestions that are specific to phenolic foam, such as pressing items directly into the foam for protection and as a dunnage material to replace shredded paper, excelsior, and other packing materials.

Of interest is some recent work at the Oak Ridge National Laboratories on the use of fiber-glass-reinforced phenolic foam as insulation for containers of fissile materials [13]. The cellular material (6 lb/ft^3) provides protection for uranium hexafluoride cylinders during transit from shock and fire damage up to 1200° C. Sufficient protection for containers loaded with a 12-in. cylinder under a 30-ft drop test with a deceleration on impact equivalent to 1400 g was provided by the foam. The termal insulation provided by these containers was also impressive. Heating the containers with 5-in. thicknesses of foam for 1 h at temperatures of up to 1150° C only caused charring to penetrate to about 2.4 in. with an average cylinder-surface temperature of 66° C. The foam also provided excellent neutron attenuation, which was probably due to the presence of boron from the boric-oxalic acid catalyst employed.

D. Miscellaneous

Phenolic foams have been used for filling cavities like boat hulls. Judging from the many patents issued to aircraft companies during the late 1950s, there was much effort expended to utilize phenolic foam as fairings and filler for aircraft. With the availability of noncorrosive phenolic foam these applications could gain further attention.

The highly effective energy-absorption (impact) characteristics of phenolic foam have led to other suggested uses in transportation [13]. For example, they have been recently recommended for crash-pad panels in automobiles. As federal safety regulations call for improved energy-absorption materials, phenolic foam could become a strong competitor for collapsible metal structures.

Phenolic foams are especially useful for the preparation of cellular carbon for refractory insulating material [135]. These carbonized foams will withstand temperatures of up to 3300° C under inert atmospheric conditions. The foam is prepared by successive heating during about 2 h up to 1230° C. The graphitelike products were at about 3-lb/ft^3 density, and the final structural dimensions were identical with those of the starting material.

REFERENCES

1. L. H. Baekeland, Ind. Eng. Chem., 1, 149 (1909).
2. L. H. Baekeland, U.S. Pat. 942,699 (1909).
3. H. Junger, Kunststoffe-Rundschau, 9, 437 (1962).
4. J. H. Rooney, G. M. Kline, J. W. C. Crawford, T. W. M. Pond, T. Love, and R. T. Richardson, Investigation of German Plastics Plants, Part 2, P.B. 25,642, Technical Information and Documents Unit, London, 1945, pp. 169-170.
5. A. Cooper, A. K. Unsworth, and A. Hill, Cellular Rubber and Plastics, P.B. 93,484, Technical Information and Documents Unit, London, 1947, pp. 62-63.

6. A. Cooper, Plastics, 29 (321), 62 (1964).
7. J. Cell. Plastics, 2, 189 (1966).
8. J. D. Nelson, SPE Journal, 9, 14 (1953).
9. A. G. Winfield, in Plastics in Building (I. Skeist, ed.), Reinhold, New York, 1966, p. 368.
10. Anon., Connais. Plast., 7 (67), 8 (1966).
11. Anon., Pensez Plast., 14 (431-432), 41 (1967).
12. International Foamed Plastic Markets and Directory, Technomic Publishing, Stamdord, Conn., 1965-1966.
13. S. J. Wheatley and A. J. Mallett, in Proc. 2nd SPI Intern. Cellular Plastics' Conf., The Society of the Plastics Industry, New York, 1968, Section 2-E, p. 1.
14. Phenolic Foams, Union Carbide Corp., Modern Plastics Encyclopedia, Vol. 45, No. 1A, McGraw-Hill Inc., New York, 1968, p. 364.
15. T. H. Ferrigno, Rigid Plastics Foams, 2nd ed., Reinhold, New York, 1967, p. 323.
16. M. N. Paul, in Handbook of Foamed Plastics (R. J. Bender, ed.), Lake Publishing, Libertyville, Ill., 1964, p. 306.
17. F. Weissenfels, Kunststoffe, 51, 698 (1961).
18. D. F. Gould, Phenolic Resins, Reinhold, New York, 1959 p. 177.
19. A. A. K. Whitehouse, E. G. K. Pritchett, and G. Barnett, Phenolic Resins, Iliffe, London, 1967.
20. N. J. S. Megson, Phenolic Resin Chemistry, Academic Press, New York, 1958.
21. R. W. Martin, Chemistry of Phenolic Resins, Wiley, New York, 1956.
22. E. Simon and F. W. Thomas (to Lockheed Aircraft), U.S. Pat. 2,728,741 (1955).
23. P. H. Rhodes (to Koppers), U.S. Pat. 2,443,197 (1942).
24. P. H. Rhodes (to Penn. Coal Products), U.S. Pat. 2,414,415 (1943).
25. E. A. Lauring (to Minnesota and Ontario Paper), U.S. Pat. 2,926,722 (1960).
26. R. A. Darrall (to Imperial Chemical Industries), British Pat. 861,156 (1958).
27. E. H. G. Sargent, British Pat. 587,833 (1945).
28. P. N. Erickson (to Evans Products), U.S. Pat. 3,300,419 (1967).
29. W. E. McCullough, Australian Pat. 25,977 (1930).
30. A. E. T. Neale (to Dunlop Rubber), British Pat. 972,094 (1964).
31. See Ref. 21, p. 150.
32. R. T. Conley and J. E. Bieron, J. Appl. Polymer Sci., 7, 103 (1963).
33. Z. Katovic, J. Appl. Polymer Sci., 11, 85 (1967).
34. R. C. Vasishth, W. C. Ainslie, and S. Y. Leong, in Abstracts of Papers of 155th National American Chemical Society Meeting, San Francisco, Calif., March 31-April 5, 1968, p. 712.

35. Bakelite Phenolic Resins BRL-2759, BRL-2760, and BRL-2761 for Rigid Foam, Phenolic Product Data Bull., Union Carbide Corp., New York, 1961.
36. G. M. R. Lorentz, H. Neises, and R. Stroh (to Farbwerke Hoechst), U.S. Pat. 3,313, 766 (1967).
37. R. W. Quarles and J. A. Baumann (to Union Carbide), U.S. Pat. 3,298,973 (1967).
38. British Pat. 1,088,056 (to Union Carbide) (1967).
39. P. N. Erickson and A. N. Erickson (to Evans Products), U. S. Pat. 3,256,216 (1966).
40. H. Junger (to Dynamit Nobel), German Pat. 1,240,274 (1967).
41. F. W. Thomas and E. Simon (to Lockheed Aircraft), U.S. Pat. 2,744,875 (1956).
42. British Pat. 1,077,423 (to Canadian Industries) (1967).
43. British Pat. 1,004,607 (to Peltex) (1965).
44. British Pat. 676,381 (to Westinghouse Electric International) (1952).
45. E. Simon and F. W. Thomas (to Lockheed Aircraft), U.S. Pat. 2,772,246 (1956).
46. W. A. LaLande and H. Green (to the Pennsylvania Salt Manuf.), U.S. Pat. 2,493,075 (1950).
47. British Pat. 571,284 (to Semtex Ltd.) (1945).
48. H. H. Enders (to Union Carbide), U.S. Pat. 3,271,331 (1966).
49. A. Khawam (to Allied Chemical), U.S. Pat. 3,063,964 (1962).
50. R. F. Sterling (to Westinghouse Electric), U.S. Pat. 2,608,536 (1952).
51. K. L. Proctor and H. L. Katz, U.S. Pat. 2,806,006 (1957).
52. T. L. Phillips (to the Distillers Co.), British Pat. 908,303 (1962).
53. R. F. Shannon and P. W. Sullivan (to Owens-Corning Fiberglas), U.S. Pat. 3,081,269 (1963).
54. R. Wong and P. W. Sullivan, U.S. Pat. 3,218,271 (1965).
55. British Pat. 1,091,238 (to Compagnie de Saint-Gobain) (1967).
56. Netherlands Pat. 6,609,096 (to Compagnie de Saint-Gobain) (1967).
57. British Pat. 1,062,850 (to Dynamit Nobel) (1967).
58. G. A. Mullen (to Westinghouse Electric), U.S. Pat. 2,933,461 (1960).
59. British Pat. 586,199 (to Semtex Ltd.) (1947).
60. W. J. D'Alessandro (to Union Carbide), U.S. Pat. 3,389,094 (1968).
61. K. F. Krebs, W. Caldwell, and P. F. Urich (to Union Carbide), U.S. Pat. 2,845,396 (1958).
62. R. D. Maitrot, French Pat. 1,519,474 (1968).
63. K. L. Schmidt, E. Muller, and H. Scheurlen (to Farbenfabriken Bayer), U.S. Pat. 3,192,170 (1965).
64. O. L. Mageli, C. S. Sheppard, and N. H. Schack (to Wallace and Tiernan, Inc.), U.S. Pat. 3,306,862 (1967).

65. A. P. Armstrong and J. B. Cameron (to the Fairey Aviation Co.), U.S. Pat. 2,828,271 (1958).
66. A. Lambert and G. H. Lang (to Imperial Chemical Industries), British Pat. 791,144 (1958).
67. R. L. Frank (to Ringwood Chemical), U.S. Pat. 2,676,928 (1954).
68. R. L. Frank (to Ringwood Chemical), U.S. Pat. 2,708,661 (1955).
69. I. L. Newell (to United Aircraft), U.S. Pat. 2,626,968 (1953).
70. R. F. Sterling (to Westinghouse Electric), U.S. Pat. 2,653,139 (1953).
71. S. P. Kish and L. C. Coe (to Kish Plastic Products), U.S. Pat. 2,733,221 (1956).
72. T. L. Phillips (to the Distillers Co.), British Pat. 937,855 (1963).
73. R. J. Brinkema (to Koppers), U.S. Pat. 2,582,228 (1952).
74. I. Tashlick and G. M. Grudus (to International Pipe and Ceramics), U.S. Pat. 3,390,107 (1968).
75. J. V. Duffy (to Atlantic Research), U.S. Pat. 3,215,648 (1962).
76. R. F. Shannon (to Owens-Corning Fiberglas), U.S. Pat. 3,207,652 (1965).
77. F. W. Thomas (to Lockheed Aircraft), U.S. Pat. 2,802,240 (1957).
78. E. G. K. Pritchett and D. Smith (to Bakelite Ltd.), British Pat. 965,218 (1964).
79. J. K. Atticks (to United Aircraft), U.S. Pat. 2,895,173 (1959).
80. E. Simon and F. W. Thomas (to Lockheed Aircraft), U.S. Pat. 2,798,054 (1957).
81. W. L. Garrett (to Dow Chemical), U.S. Pat. 3,336,243 (1967).
82. E. B. McMillan and A. R. Olson (to United Shoe Machinery), U.S. Pat. 2,678,293 (1954).
83. J. D. Nelson and P. V. Steenstrup (to General Electric), U.S. Pat. 2,446,429 (1948).
84. British Pat. 638,416 (to Dunlop Rubber), (1950).
85. M. Goldstaub (to Dunlop Rubber), British Pat. 618,197 (1946).
86. W. L. Garrett (to Dow Chemical), U.S. Pat. 3,389,095 (1968).
87. C. G. Harford (to Arthur D. Little, Inc.), British Pat. 887,078 (1957).
88. British Pat. 973,835 (to Dynamit Nobel) (1964).
89. Netherlands Pat. 6,702,335 (to Compagnie de Saint-Gobain) (1967).
90. P. Dorier and B. Robinet (to Plastigel S.A.), Netherlands Pat. 6,705,248 (1967).
91. P. I. Vidal (to Rocma Anstalt), U.S. Pat. 3,400,183 (1968).
92. W. A. Miller (to Union Carbide), U.S. Pat. 2,649,620 (1953).
93. M. Raymond (to Maitrot), French Pat. 1,517,757 (1968).
94. P. I. Vidal (to Rocma Anstalt), U.S. Pat. 3,214,793 (1965).
95. M. Paul and P. I. Vidal, French Pat. 1,272,193 (1961).
96. Nonflammable Insulating Material. Old Material with New Properties, Plastugil, Plastics, and Elastomers Ugine-Pragil Co., Paris, France, 1967.

97. P. I. Vidal (to Rocma Anstalt), U.S. Pat. 3,340,909 (1967).
98. D. P. Cook (to Union Carbide), U.S. Pat. 3,122,326 (1964).
99. A. J. Papa and W. R. Proops, Union Carbide Corp., South Charleston, W. Va., unpublished results.
100. Thermal Insulation for Industrial Plants, Catalog R-8, Riva and Mariana, Milan, Italy.
101. D. P. Cook, Union Carbide Corp., Bound Brook, N.J., private communication.
102. P. Pinten (to Dynamit Nobel), German Pat. 1,073,199 (1960).
103. P. Dorier and J. Potier (to Plastugil), French Pat. 1,468,907 (1967).
104. W. L. Morgan and C. A. Matuszak (to Owens-Corning Fiberglas), U.S. Pat. 3,138,563 (1964).
105. R. H. Harding, J. Cell. Plastics, 1, 2 (1965).
106. H. W. Lochte, E. L. Strauss, and R. T. Conley, J. Appl. Polymer Sci., 9, 2799 (1965).
107. W. M. Jackson and R. T. Conley, J. Appl. Polymer Sci., 8, 2163 (1964).
108. R. T. Conley and J. F. Bieron, J. Appl. Polymer Sci., 7, 103 (1963).
109. R. T. Conley and J. F. Bieron, J. Appl. Polymer Sci., 7, 171 (1963).
110. R. E. Horste (to IIT Research Institute), U.S. Pat. 3,267,048 (1962).
111. H. D. Stuck (to J. W. Bolton and Sons, Inc.), U.S. Pat. 2,561,999 (1946).
112. A. R. Olson (to U.S. Shoe Machinery), U.S. Pat. 2,632,210 (1950).
113. T. R. Korkatti (to U.S. Shoe Machinery), U.S. Pat. 2,602,193 (1949).
114. W. L. Morgan (to Owens-Corning Fiberglas), U.S. Pat. 3,062,682 (1957).
115. R. F. Shannon (to Owens-Corning Fiberglas), U.S. Pat. 2,979,469 (1957).
116. I. L. Newell (to United Aircraft), U.S. Pat. 2,683,697 (1951).
117. L. P. Biefeld (to Owens-Corning Fiberglas), U.S. Pat. 2,673,825 (1949).
118. L. P. Biefeld (to Owens-Corning Fiberglas), U.S. Pat. 2,673,824 (1949).
119. L. P. Biefeld (to Owens-Corning Fiberglas), U.S. Pat. 2,673,823 (1949).
120. R. A. Barkhuff (to Monsanto), U.S. Pat. 2,513,274 (1947).
121. O. H. Huchler (to Badische Anilin- and Soda-Fabrik), Canadian Pat. 718,474 (1961).
122. M. J. Hiler (to Midland Chemical), U.S. Pat. 2,862,834 (1954).
123. M. J. Hiler (to Midland Chemical), British Pat. 855,152 (1958).

124. J. G. Hendricks (to Stowe-Woodward), U.S. Pat. 3,206,201 (1961).
125. W. F. Dermody (to Electric Storage Battery), U.S. Pat. 3,062,760 (1959).
126. G. H. Billue, U.S. Pat. 2,761,511 (1953).
127. J. W. Tijen (to N. V. Phillips Gloeilampen-fabrieken), Australian Pat. 233,380 (1958).
128. A. W. Edmunds (to Dow Chemical), Canadian Pat. 566,389 (1956).
129. W. Engel, U.S. Pat. 2,189,889 (1936).
130. French Pat. 1,387,941 (to Societe Le Carbone-Lorraine) (1965).
131. Advertisement, Riva and Mariana Insulating Contractors, Chem. Age, Dec. 16, 1967, p. 25.
132. D. H. Way and C. J. Hilado, J. Cell. Plastics, 4, 1 (1968).
133. J. A. Baumann, Union Carbide Corp., Bound Brook, N.J., unpublished data.
134. R. E. Skochdopole, Wayne State University Polymer Conference Series, May 24-28, 1965.
135. W. D. Ford (to Union Carbide), U.S. Pat. 3,121,050 (1964).

Chapter 12

UREA-FORMALDEHYDE FOAMS

K. C. Frisch

Polymer Institute
University of Detroit
Detroit, Michigan

I. INTRODUCTION

Urea-formaldehyde (U-F) foams were developed as early as 1933 by investigators at I. G. Farbenindustrie [1]. They were first developed for use as insulation and were introduced commercially in Germany under the trade names Iporka (I. G. Farbenindustrie) and Isoschaum (Badische Anilin und Soda Fabrik AG (BASF) [2]. Later on modified urea-formaldehyde foam was utilized as agricultural mulch under the trademark Hygromull (BASF) and other uses for U-F foams in the horticultural field were commercialized in the United States by a number of companies. In England Imperial Chemical Industries Ltd. contributed greatly toward the introduction of U-F foams (Ufoam) in the construction industry. Thermalon Ltd. and more recently Ciba (foamed Aerolite resins) are companies actively engaged in the marketing of U-F foams in the United Kingdom. In the United States a number of companies, including Filfast Corp., U. F. Chemical Corp., Brand Industrial Services, Inc., and A. L. Randall Co., are offering U-F foams for a variety of applications [3].

II. CHEMISTRY

Urea-formaldehyde foams are generally prepared from an aqueous solution of urea-formaldehyde resin and foaming agent (surfactant) in an aqueous solution. Acid catalysts are usually employed for the curing of these foams.

Urea-formaldehyde resins in a mole ratio of 1:1.6 to 1:2 are suitable for foam production. Urea and formaldehyde solutions are condensed by refluxing at pH 5 to 6 and 80 to 100° C. During this process methylolureas and methyleneureas are formed, together with some low-molecular-weight condensation polymers containing methylol and methylene groups. After neutralization to a pH of 7 to 9, the resin solution is dehydrated to a resin-solids content of 50 to 70%. In this form the resin can be stored for months at ambient temperature. Spray drying of the resins can also be employed for products with a long shelf life. These resins can subsequently be cured with acid catalysts.

The crosslinking reaction in dilute aqueous acid solution consists of the conversion of methylol groups present in methylolurea or in low-molecular-weight urea-formaldehyde condensation products to methylene linkages. The mechanism proposed by De Jong and De Jonge [4] as well as by others is as follows:

$$-\underset{|}{N}-CH_2OH + HA \longrightarrow -\underset{|}{N}-\overset{+}{C}H_2 + H_2O + A^-, \qquad (1)$$

$$-\underset{|}{N}-\overset{+}{CH_2} + H-\underset{|}{N}- \longrightarrow -\underset{|}{N}-CH_2-\underset{|}{\overset{+}{N}}H, \tag{2}$$

$$-\underset{|}{N}-CH_2-\underset{|}{\overset{+}{N}}H + A^- \longrightarrow -\underset{|}{N}-CH_2-\underset{|}{N}- + HA. \tag{3}$$

The curing of the foam depends on the following factors:

1. The degree of condensation of the U-F resin.
2. The concentration of the foam components.
3. The temperature of the system.
4. The "buffering effect" of the foaming agent (surfactant).

A commonly used foaming agent is an alkyl aryl sulfonate, for example, a salt of a butylated sulfonic acid, sodium alkylbenzene sulfonate, or an alkylnaphthalene sulfonate [5]. However, many other foaming agents can be used.

III. FOAM PREPARATION

A number of procedures have been described for the preparation of U-F foams. The liquid U-F resin can be foamed by the dispersion of air or another gas employing high-speed agitation, followed by the addition of an acid catalyst to cure the foam.

In another procedure gas is dissolved under pressure in the liquid resin, followed by foaming of the froth by the sudden release of pressure [6].

In addition to foam processes utilizing mechanical expansion of the U-F resin, foams can also be produced by chemical expansion, such as the in situ generation of such gases as carbon dioxide produced by the reaction of inorganic carbonates or bicarbonates with the acidic curing agents.

Another technique consists in the expansion of the U-F resin by the evaporation of such low-boiling-point solvents as fluorocarbons or pentane, caused by the heat of the exothermic reaction of the U-F condensation crosslinking in the presence of an acidic catalyst, such as phosphoric acid [7, 8]. Foams made by this technique exhibit a predominantly closed-cell structure, whereas foams prepared by other methods are usually open-celled ones.

A number of processes have been described for the continuous manufacture of U-F foams [6, 9-13].

Vieli and Kreidl [6, 9] described a process employing frothing under pressure of the aqueous solution of the foaming agent (alkyl sulfonate) and hardening agent (strong acid). The U-F resin solution is added while subjecting the foam to centrifugal forces of 500 to 600 rpm during the pressure relief. This process is very effective in producing a foam with a very fine cell structure.

Shriver et al. [10] disclosed a process for the continuous production of closed-cell U-F foam. The apparatus for the preparation of this foam is shown in Fig. 1 [10]. It consists of a cylindrical foaming chamber that has at its lower end a liquid inlet and a gas-atomizing inlet, a cylindrical blending chamber terminating in a restricted throat, and a transfer line connecting the top of the foaming chamber with the blending head. A resin-inlet pipe extends into the blending head and terminates in the throat. A cylindrical cure tube extends downward from the throat. A sump pump allows incompletely frothed liquid to drain from the foaming agent-catalyst solution before the addition of the U-F resin solution.

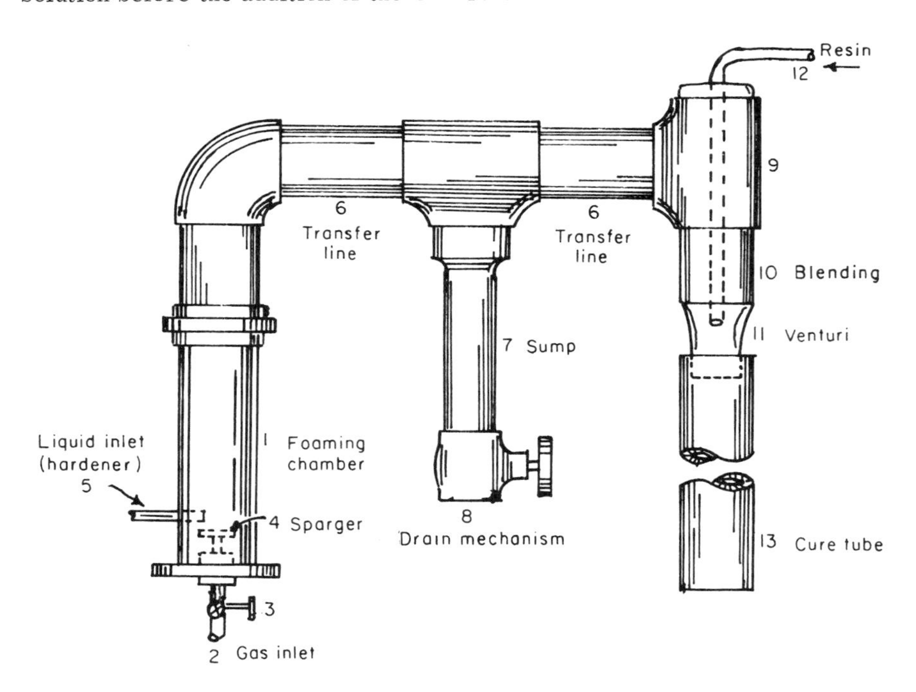

FIG. 1. Continuous foam apparatus. Reprinted from Ref. [10].

Pressurized air is fed into the foaming chamber through a sparger, and the liquid acidic foaming agent is injected into the chamber and foamed by the action of the sparged air. The stream of the aqueous acidic foam is blended with the aqueous U-F resin solution by means of Venturi action, producing a partially hardened foam that will flow sufficiently to fill a mold.

Unterstenhöfer [11] described four different types of equipment used for the production of U-F foam made by the dispersion process. Figure 2a [11] shows an apparatus equipped with an agitator into which the aqueous solution of the foaming agent is fed under pressurized air and is converted into foam by means of vigorous agitation. Figure 2b shows a reaction vessel for the production of foam, containing a frit through which pressurized air is forced into the aqueous solution of the foaming agent.

In another variation of the dispersion process the aqueous foaming agent is fed under pressure into a vessel filled with glass beads, which cause initimate mixing and generation of foam (see Fig. 2c). Figure 2d shows the "swirl foamer" employed by BASF in the production of Hygromull foam. In this procedure air is fed trangentially at high speed into a rotating prefoamer, consisting of a cylindrical vessel with an exit pipe provided with a coarse sieve, and forms a vortex. The foaming solution enters the prefoamer through an inlet tube and through the action of the turbulent air is converted into a coarse foam. By selecting the appropriate diameter-to-length ratio of the exit pipe, the coarse foam is changed into a fine-celled one.

Equipment for the continuous production of U-F foam has also been described by Scheuermann [12] and Graf [13] (see Fig. 3 [11]), who employed a vertically stirred reaction vessel for the blending of the resin and the foaming agent. The resin solution is introduced in the lower part, and the foaming agent is added in the upper part of the reaction vessel. Both resin and foaming agent are fed continuously with air, allowing for the foam to be discharged on a rubberized conveyor. The foam hardens sufficiently in about 0.5 to 1 min that it can be handled and stored for completion of cure, which takes about 8 h.

The BASF process for the production of Isoschaum is shown schematically in Fig. 4 [11]. The resin solution and the aqueous foaming-agent solution are mixed in a tubular reactor,. The air pressure forces the expanding resin mass through a flexible line to the point of application, where the foam solidifies and cures. This process can be employed for the foam filling of not readily accessible cavities and crevices in buildings.

The foam machine that was recently developed by BASF and produced by Carl Platz for the production of Hygromull, the U-F foam for agricultural applications, is shown in Fig. 5 [11]. In this process the resin solution is premixed with compressed air and blended with the aqueous foam in a special mixing head, called a homogenizer. The latter, shown

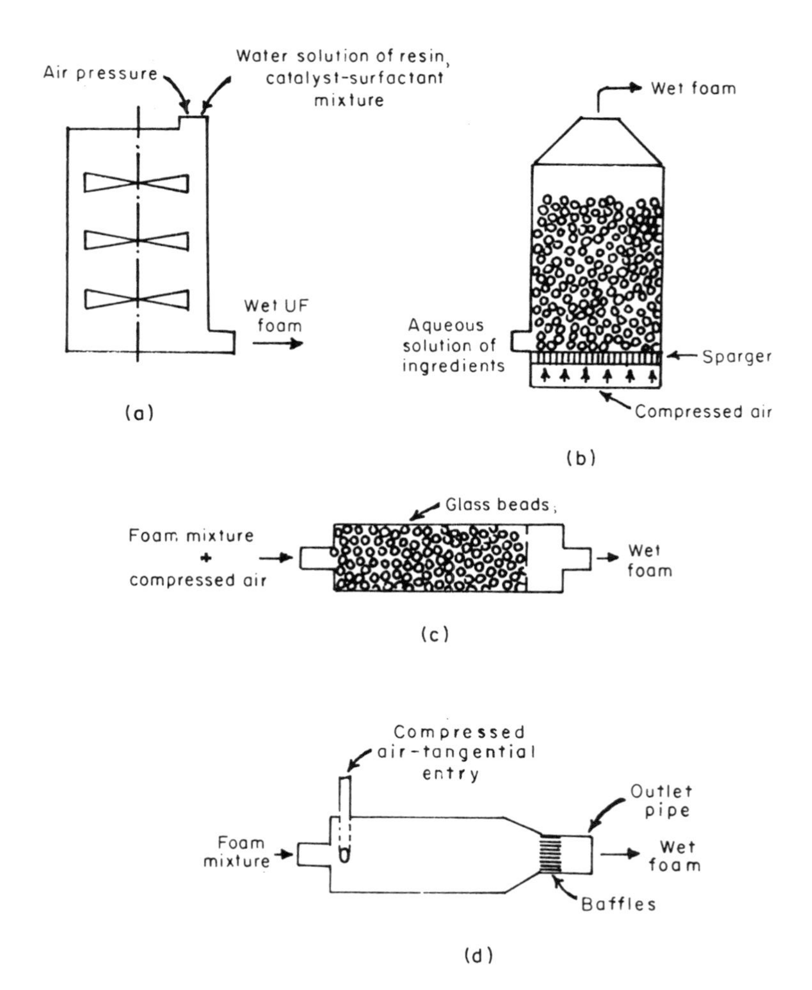

FIG. 2. Apparatus for the production of U-F aqueous foam: (a) stirred reactor; (b) porous-bottom, or sparger, reactor; (c) spray gun with packing; (d) swirl foamer for Hygromull production. Reprinted from Ref. [11] by courtesy of Kunststoffe.

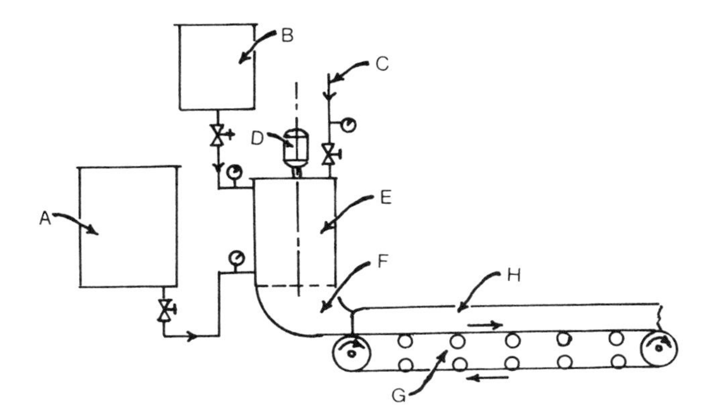

FIG. 3. Continuous foam production with stirred reactor and conveyor: A, resin solution; B, surfactant solution; C, compressed air; D, stirrer; E, vessel; F, adjustable value; G, conveyor; H, foam plastic. Reprinted from Ref. [11] by courtesy of Kunststoffe.

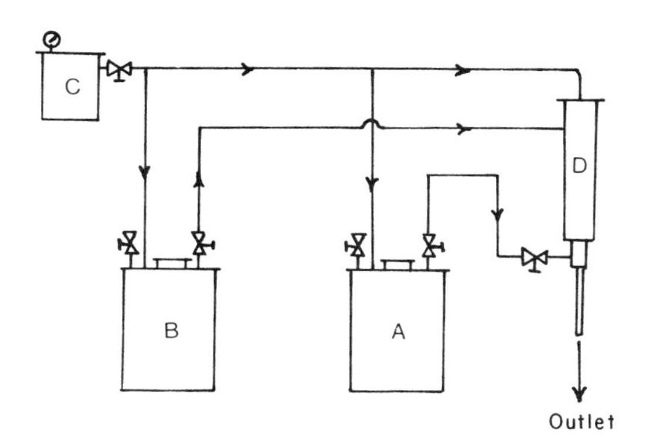

FIG. 4. Preparation of U-F foam by the Isoschaum process: A, resin solution; B, foaming-agent solution; C, air supply; D, tubular reactor for foam plastic production. Reprinted from Ref. [11] by courtesy of Kunststoffe.

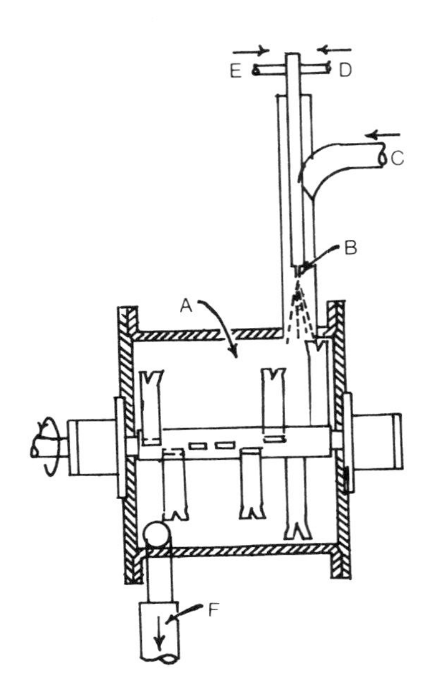

FIG. 5. Apparatus for Hygromull production: A, aqueous foam-resin solution; B, aqueous foaming-agent solution; C, compressed air; D, water; E, flow meter; F, swirl foamer; G, mixing section; H, homogenizer; I, Hygromull outlet. Reprinted from Ref. [11] by courtesy of Kunststoffe.

in Fig. 6 [11], consists of a cylindrical reaction vessel containing a rotating mixing device with side arms to provide for intimate mixing.

The foam, consisting of the mixture of the resin solution under air pressure and aqueous foam, is fed through the premix pipe into the mixing chamber of the homogenizer. The premixed foam is continuously agitated to prevent any settling of the components. A dwell time of 2 to 5 sec in the homogenizer is generally sufficient to result in a uniform plastic foam. The mixing speed and the dimensions of the homogenizer have a pronounced effect on the quality and uniformity of the foam. The U-F foam is stored for at least 24 h. before further processing.

For the production of the agricultural foam Hygromull, which involves the addition of considerable amounts of garden soil to U-F foam, it is advantageous to mount the foam machine on a truck along with the storage containers for the foaming-agent solution and foam-resin solution. In addition, the truck must carry an air compressor, pumps, drivers, and

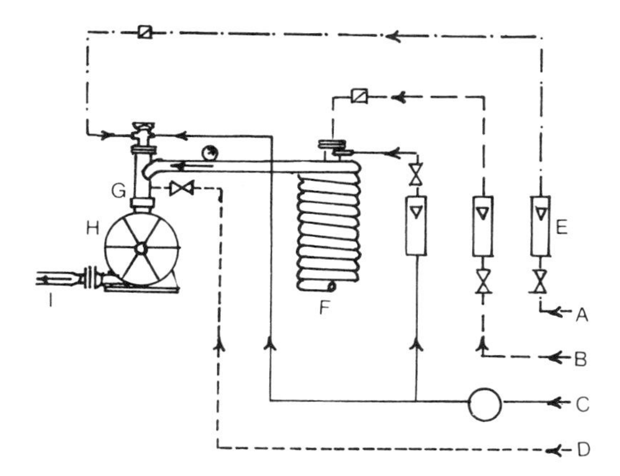

FIG. 6. Homogenizer of Hygromull production apparatus (see Fig. 5): A, homogenizer; B, premix pipe; C, aqueous foam; D, resin solution; E, compressed air; F, delivery tube. Reprinted from Ref. 11 by courtesy of Kunststoffe.

the homogenizer. The foam machine is also equipped with a flexible hose for dispensing the foam. For best results the temperature of the foaming components should be about 20° C and not fall below 15° C.

Imperial Chemical Industries Ltd. has developed a mobile foam-dispensing unit for the insulation of buildings with U-F foam [5]. This system is based on the injection of a mixture of U-F resin, a foaming agent, and a hardener through a foaming head into the cavity wall.

The equipment used in field applications consists essentially of two storage vessels for the liquid components and a foaming head, as seen in Fig. 7 [5]. One vessel contains a diluted U-F resin, the other a mineral acid. The foaming agent is added to one of these solutions which is then pressure fed or pumped to the rear of the foaming head in which it is aerated to a foam. The foam is then stabilized with the other component, which is also pressure fed or pumped to the rear of the foaming head where intimate mixing of the components takes place.

Fig. 8 [5] shows a mobile foam unit mounted on a truck. The injection of the foam is carried out through small holes, 3/4 in. in diameter and about 3 ft apart, drilled through the mortar joints into the outside wall of buildings that are already occupied. In the case of new buildings before plaster has been applied or buildings under construction without any roofing, foam is applied through the open top of the cavity, as shown in Fig. 9 [14]. The foam has an initial consistency similar to shaving-cream lather, which enables it to fill crevices and not readily accessible gaps. It dries out completely within a few days and shrinks slightly on drying. Shrinkage depends on a number of factors, such as density and ambient

FIG. 7. Mobile foam-dispensing unit developed by Imperial Chemical Industries Ltd. for insulating buildings with U-F foam. Reprinted from Ref. [5] by courtesy of Imperial Chemical Industries, Ltd.

FIG. 8. Mobile foam unit mounted on a truck. Reprinted from Ref. [5] by courtesy of Imperial Chemical Industries, Ltd.

FIG. 9. Application of U-F foam insulation in buildings under construction. Reprinted from Ref. [14] by courtesy of Ciba Ltd.

temperature [15]. The addition of a small amount of plasticizer or humectant to the U-F resin prior to foaming has been found to reduce linear shrinkage of the foam [5].

IV. MODIFICATIONS OF U-F FOAMS

Many chemical modifications of U-F foams have been made to overcome certain apparent weaknesses of unmodified U-F foams, such as friability, high water absorption, and lack of dimensional stability.

A series of plasticizers have been used to reduce the friability of U-F foams. Specific examples include N, N, N^1, N^1-tetrakis(2-hydroxethyl) ethylenediamine tetraacetate and higher homologs [16]. Addition products of ethylene oxide to hexadecyl alcohol have also been cited for this purpose [17].

Polyethylene glycols have been reported to increase the toughness of U-F foams [18]. In particular, polyethylene glycols with 5 to 25 oxyethylene groups per molecule exhibit good plasticizing action. In general 10 to 35% of these polyethylene glycols is added to the U-F resin, containing high proportions of formaldehyde to urea (5:1). The hydroxyl groups of the glycols react with the formaldehyde to become an integral part of the resin network.

The addition of emulsions of thermoplastic resins has been shown to improve resistance to air and moisture. Daly [19] added emulsions of acrylic resin and polyvinyl acetate to a U-F foam made by frothing techniques, yielding a low-density foam (2 lb/ft^3) that had a K-factor of 0.30 to 38° C and was impervious to moisture and air. A similar technique was used by Marcora [20], who observed an increase in the flexibility as well as the sound absorption of the resulting foams. These effects were presumably due to a change in the cell structure of these foams.

Urea-formaldehyde foams can also be made more flexible by the addition of nitrogen derivatives of polyacrylic acid and its copolymers with acrylic esters, acrylonitrile, acrylamide, vinyl esters, or partially saponified copolymers of methyl acrylate and acrylonitrile [21]. Urea-formaldehyde foams are capable of absorbing a variety of liquids. By using solutions of elastomeric polymers, U-F foams can be impregnated, and, on evaporation of the solvent, a uniform coating of the elastomer is obtained [22]. By crushing the original foam structure one obtains a soft and resilient foam that can be used as an absorbent for medical uses.

Eberl and Coppick [23] reported that by uniform compression and carefully controlled processing conditions (e.g., resin composition, temperature, humidity, amount and duration of compression), on release of the compressive force one can obtain U-F foams that exhibit a certain degree of resilience and softness.

Furfuryl alcohol has also been reported to reduce the friability of U-F foams at low densities [24]. However, the improvement was said to be relatively small.

Various approaches have been described to reduce the tendency of U-F foam to shrink and crack on drying [25]. One method [26] consists of treating the partially cured molded foam with ammonia, thereby reducing its acidity. This process yields dimensionally stable foams.

The use of ammonia for the production of dimensionally stable foams is also reported in another patent [27]. However, in this case ammonia is added to the reaction mixture before foaming.

The shrinkage of U-F foams has been described by Baumann 2 (see Table 1). It can be noted that linear shrinkage and weight loss increase with increasing temperature.

TABLE 1

Weight Loss and Linear Shrinkage of U-F Foam (Isoschaum) in 3-h Test[a]

Temperature (° C)	Average weight loss (%)	Average linear shrinkage (%)
40	7.7	1.8
60	9.8	3.8
80	10.5	3.8
100	13.2	4.9
120	14.4	5.2
140	16.9	5.2
160	19.1	5.1
180	19.4	5.0
200	19.8	5.8
220	22.0	7.3

[a]Data from Ref. [2].

In addition to the use of ammonia to reduce shrinkage, other techniques have been described: the addition of fillers (e.g., wood flour) and the blending of thermoplastic emulsions with the U-F foam reactants before foaming [28].

The modification of U-F foams with α, ω-bisepoxies has been reported to yield foams with lower friability and greater impact strength at low densities [29].

A number of methods have been described to reduce the odor of formaldehyde that is present during the foaming process and may also persist for some time after foaming is completed. The odor is ascribed to both free formaldehyde and formaldehyde in the form of methylol groups.

Justice [30] reported a technique for reducing the odor of U-F foams by adding between 0.5 and 8 parts by weight of ammonium carbonate or ammonium bicarbonate (per 100 parts of U-F resin solids) to a partially resinified, aqueous U-F resin solution. The resin solution is allowed to stand prior to the addition of the dilute acid hardener until the odor is substantially reduced.

TABLE 2

Properties of U-F Foams[a]

Property	ASTM test method	Unmodified		Modified[b]	
		Parallel[c]	Perpendicular[c]	Parallel[c]	Perpendicular[c]
Density, lb/ft^3		1.47	1.47	1.47	1.47
Compressive modulus, psi:					
Initial	D-790	650.0	270.0	25.4	146.0
Ultimate	--	--	--	3.9	--
Compressive strength, psi	D-695	12.2	8.30	No failure	4.08[d]
Impact strength, film	(e)	0.0043	0.0062	0.0144	0.0108
Friability				Not readily friable	

[a]Reprinted from Ref. [34] by courtesy of Wiley-Interscience.

[b]Modified with 14% polyethylene glycol.

[c]Relative to the direction of foam rise.

[d]Slight fracture; severe fracture; occurs with unmodified foams.

[e]Relative values, estimated by the same empirical test, not comparable to values for unfoamed resin.

In another process, described by Kelly and Wells [31], at least 4% by weight, based on the acid hardener solution, of urea is added to the acidic foam (prepared by foaming an acid hardener solution of a surfactant) prior to mixing with the U-F resin solution. Foams made by this method are reported to be free of formaldehyde odor during manufacture as well as on storage of the resulting foam. Presumably urea acts as a scavenger for both the unreacted formaldehyde and the methylol groups to form methyleneurea linkages.

Other types of modified U-F foams that have been prepared are based on cocondensates of U-F resins with phenol [8]; melamine [32]; melamine, thiourea, and phenol [33]; and with furan resins [24].

V. PROPERTIES

A. Mechanical Strength

Unmodified U-F foams have very little mechanical strength. As has already been mentioned (see Section IV), the addition of certain plasticizers or modifiers, such as polyethylene glycols [18], can result in an improvement in mechanical properties. Table 2 [34] shows the effect of polyethylene glycol on the properties of U-F foams.

Another method to improve the compressive strength of U-F foams consist of adding various amounts of urea (5-60%) to a U-F condensation product [35]. The resulting rigid foams had a density of 0.7 to 2.5 lb/ft^3 and a compressive strength of 6 to 8 psi.

Various fillers have been used to enhance the strength of U-F foams [12, 36, 37]. In one investigation [36] wood flour, asbestos, and glass powder were used, wood flour producing the greatest increase in strength. Other fillers, such as gypsum [37], have been reported to have a similar effect. Scheuermann [12] has described the use of inorganic fibers to make considerable improvements in the strength of U-F foams.

B. Heat Resistance

Urea-formaldehyde foam in the form of 10-cm^3 cubes was exposed for 1 year alternately for 12 h at 130° C and 12 h at -30° C [2]. After the heat cycle was completed, no change was observed in the foam.

The actual decomposition temperature of U-F foam (Isoschaum), as determined in a melting-point apparatus, was 220° C, although some variation in the decomposition temperature was observed (218-232° C) [2].

C. Flammability

When exposed to a Bunsen-burner flame at 1000 to 1500° C, U-F foam decomposes immediately [2]. With large foam samples severe shrinkage of the foam takes place at the edges, where the foam is in direct contact with the flame. However, the foam does not exhibit any afterglow ("punking").

In general U-F foam must be rated as a material that burns with difficulty, although there may be considerable differences in the degree of flammability, depending on whether the foam is unmodified or contains additional flame-retarding ingredients.

Baumann [2] reported a number of flammability tests carried out with U-F foam (Isoschaum):

1. Foam samples containing 5 and 28% absorbed moisture, respectively, were suspended above a wood fire in close proximity to the flame. After removal of the foam sample from the flame, the foam did not continue to burn and no afterglow was observed.
2. In another flammability test a slow and continuous charring of the surface of the foam occurred, with simultaneous formation of wrinkles and bubbles.
3. A layer of foam was placed between two steel plates, and one of the plates was heated with a blow torch. After 1 min, a small flame appeared, but it extinguished itself after 15 sec without any afterglow.
4. After heating the foam to 600° C, no combustible decomposition products were found. An electric resistance-heating unit, placed between foam layers of 5-cm thickness, when heated to red heat, did not cause a fire in the foam.

A number of methods have been disclosed to improve the flame resistance of U-F foams. Among these are the incorporation of such phosphorus-containing compounds as phosphonates [38] and the use of furfuryl alcohol and ethylene glycol [39].

Urea-formaldehyde foaming resins were also incorporated into intumescent coatings for the protection of wood and other materials of construction. These types of coatings swell under the influence of heat and are capable of forming a carbonaceous foam that protects the underlying substrate. Many formulations have been developed containing U-F resins, ammonium phosphates (or other phosphorus-containing compounds, such as polyphosphorylamide), polyols, starch or dextrin, and in some cases inorganic fillers like titanium dioxide [40, 41].

D. Water Absorption and Water-Vapor Transmission

The water absorption of U-F foam (Isoschaum) at 20° C and 30, 50, 70, and 95% relative humidity is shown in Fig. 10 [2].

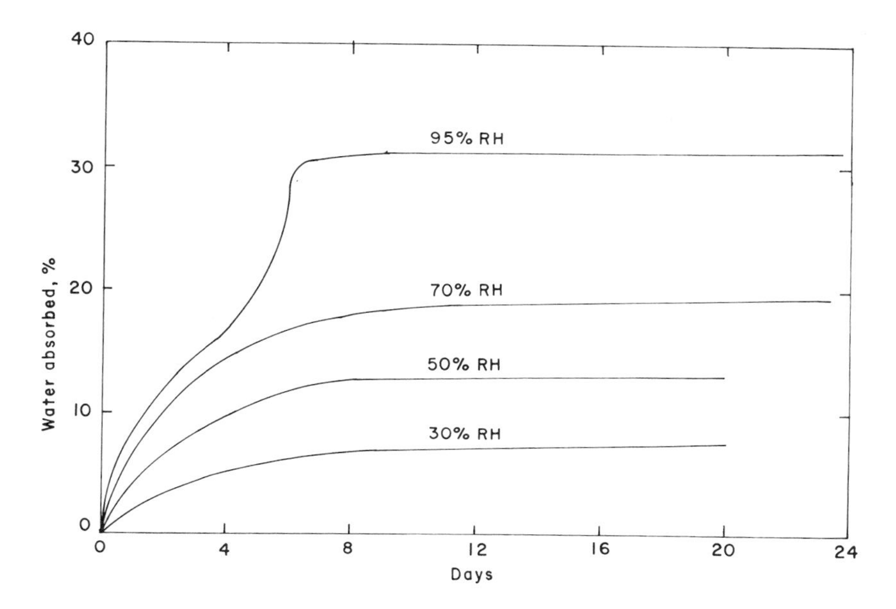

FIG. 10. Water absorption of U-F foam (Isoschaum). Reprinted from Ref. [2] by courtesy of Kunststoffe.

Urea-formaldehyde foam that has been submerged in water for 8 years was allowed to dry at room temperature for 24 h [2]. The cell structure was found to be unchanged, and a microscopic examination of the foam revealed no differences from a freshly prepared foam of the same composition.

The water-vapor transmission of U-F foam was reported to be 35 to 40 perms at 38° C and 90% relative humidity [14].

E. Chemical Resistance

Urea-formaldehyde foam (Isoschaum) did not exhibit any changes on storage in the following chemicals at 20° C for 21 days [2]:

Ammonium chloride (25%)	Acetic acid
Sodium carbonate	Propionic acid
Sodium chloride	Butyric acid
Acetaldehyde	Caproic acid
Formaldehyde	Petroleum ether
Acetone	Methanol

Diethyl ether
Ethanol
Ethyl acetate
Pelargonic acid
Tartaric acid
Benzene
Chloroform
Dibutyl phosphate

Concentrated and dilute caustic solutions and inorganic acids decompose U-F foam. The rate of decomposition increases with increasing concentration of the base or inorganic acid [2]. Among organic acids, concentrated formic acid decomposes the foam rapidly, whereas 50% solutions of formic and lactic acids bring about a slower decomposition.

F. Thermal Conductivity

Baumann [2] reported the thermal conductivity of U-F foams (Isoschaum) of different densities at temperatures of 25, 35, and 45° C. These data are shown in Table 3.

TABLE 3

Thermal Conductivity of U-F Foams (Isoschaum) at Various Temperatures[a]

Temperature (° C)	Density (lb/ft^3)	Thermal conductivity ($Btu/(sec)(ft^2)(° F/in.)$)
25	0.62	0.185
35	0.62	0.218
45	0.62	0.250
25	0.70	0.177
35	0.70	0.234
45	0.70	0.250
25	1.12	0.202
35	1.12	0.250
45	1.12	0.266

[a]Data from Ref. [2].

G. Sound Absorption

The sound absorption of perforated U-F foam (Isoschaum) of different foam thickness is given in Table 4 [2]. It has been shown that perforation (about 3%) of the foam increases the sound absorption, presumably due to an increase in the effective surface area. It can be observed from Table 4 that maximum absorption (70-95%) takes place at frequencies of 800 to 3200 Hz. The sound absorption of perforated U-F foam of 3-cm thickness, provided with air cushions of varoius thickness, is shown in Table 5 [2].

H. Resistance to Microorganisms

Under normal conditions of temperature and humidity U-F foam does not support fungus growth. Cultures of fungi, such as Mucor, Alternaria, and Aspergillus, when injected into U-F foam, died after a short time, indicating that U-F exerted some fungicidal action [2].

I. Radiation Resistance

Baumann [2] reported the resistance of U-F foam toward both soft and hard radiation. Foams containing lead compounds exhibit increased protection against α - and β -radiation as the lead content in the foam increases.

The penetration of U-F foam by γ rays from cobalt-60 was also investigated and is shown in Fig. 11 [2]. Foam samples 1 through 3 contained lead, with sample 1 possessing the highest lead content ($\sim$ 10%) and sample 3 containing the lowest lead content; sample 4 contained a cadmium compound. It can be seen that the foam sample containing the highest amount of lead and the foam containing the cadmium compound were most effective against γ radiation. Comparative data for pure lead are also included in Fig. 11. Table 6 [2] presents data on the absorption of γ rays as a function of foam thickness. The filled U-F foams used in these measurements were identical with the samples used to obtain the data of Fig. 11.

VI. APPLICATIONS

Although U-F foams have been replaced in many applications by polystyrene or urethane foams, they have found uses in widely different fields where their low cost, low density, and combination of properties have made them quite suitable. Nevertheless, it should be recognized

TABLE 4

Sound Absorption of Perforated U-F Foam (Isoschaum)[a]

Foam thickness (cm)	Sound absorption (%)						
	Frequency (Hz)						
	100	200	400	800	1600	3200	6400
3	9	23	58	70	89	78	77
4	10	34	78	85	93	86	79
5	12	44	83	92	95	92	83

[a]Data from Ref. [2].

TABLE 5

Sound Absorption of Perforated U-F Foam[a] with Air Cushions of Various Thicknesses[b]

Thickness of air cushion (cm)	Sound absorption (%)						
	Frequency (Hz)						
	100	200	400	800	1600	3200	6400
5	18	28	68	60	70	80	60
10	27	38	72	57	78	79	72
15	38	52	63	53	80	77	60
20	53	50	57	57	69	78	62
25	57	62	58	59	83	78	50
30	68	65	54	69	80	80	72

[a]Isoschaum foam, 3 cm thick.

[b]Data from Ref. [2].

TABLE 6

Absorption of γ Rays by Lead and U-F Foams (Isoschaum) Containing Lead and Cadmium[a]

Foam sample[b]	Foam thickness (m)			
	γ-Ray absorption			
	3%	6%	25%	50%
1	0.8	1.6	3	4
2	2.0	4.0	8	10
3	3.2	6.4	13	16
4	1.2	2.4	5	6
Lead	0.014	0.028	0.056	0.070

[a]Data from Ref. [2].

[b]Samples 1 through 3 contained lead, with sample 1 possessing the highest lead content (~10%); sample 4 contained a cadmium compound.

that the major disadvantages of U-F foams are their relatively low strength, their tendency toward brittleness, and water absorption. Methods to overcome these deficiences have been discussed in Sections IV and V.

A major application of U-F foams in Europe, notably England and Germany, has been in cavity wall insulation [5, 14, 42]. Urea-formaldehyde foam is being used for the insulation of private homes, apartment houses, industrial and agricultural buildings, hospitals, churches, schools, offices, hotels, and public buildings [14]. It can also be cast into blocks from which slabs can be cut to size for use in new buildings or between wall battens in old houses.

Urea-formaldehyde foams have also found application as pipe insulation [5]. Small pellets of U-F foam of low density (2% by volume solids) and very small cell size ($10^6/cm^3$) is used as a filler for insulation in construction applications [43]. The pellets are produced by centrifugal pelletizing of a fast-curing U-F foam. The cell walls of the foam are destroyed by air jets, forming extremely small cellular pellets.

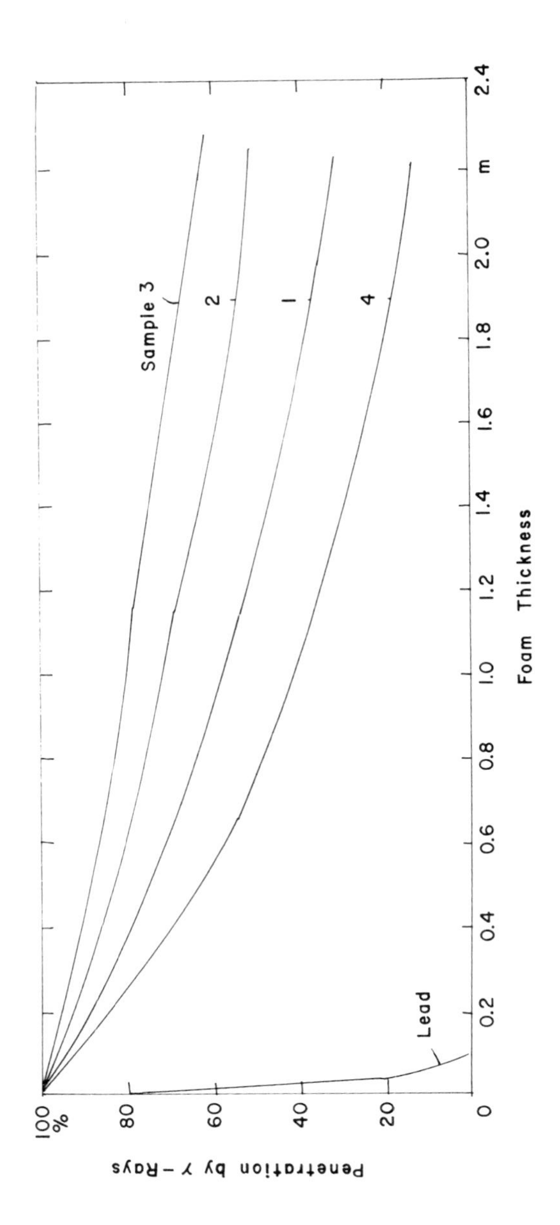

FIG. 11. Penetration of filled U-F foams by γ rays from cobalt-60. Samples 1, 2, and 3 contained lead (highest concentration, in sample 1, of about 10%) and sample 4 contained a cadmium compound. Reprinted from Ref. [2] by courtesy of <u>Kunststoffe</u>.

Although the ability of U-F foam to absorb water is a liability in insulation applications, its capacity to absorb large amounts of liquids has been put to use in a number of applications. Since soil bacteria decompose U-F resins with release of nutrient nitrogen, U-F foams, unmodified and modified, have been used for a variety of agricultural and horticultural applications [5, 11, 44-47]. The foam can absorb nutrient solutions and various other agricultural preparations that are retained and subsequently released at a slow rate, thus acting as a combination of carrier and fertilizer or plant food. Heavy rains do not wash away the various consitituents necessary for plant growth, as with the customary fertilizers. The foam prevents excessive drainage and supplies moisture when ordinary soils would become dry. The use of the modified U-F resin, Hygromull (BASF) as an agricultural mulch and the method of applying it have already been mentioned [11]. Mixed with water, Hygromull has also been employed for the conversion of desert land into plant-growing areas [46].

Wilmsen [45] reported the incorporation of seeds, bulbs, fertilizers, insecticides, and trace elements in U-F foam for horticultural applications.

Urea-formaldehyde foams have also found use in the shipping and display of blossoms and cut flowers [44, 47]. By soaking the foam in water, the moisture will be retained and will keep the flowers fresh for several days. In addition, nutrients that are added to the foam permit the flowers to flourish under novel display conditions. The U-F foam also provides nondripping shipping containers for flowers and delicate blossoms.

Molded U-F flowerpots combine light weight with water permeability and water retention, and have been shown to have a beneficial effect on plant growth [48].

Other applications that make use of the good absorption of liquids by U-F foam include surgical sponges and dressings after sterilization of the plasticized foam [47].

An interesting application of U-F foams is a partial replacement for wood pulp in the formation of paper composites, developed by investigators of the Scott Paper Co. [49-52]. After the U-F foam has been prepared, it is compressed and then disintegrated in aqueous suspension by means of a high-speed cutter. It is then dewatered to remove residual acid curing agent and centrifuged to separate large and granular particles. The aqueous suspension of U-F foam fragments (48-150 mesh) can be combined with wood pulp and sheets formed in conventional papermaking equipment. The resulting sheets are softer and higher in bulk and exhibit 4 to 5.5 times greater water absorption than sheets derived solely from wood pulp. Furthermore, wetting of these sheets occurs several hundred times faster than with ordinary wood-pulp sheets [49]. Because of these properties these composites find application in highly absorbent bandages.

Sheets obtained from mixtures of uniaxial cellulosic fibers and multiaxial fiber assemblies from disintegrated U-F foam exhibit in addition to greater bulk an increase in opacity and brightness as well as improved appearance [51].

One process for increasing the water absorption of U-F foam consists of incorporating into the foam formulation a soluble compound that is later removed by extraction [53]. This treatment increases the surface area, thus enhancing the hydrophilicity of the foam.

Another method for increasing hydrophilicity is treatment of cocondensates of U-F and phenolic foams with ammonia [33]. Highly absorbent fibers are produced by hydropulping these foams and subsequently forming sheets that can be used as absorbent pads for surgical dressings, diapers, and sanitary napkins.

As described in Section IV, U-F foam can be made quite soft and resilient by subjecting it to uniform compression-relaxation cycles under carefully controlled conditions [23]. The resulting foams are utilized as absorbents for surgical and catamenial devices.

Urea-formaldehyde foams have also been found useful as filters for dry-cleaning solvents [44].

Urea-formaldehyde foams have been employed in the preparation of foamed adhesives in the bonding of wood (e.g., particle board), paper, synthetic fibers, canvas, and other materials [54, 55]. In a process described by Klug [54], the adhesive binder is formed from a 50:50 mixture of a U-F resin (50% aqueous solution) and concentrated sulfite waste liquor (50%). In addition to these components a typical foam formulation includes calcium carbonate (chalk) as blowing agent and formic acid as a catalyst. Sawdust or wood shavings and smaller amounts of shredded and air-dried peat are included as fillers. By varying the concentration of the acid hardener and the blowing agent, either medium-density (e.g., 24-lb/ft^3) or low-density (e.g., 3-lb/ft^3) wood composites can be formed. The medium density wood composites are suitable for such applications as wall panels and curtain wall ceilings.

Other applications of U-F foam include its use as artificial snow for the film industry and arrestor pads for airport runways [56].

REFERENCES

1. German Pat. 636,658 (to I.G. Farbenindustrie) (1933).
2. H. Baumann, Kunststoffe, 48, 362, 406 (1958).
3. U.S. Foam Plastic Markets and Directory, Technomic Publishing, Stamford, Conn., 1970.
4. J. L. DeJong and J. DeJonge, Rec. Trav. Chim., 27, 139 (1953).
5. C. A. Schutz, J. Cell. Plastics, 3, 37 (1968).

6. O. A. Vieli (to W. H. Kreidl), U.S. Pat. 3,150,108 (1964).
7. P. E. Lindvig (to DuPont), U.S. Pat. 2,789,095 (1957) .
8. L. Rechner, French Pat. 1,399,642 (1965) .
9. O. A. Vieli and W. H. Kreidl (to Vereinigte Chemische Fabriken, Kreidl, Rutter and Co.), U.S. Pat. 2,970,120 (1961).
10. D. S. Shriver, R. R. MacGregor, and W. P. Moore (to Allied Chemical), U.S. Pats. 3,186,959 (1965), and 3,256,067 (1966).
11. L. Unterstenhöfer, Kunststoffe, 57, 850 (1967).
12. H. Scheuermann (to BASF), German Pats. 800,704 (1948), and 802,890 (1951).
13. F. Graf, Baugewerbe, 39 (8), 314 (1959).
14. Thermal Insulation with Foamed Aerolite® Resins, Ciba (A.R.L.) Ltd., Publication No. FM. 1a, June 1968.
15. Imperial Chemical Industries Ltd., Plastics Division Report No. PLI/1371B.
16. P. Blackman (to ICI/Organics), U.S. Pat. 3,142,653 (1964).
17. P. Blackman and J. P. Conbere (to ICI/Organics), U.S. Pat. 3,174,943 (1965).
18. N. Brown (to DuPont), U.S. Pat. 2,807,595 (1957).
19. L. E. Daly (to U.S. Rubber), U.S. Pat. 2,432,389 (1947).
20. A. Marcora, Italian Pat. 679,248 (1964).
21. J. Lentz and H. Scheuermann (to BASF), German Pat. 875,557 (1953).
22. I. D. Roche (to DuPont), U.S. Pat. 2,805,208 (1957).
23. J. J. Eberl and S. Coppick (to Scott Paper), U.S. Pat. 3,063,953 (1962).
24. G. B. Sunderland (to American Cyanamid), U.S. Pat. 3,006,871 (1961).
25. R. J. Bender, Handbook of Foamed Plastics, Lake Publishing, Libertyville, Ill., 1965, p. 330.
26. F. Brasco and P. R. Temple (to A. D. Little, Inc.), U.S. Pat. 3,284,379 (1966).
27. Grisopat Anstalt für Patentwertung und Forschung, Swiss Pat. 400,562 (1966).
28. C. J. Benning, Plastic Foams, Vol. I, Wiley-Interscience, New York, 1969, p. 466.
29. T. Hairida and Y. Miyazu (to Riken Synthetic Resins), Japanese Pat. 2783 (1963).
30. G. H. Justice (to Allied Chemical), U.S. Pat. 3,306,861 (1967).
31. F. L. Kelly and R. I. Wells (to Allied Chemical), U.S. Pat. 3,231,525 (1966).
32. W. Ulbricht, Belgian Pat. 630,241 (1963).
33. S. Coppick and R. L. Beal (to Scott Paper), U.S. Pat. 3,189,479 (1965).

34. C. J. Benning, Plastic Foams, Vol. I, Wiley-Interscience, New York, 1969, p. 463.
35. L. S. Meyer (to L.O.F. Glass), U.S. Pat. 2,559,891 (1951).
36. V. I. Kabaivanov et al., Godishnik Khim.-Tekhnol. Inst., 3 (1), 47 (1956).
37. S. Maciaszek and L. Horyl, Polish Pat. 46,347 (1962).
38. Allied Chemical Corp., British Pat. 1,021,248 (1966).
39. P. J. Mason (to Allied Chemical), French Pat. 1,432,889 (1966).
40. R. H. G. Fenner and R. Thompson (to Intubloc), British Pat. 978,623 (1964).
41. H. L. Vandersall, paper presented at Polymer Conference, University of Utah, Salt Lake City, Utah, June 1970.
42. "Ufoam" Insulation, Imperial Chemical Industries Data Sheet, Nov. 1969.
43. British Plaster Board Holdings Ltd., Belgian Pat. 639,416 (1964).
44. T. H. Ferrigno, Rigid Plastics Foams, Reinhold, New York, 1967.
45. H. Wilmsen, Belgian Pat. 660,198 (1964).
46. M. Fischer, Garten und Landschaft, No. 12 (1966).
47. H. Scherr, A. Gottfurcht, and R. W. Stenzel, Plastics, 8 (11), 8 (1949).
48. D. C. Frysinger and O. R. Odhner, U.S. Pat. 3,282,868 (1966).
49. J. J. Eberl and S. Coppick (to Scott Paper), U.S. Pat. 3,164,559 (1965).
50. H. W. Steinmann (to Scott Paper), U.S. Pat. 3,407,538 (1962).
51. J. J. Eberl and S. Coppick (to Scott Paper), U.S. Pat. 3,004,884 (1961).
52. J. J. Eberl and S. Coppick (to Scott Paper), U.S. Pat. 3,210,239 (1965).
53. H. Baumann and F. Graf (to BASF), German Pat. 1,097,669 (1961).
54. O. W. H. Klug, British Pat. 989,799 (1965).
55. C. J. Benning, Plastic Foams, Vol. I, Wiley-Interscience, New York, 1969, pp. 474-475.
56. Imperial Chemical Industries Ltd., private communications, Sept. 4, 1970.

Chapter 13

EPOXY-RESIN FOAMS

Henry Lee
and
Kris Neville

Lee Pharmaceuticals
South El Monte, California

I. INTRODUCTION

Interest in epoxy-resin foams was most pronounced during the late 1950s and early 1960s. By 1964 estimates of epoxy-foam production in the United States were given at about 2% of the total production of rigid and semirigid foams [1], although this figure appeared then to be somewhat high. There is no evidence that epoxy foams have subsequently increased their share of the market. In the late 1960s, aside from the continuing patent literature issuing from earlier years, epoxy foams were mentioned only in review articles, where data on them were given for comparative purposes. Probably the largest single application, entering the 1970s, is for syntactic foams as core materials for laminates, and here volume usage is relatively small.

The present epoxy resins, though they can be flexibilized, are not truly elastomeric and thus can compete only in the rigid- and semirigid-foam markets. The basic properties of rigid foams are not sufficiently outstanding to warrant the development expenses and the higher formulation costs involved in their successful use for potential volume applications. Strength properties are no better than those of the rigid polyurethanes, and marginal improvements in chemical resistance are of little consequence in most commercial applications. Epoxy foams offer some advantage with regard to water sensitivity and aging properties [2]. By proper formulation, systems with superior high-temperature resistance can be developed. However, in total they have been and remain of interest in only limited and specialized areas.

Epoxy foams are not as satisfactory as polyurethanes for foaming-in-place applications because it is more difficult to control their chemical reactions [3]. From the beginning of the technology, therefore, emphasis was placed on syntactic, rather than chemical, systems. Syntactic foams are obtained by adding hollow spheres or other low-density fillers to the formulation. Syntactic foams are inherently less versatile than the chemical foams, which further restricts their overall usefulness.

II. RAW MATERIALS

Epoxy-resin-foam formulations contain the resin itself; a curing agent selected for the desired handling and cured properties, such other modifiers as are dictated by the application to produce flame resistance, lower cost, modified electrical properties, and the like; and specialized ingredients required to produce foam of the proper consistency.

The basic resin suppliers have, from time to time, suggested starting foam formulations and, in at least one case, offered epoxy-resin foams on an experimental basis. Today, however, the foams are available only through specialized epoxy-resin formulators and specialized producers of prefoamed blocks. Such literature as exists, therefore, reports data on proprietary formulations, the ingredients of which are not disclosed.

A. Epoxy Resins

Over the years a large number of distinct epoxy resins have been offered commercially and experimentally from the basic suppliers. These resins have been the subject of a number of books [4-9], the most comprehensive of which is the Handbook of Epoxy Resins [4].

The most useful epoxy resins are those prepared by the reaction of epichlorohydrin (1) with a phenolic or alcoholic compound followed by dehydrohalogenation. The resultant structure, the epoxy group (2), is common to all the epoxy resins.

$$ClCH_2\overset{O}{\overbrace{CH-CH_2}} + HO-R \longrightarrow R-OCH_2\overset{O}{\overbrace{CH-CH_2}} \qquad (1)$$

(1) (2)

Other techniques for obtaining the epoxy group are available, one of which involves adding oxygen to double bonds along aliphatic or cycloaliphatic chains. Peroxyacetic acid is the oxygen-bearing donor compound, and the resins are thus often referred to as "peroxyacetic acid epoxies." A trememdous variety of components can be epoxidized with peroxyacetic acid, although only a few of them have warranted commercial marketing, and, in total, excluding the use of such compounds as stabilizers for vinyl-resin formulations, they contributed very little to the overall volume of epoxy resins produced.

Although some of the peroxyacetic acid epoxies have been used to produce foams, they have not been shown to offer evident advantages. The epoxy resins of choice are the diglycidyl ethers of bisphenol A (DGEBA) of the generalized formula (3).

When n is zero, this resin is a liquid species with the viscosity of light machine oil at room temperature. It is straw colored and has a mild, not unpleasant, characteristic odor. A number of commercial versions are available, some of which have improved reactivity, some contain a variety of monoepoxy diluents to provide lower viscosities, some have other properties tightly controlled, including color and defoam time under vacuum. Typical commercial materials are listed in Table 1.

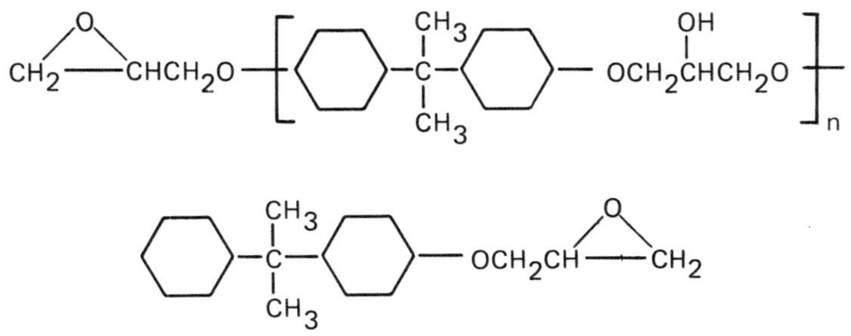

$$CH_2\overset{O}{—}CHCH_2O\left[C_6H_{10}—\underset{CH_3}{\overset{CH_3}{C}}—C_6H_{10}—OCH_2\overset{OH}{C}HCH_2O\right]_n$$

$$C_6H_{10}—\underset{CH_3}{\overset{CH_3}{C}}—C_6H_{10}—OCH_2CH\overset{O}{—}CH_2$$

(3)

TABLE 1

Typical Commercial Epoxy Resins

Resin	Manufacturer
DGEBA[a] (n = 0):	
Araldite 6010	Ciba Products Co.
DER 331	Dow Chemical Co.
Epi-Rez 510	Jones-Dabney Co.
Epon 828	Shell Chemical Co.
Epotuf 37-140	Reichhold Chem., Inc.
ERL 2794	Union Carbide Corp.
DGEBA, diluted:	
Araldite 502	Ciba Products Co.
DER 334	Dow Chemical Co.
Epi-Rez 504	Jones-Dabney Co.
Epon 815	Shell Chemical Co.
Epotuf 37-135	Reichhold Chem., Inc.
ERL 2795	Union Carbide Corp.
Higher functionality resins:	
Araldite EPN 1138	Ciba Products Co.
DEN 431	Dow Chemical Co.
Epi-Rez 5155	Jones-Dabney Co.
Epon 1031	Shell Chemical Co.
Epotuf 37-170	Richhold Chem., Inc.
ERLB 0447	Union Carbide Corp.

TABLE 1 (Continued)

Resin	Manufacturer
Chlorinated or Brominated resins:	
Araldite 8011	Ciba Products Co.
DER 511	Dow Chemical Co.
Epi-Rez 5161	Jones-Dabney Co.
Epon H-45	Shell Chemical Co.
Epoluf 37-200	Reichhold Chem., Inc.
ERLA 0625	Union Carbide Corp.

[a] See structural formula (3).

B. Curing Agents

When epoxy resin is intimately admixed with a stoichiometric amount of a curing agent containing labile hydrogen atoms, the epoxy ring opens as follows:

$$R—CH\overset{O}{\frown}CH_2 + H—R' \longrightarrow R—\underset{}{\overset{OH}{|}}{CH}CH_2R'$$

The most commonly used curing agents are primary and secondary amines and carboxylic acid anhydrides. In the former case the reaction is straightforward, through the amine hydrogens, and for thermoset structures to occur, the amine should contain at least three labile hydrogens for the diepoxy resins and at least two labile hydrogens for the more functional species. In the case of the anhydrides the nascent hydroxyls enter into the reaction to open the anhydride ring, thereby increasing the system functionality to some value in excess of 2 with a monoanhydride and a diepoxy, and permitting crosslinking. In both cases optimum properties require the intimate admixture of just the proper amounts of the two ingredients.

A second general method is available for effecting conversion of the epoxy resin to the thermoset state. By it a curing agent, either a tertiary amine or a Lewis acid, is employed in relatively small amounts. The curing agent serves to trigger a reaction between epoxy groups, the reaction continuing in the presence of hydroxyls but not necessarily incorporating the hydroxyl-containing species directly into the cured network. These reactions are somewhat less straightforward.

When the curing agent contains labile hydrogens and is directly incorporated into the cured structure, the final properties partake of the properties of both the initial reactants. When a Lewis acid or tertiary amine is used as curing agent without diamine or anhydride, the cured properties are more directly related to the properties of the starting epoxy resin and the system becomes essentially a homopolymer.

As is evident, the final properties of the cured network can be varied over quite a wide range by selection of curing agents and/or specific resins. The cured products of commercially useful systems are characterized by good toughness, outstanding resistance to solvents and most bases and acids, good physical strength, outstanding adhesion to properly prepared surfaces, and low final shrinkage. Cure times can be regulated by regulating the temperature at which the reaction is conducted or by varying the reactivity of the curing agent.

As a general rule, aliphatic primary amines react exothermically to produce thermoset structures in 15 to 20 min at room temperature, provided the mass being reacted is sufficiently large to permit the exotherm to build up to about 125° C or more. When it is not accelerated by exotherm or the application of external heat, the reaction at room temperature will not be completed for a number of hours as is the actual case with thin films.

The aromatic primary amines react quite rapidly at 120 to 150° C, but only very sluggishly at room temperature, even in larger quantities.

The anhydride curing agents require even higher reaction temperatures or longer reaction times. Moreover, as a general rule they are only slightly exothermic and do not show as sharp a gel point as the amine-cured systems, tending instead to thicken progressively.

Tertiary amines provide about the same order of reactivity as the alphatic primary amines, whereas the Lewis acids, especially selected for use with epoxy resins, provide extremely fast reactions at room temperature, with correspondingly high exotherms, even in small masses. Reactions are often complete in less than 60 sec.

The most frequently used curing agents are those listed in Table 2. The aliphatic primary amines, the tertiary amines, and the Lewis acids provide cured systems that begin to soften at 100 to 120° C and thus inherently offer limited heat resistance. The aromatic primary amines provide somewhat better heat resistance up to about 150°C. Special dianhydride formulations offer excellent long-term stability above 200° C and are good for short-time use at temperatures around 400°C.

The Lewis acids, most usually boron trifluoride complexes with a polyol, because of fast reaction time, are most suitable for use with

TABLE 2

Typical Epoxy-Resin Curing Agents

Aliphatic Primary Amines	
Diethylenetriamine	Dimethylaminopropylamine
Triethylenetetramine	Diethylaminopropylamine
Tetraethylenepentamine	Aminoethylethanolamine
Alicylic Polyamines	
Menthanediamine	N-Aminoethylpiperazine
Aromatic Polyamines	
m-Phenylenediamine	Diaminodiphenylsulfone
4,4'-Methylenedianiline	Aniline-formaldehyde resins
Tertiary Amines	
2, 4, 6-Tris(dimethylaminomethyl) phenol	Benzyldimethylamine
Lewis Acids	
Boron trifluoride etherates	Boron trifluoride-monoethylamine complexes
Anhydrides	
Phthalic anhydride	Chlorendic anhydride
Maleic anhydride	Pyromellitic dianhydride

room-temperature-curing formulations provided only small masses are to be foamed. With larger masses the exothermic heat is sufficiently great to char the curing structure. The material is covered in the patent literature when used as the sole curing agent [10, 11] or when used as an accelerator for the slower amine-cured systems [12]. The reaction usually takes place within 30 sec, so that additives may be used to slow it down somewhat. Ammonium halides will extend the time to about 70 sec [13].

With primary aliphatic amines reactions are considerably slower, but exothermic considerations dictate careful regulation of masses being cast and require relatively good control of the curing temperature [14]. Because they do not offer improved properties commensurate with the inconvenience of their handling, such systems have generally been abandoned.

The aromatic primary amines, because of somewhat higher heat resistance, have been investigated in more detail, not only in the United States but also abroad [15]. m-Phenylenediamine is typical of these amines.

A system of improved high-temperature properties has been produced by substituting diaminodiphenylsulfone for m-phenylenediamine, with further improvements being obtained by the use of a more highly functional solid epoxy resin [16].

Although anhydride curing agents have been used experimentally [17], they are most suited for syntactic foams.

Polyisocyanates have been suggested as cocuring agents for anhydrides [18] and for use as curing agents and coreactants [19, 20], although they appear to offer no advantage.

C. Blowing Agents and Surfactants

A number of compounds capable of reacting with epoxy resins to provide cure simultaneously liberate gas, thereby producing foamed products of indifferent value. Among them are aconitic acid [21], trimethoxyboroxine [22, 23], phosphitoborohydrides [24], and polyamine carbonates [25, 26]. Diisobutylaluminum-o-acetyl phenoxide is reported as foaming peroxyacetic acid epoxies [27]. Other agents are capable of reacting with the curing agents to produce foams. Among these are N, N'-dinitroso-N, N'-dimethyl terephthalamide [28] and hydrogen peroxide [29].

Water has been reported as a foaming agent [30], as has balsa wood and cork when used as a powder, the foam being produced from water contained in the additive [31].

The fluorocarbons have been suggested more often [32], as well as commercial materials from the more general foam technology, such as Celogen and Unicel ND [33] and toluene [34, 35]. Figures 1 and 2 give typical blowing-agent parameters for DGEBA cured with an aliphatic diamine like diethylenetriamine.

The chemical blowing agents follow the wider commercial practice and are selected in terms of decomposition temperature with relation to the specific formulation involved.

Surface-active agents, such as Tween 20 or silicone oils, are also used, and a variety of these have been suggested in the patent literature [36].

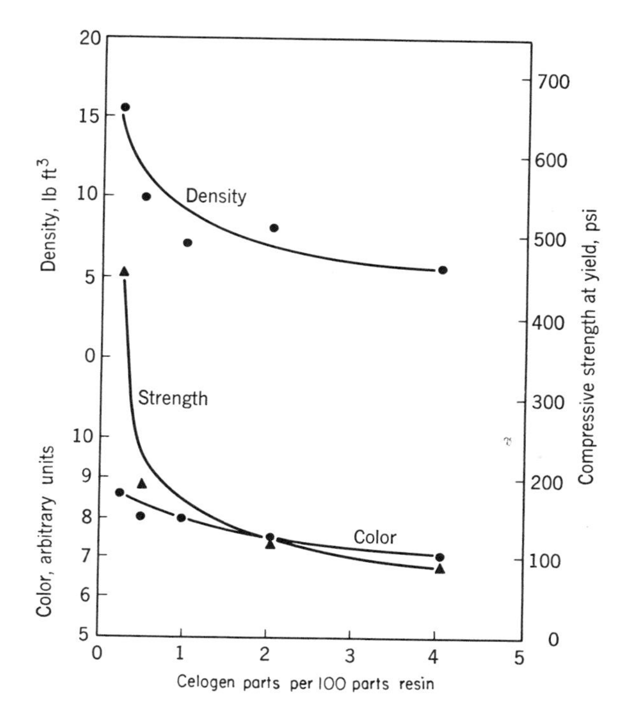

FIG. 1. Effect of blowing-agent concentration on the properties of epoxy-resin foam. Reprinted from Ref. [4] by courtesy of the McGraw-Hill Book Co.

D. Fillers

Fillers are conventionally used with epoxy-resin formulations for a variety of reasons, including lowering cost, reducing exotherms, reducing shrinkage rates, modifying surface characteristics, improving chemical resistance, and increasing adhesion. Finely divided fillers, such as talc and silica, have been used experimentally in chemical foam formulations, but as the density of the foam decreases, the amount of filler that can be tolerated without sacrifice of strength likewise decreases, until at lower densities only a few parts per hundred can be tolerated [8].

Small amounts of filler can be used for pigmentation, and special fillers like antimony oxide can be used to improve the flame resistance of formulations containing chlorinated resins, curing agents, or additives [37].

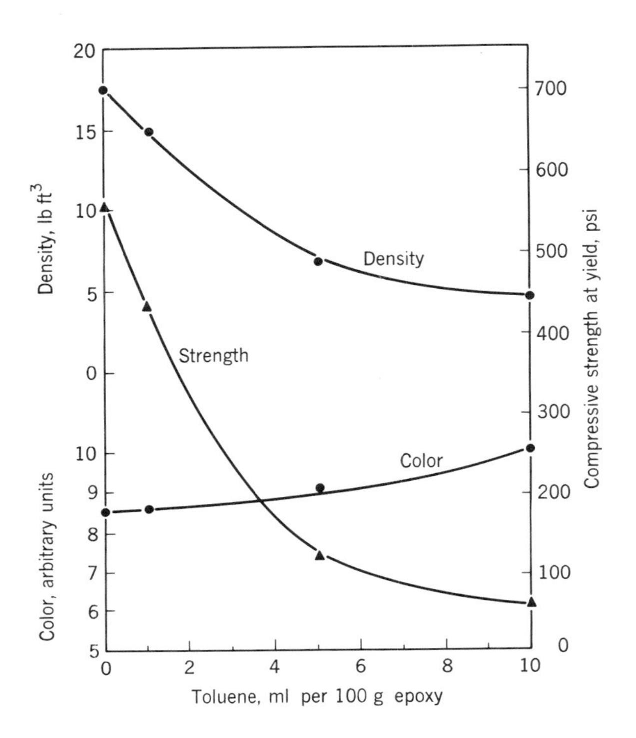

FIG. 2. Effect of toluene content on the properties of epoxy-resin foam. Reprinted from Ref. [4] by courtesy of the McGraw-Hill Book Co.

Commercially, metallic fillers are more significant, being used to control dielectric constants [38]. Low-cost fillers may be added in cases where full weight reduction is not required. Cement has been suggested in one case [39].

A variety of low-density fillers--so-called microballoons--are used to provide syntactic foams (Table 3), and these are of more general interest.

1. Microballoons

The influence of microballoons on epoxy-resin formulations is not uniquely different from their influence on other plastic formulations. Physical properties decrease in relationship to the loading volume of the filler, and the system viscosity increases proportionally.

TABLE 3

Low-Density Fillers for Epoxy Foams[a]

Name	Description	Particle size	Bulk density (g/ml)
Kanamite grade 200	Unicellular hollow clay spheres, primarily aluminum silicate	90% 35-60 mesh 10% 20-100 mesh	0.9
Colfoam micro-balloons	Urea-formaldehyde spheres	90% 20-100 mesh 10% through 200 mesh	0.2-0.5
Microballoons, BJOA-0840	Hollow phenolic spheres	10% 40-100 mesh 74% 40-200 mesh 16% through 200 mesh	0.3-0.4
Glass micro-ballons, CPR 2077	Hollow glass microspheres	89% 60-325 mesh 11% through 325 mesh	0.26

[a]Reprinted from Ref. [40] by courtesy of the American Chemical Society.

As a general rule the best syntactic foams are obtained with the maximum amount of the microballoons and the minimum amount of resinous matrix. Lightly filled systems tend to create filler-settling problems during cure, which require the addition of thixotropic agents with the attendant increase in viscosity that their use entails. At densities below 35 lb/ft^3 in any event, the systems are not pourable [41], and generally must be applied by packing-in-place techniques. Mixing the heavily loaded formulations is somewhat difficult, and with organic microballoons excessive mixing will tend to break down the delicate cell walls of the fillers [40].

In order to remove the entrapped air from the heavily filled systems, special mixing techniques may be required. One involves vibrating the mixture as a film and passing hot air over the surface to break the bubbles [42].

When very low densities are the aim, microballoons may be coated with the epoxy formulation and the product supplied as free-flowing powders that can subsequently be cured with heat and pressure to produce the lightweight structure [43].

Both organic and inorganic microballoons have been suggested. The organic materials offer somewhat lighter weight at some sacrifice in strength properties. However, at the high loading volumes usually employed, they, rather than the specific matrix binder, govern the main parameters of the foams, and therefore the use of epoxy resins, in preference to lower cost matrices, does not appear to be warranted in most cases. Of the organics, the phenolic microballoons appear to be preferred [44], as they offer better thermal shock resistance [40].

Polystyrene beads can be expanded during cure [45] to form positive filling actions at pressures from 10 to 60 psi, but they have limited storage life and should not be used in applications in which the exotherm cannot be carefully controlled [46].

Inorganic microballoons are of greater usefulness, because of the improved strengths and better heat resistance they impart to the systems and also because of somewhat better stabilities [47]. Such syntactic foams provide relatively smooth surface finishes [48].

2. Porous Laminates

It has been suggested that, for some applications, microballoons be added to the laminating resin to reduce the weight of the laminate [49]. More interestingly, and more importantly, controlled amounts of solvents have been employed during the layup to produce porous structures in which the glass can be considered to be the filler.

III. FOAMING PROCESSES

As is evident from the preceding discussion, epoxy-resin chemical foams are produced under tightly controlled conditions with special molds designed with the specific system being cast in mind. Only a limited adjustment in the amount of curing agent can be used to control exotherm without severely sacrificing the cured properties. Thus equipment that is designed for casting a specific mass in a specific configuration at a specific temperature will not necessarily be suitable when the mass or temperature is either increased or decreased, or the configuration is changed drastically. With properly designed molds, one can develop prefoamed blocks of respectable size that are suitable for further fabrication with woodworking tools. In view of the critical nature of the formulations, the empirical approach is best suited to developing handling procedures.

For systems based on boron trifluoride etherates and the like, it has been reported that processing systems useful with the polyurethanes can be used, but it is not believed that the technology has reached the degree of industrial acceptance to warrant the extensive developmental work required to refine and commercialize such processing systems. Because of the extremely short reaction time, the temperature at which such systems are mixed requires fairly careful control. Mixing temperatures of about 20° C are preferred, the resin being nucleated by stirring air into it until it is creamy white before the addition of the curing agent. A 10-sec high-speed mix is then achieved prior to the pour. A typical formulation might consist of a curing agent from boron trifluoride mixed with polyethylene glycol (molecular weight 400) combined with orthophosphoric acid. This is then used with a mixture of DGEBA compounds (n = 0 and 1), together with an aliphatic diepoxide, with a fluorocarbon as the blowing agent [50].

With primary aliphatic amines, the foaming reaction (at room temperature is initiated 10 to 20 min after mixing, so that the mixing operation itself can be conducted in a more leisurely manner and does not require sophisticated or high-speed equipment. After thorough mixing the formulation may be poured into preheated molds. A typical formulation for a room-temperature foam-in-place system is shown in Table 4.

TABLE 4

Typical Formulation for a Room-Temperature Foam-in-Place System

Component	Parts by weight
DGEBA (n = 1)[a]	100
Toluene	5
Tween 20	2 drops
Diethylenetriamine	6
Celogen (blowing agent)	1

[a] See structural formula (3).

The system is best used when the reactants have been prewarmed to a convenient temperature (40-60° C), so that the gelation reaction is sharp. Because of exothermic considerations, the formulation is not suited to casting larger masses.

Aromatic amines are, in the main, solid at room temperature, and two general techniques have been advanced for their use in foam formulations. The amine and resin may be mixed and partially cured (B-staged) with a solid blowing agent incorporated. The friable, partially reacted mass is then powdered and the powder poured into a heated mold to liquefy the system and simultaneously crosslink and foam it. Alternatively the two compounds may be precooked, with an accelerator being added just before the temperature is increased to the final curing temperature [51]. A typical aromatic amine formulation is shown in Table 5. This mixture will foam on heating to about 100° C.

TABLE 5

Typical Aromatic Amine Formulation

Component	Parts by weight
DGEBA (n = 0)[a]	80
m-Phenylenediamine	28
Celogen (blowing agent)	2
Plyophen 5023 (phenolic modifier)	20
Pluronic L-64 (surfactant)	0.1

[a]See structural formula (3).

The syntactic foams are either poured or tamped in place and cured in ovens or allowed to cure at room temperature, depending on the specific formulation. In these cases careful temperature control is not usually required.

IV. PROPERTIES

It is appropriate to consider the properties of the chemical foams separately from the properties of the syntactics.

A. Chemical Foams

The physical properties of the chemical foams, at lower densities, are controlled by the foam structure throughout the operating temperature range. Typical curves for compressive properties are shown in Fig. 3. Thermal expansion rates are somewhat lower than those for the parent unfoamed formulations (e.g., 3×10^{-5} versus 6×10^{-5} in./in.-°C). The foams can be made self-extinguishing by selecting special chlorinated or brominated resins or curing agents.

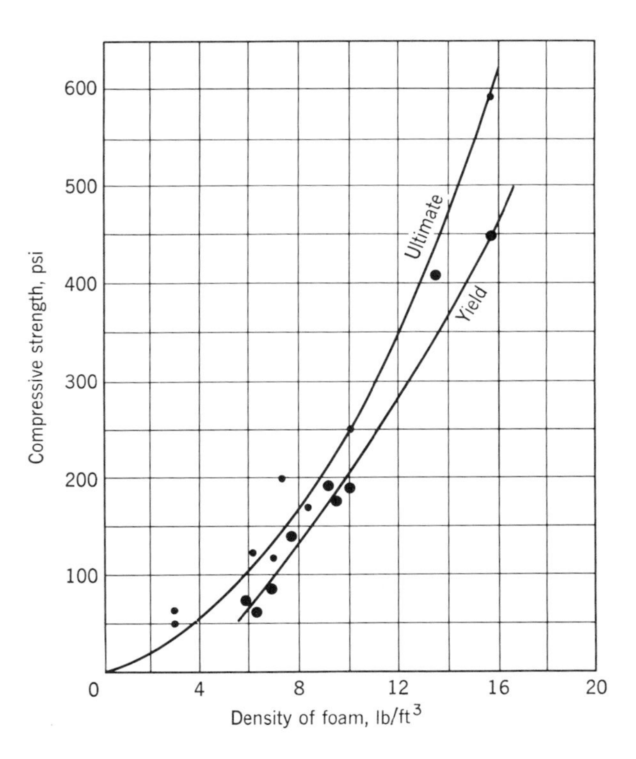

FIG. 3. Relationship between compressive strength and density of DGEBA foam cured with an aliphatic amine. Reprinted from Ref. [4] by courtesy of the McGraw-Hill Book Co.

The thermal properties of the foam are also a function of density. Thermal conductivity ranges upward from that of air to that of the parent formulation, which is a faily good thermal insulator, and is a function of cell size and density (Figs. 4 and 5). At elevated temperatures the nature of the formulation becomes more and more critical, but in general, for the chemical foams, performance is limited to about 150° C maximum. Tensile strengths at cryogenic temperatures are lower than they are at room temperature, presumably because irregularities in the structure create notch effects where cell walls join. Modulus, rigidity, and stress-strain curves at low temperatures are shown in Figs. 6 and 7.

The electrical properties are unremarkable [54]. The dielectric constant runs from about 1.1 for low-density foams up to about 1.6 for densities of about 20 lb/ft^3. The dielectric strength is about 130 V/mil, and the dissipation factor is little changed from that of the unfoamed formulations, although these properties are somewhat formulation dependent.

The chemical resistance is about what is expected of the unfoamed formulations: excellent to alkalies and fair to excellent to solvents and acids. In this respect epoxy-resin foams offer some advantages over the competitive materials.

A chemical foam formulation, based on boron trifluoride etherate and suggested for insulation of storage tanks, gives the properties indicated in Table 6.

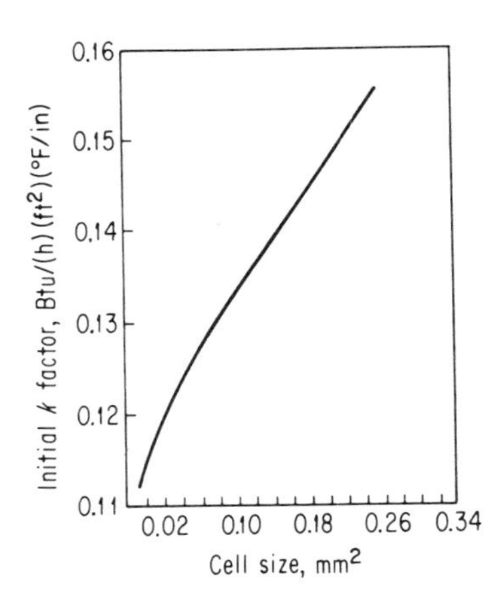

FIG. 4. Initial thermal conductivity as a function of cell size for DGEBA foam. Reprinted from Ref. [52] by courtesy of the American Institute of Chemical Engineers.

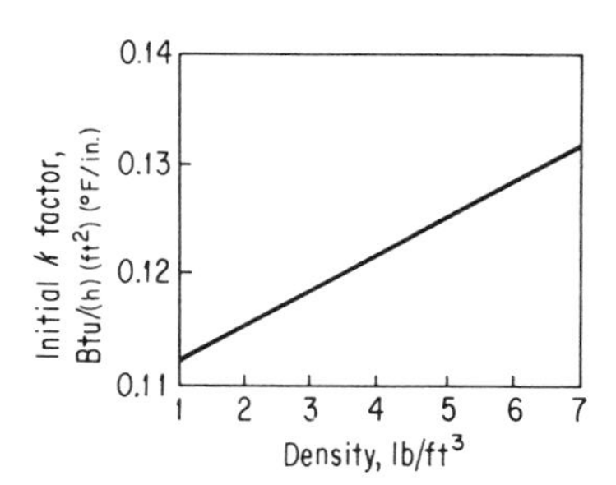

FIG. 5. Initial thermal conductivity as a function of density for DGEBA foam. Reprinted from Ref. [52] by courtesy of the American Institute of Chemical Engineers.

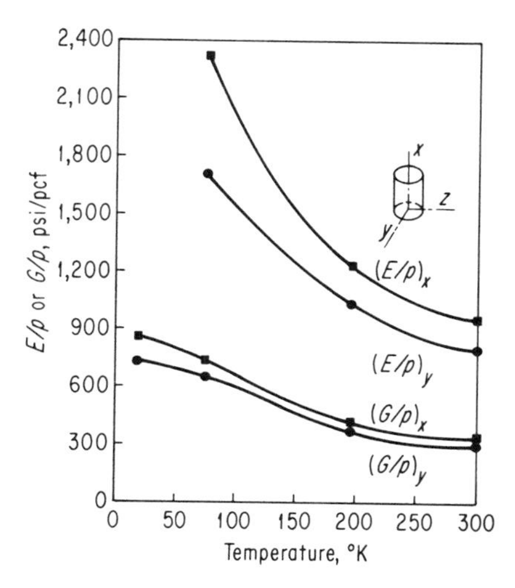

FIG. 6. Modulus of elasticity (E/p) and rigidity (G/p) as a function of temperature for DGEBA foam (density 5.5 lb/ft^3). Reprinted from Ref. [53] by courtesy of the Society of Plastic Engineers.

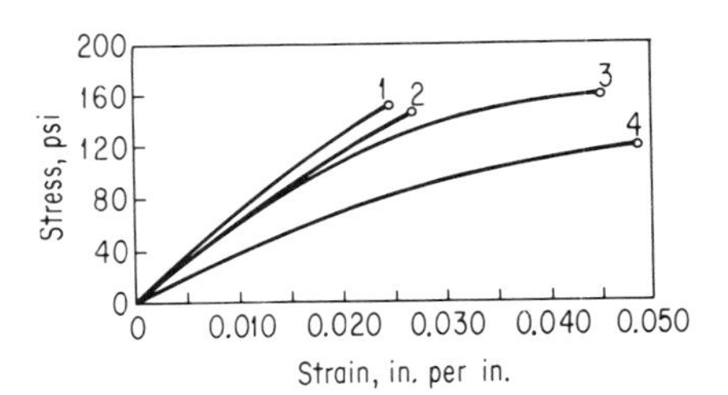

FIG. 7. Stress-strain curves for DGEBA foam. Reprinted from Ref. [53] by courtesy of the Society of Plastic Engineers. The experimental parameters are as follows:

Specimen No.	Temperature (° K)	Density (lb/ft^3)	Foam-growth direction
1	195	5.43	Parallel
2	195	5.62	Perpendicular
3	300	6.06	Parallel
4	300	5.06	Perpendicular

TABLE 6

Properties of Spray-on Epoxy Foams[a]

Property	Value
Density, lb/ft^3:	
Overall	1.8 to 2.0
Core	1.7 to 1.9
Thermal conductivity at 23° C, Btu/(h)(ft^2)(° F/in.):	
Initial k factor	0.11 to 0.12
Aged k factor[b]	0.16 to 0.17
Tensile strength, psi	26 to 31
Elongation, %	34 to 44
Compressive strength, psi	13 to 17

TABLE 6 (continued)

Properties of Spray-on Epoxy Foams[a]

Property	Value
Adhesive strength, psi:	
Of cold-rolled steel	27
Of primed steel	29
Of stainless steel	32
Of aluminum	27
Of sandblasted polystyrene	31
Water-vapor transmission at 23° C and 50% R. H., perm in.	0.9 to 1.2
Water absorption after 48 h under 10-ft head, lb/ft^2	0.03 to 0.05
Dimensional stability:	
At 160° C after 72-h total immersion in liquid nitrogen:	
Volume change, %	+0.6 to -0.8
Weight change, %	+0.2 to -0.4
At 60° C and ambient humidity after 72 h:	
Volume change, %	+2.9 to -3.7
Weight change, %	+0.2 to -0.6
Unicellularity of closed cells, %	83 to 90
Resistance to solvents:	
Aliphatic hydrocarbons	Excellent
Aromatic hydrocarbons	Fair
Low-boiling ketones	Poor
Resistance to chemical agents:	
Acid	Excellent
Alkali	Good

[a]Reprinted from Ref. [55] by courtesy of the Society of Plastic Engineers.
[b]Equilibrium value reached at ambient temperature.

B. Syntactic Foams

Cured properties of formulations with microballoons and with silica are compared in Table 7. Further comparisons are given in Tables 8 and 9.

The low conductivity of foamed formulations generally operates against their usefulness in electronic embedding applications, since thermal barriers can produce hot spots. However, with suitable formulations, it is possible to produce medium- and high-density foams that will serve satisfactorily [56], as indicated in Table 10.

Typically, formulations with microballoons may employ diluents to reduce viscosities of the vehicle to the point where larger loading volumes of fillers may be accomodated without introducing severe handling difficulties.

The better high-temperature syntactic foams give properties in the range of those indicated in Table 11 and Fig. 8.

Laminates incorporating syntactic foams as core materials (Fig. 9) give properties as indicated in Fig. 10 and Table 12. The specific formulation is selected in terms of the pressures the finished structure is required to withstand.

TABLE 7

Properties of Cured Epoxy Resin at Equivalent Loadings of Filler[a]

Property	Microballoons at 33%	Silica, 200-mesh, at 33%
Specific gravity	0.6	1.7
Flexural strength, psi	4400	4200
Linear expansion, in./(in.)(°C) x 10^6	17	16
Dielectric constant at 10^{10} Hz	1.9	3.6
Dissipation factor at 10^{10} Hz	0.015	0.028
Resistivity, ohm-em x 10^{-12}	1	4

[a]Reprinted from Ref. [48] by courtesy of the McGraw-Hill Book Co.

TABLE 8

Effect of Fillers on the Physical Properties of Epoxy Resin Cured with Tertiary Amine Salt[a]

Filler[b]	Viscosity at 25° C (cP)	Linear shrinkage (%)	Hardness, Shore D	Density at 21° C (lb/ft^3)	Tensile strength (psi)	Thermal conductivity (Btu/(h)(ft^2)(° F/in.)	Thermal expansion at 25 to 100° C (in./(in.)(° C) x 10^5)	Weight reduction over silica-filled compounds (%)
None	13,500-19,500	0.12	80-85	1.17	8000	2.68	8.7	26.5
Silica, 325 mesh, 100 phr	43,000-48,000	0.08	80-85	1.59	5500	6.38	8.6	0
Phenolic spheres, 15 phr	34,000-38,500	0.14	80-84	0.86	3300	1.91	8.2	46.0
Kanamite, 34 phr	34,000-39,000	0.06	75-80	1.01	2000	2.47	6.7	36.5
Colfoam, 4 phr	45,000-48,000	0.17	80-85	1.01	4050	4.15	8.6	36.5
Glass spheres, 14 phr	44,000-47,000	0.25	80-85	0.95	4200	4.56	8.2	41.0

[a]Reprinted from Ref. [40] by courtesy of the Americal Chemical Society.

[b]Quantities in parts per 100 parts resin (phr)

TABLE 9

Effect of Fillers on the Electrical Properties of Epoxy Resin Cured with Tertiary Amine Salt[a]

Filler[b]	Dielectric constant at 25° C		Power factor at 25° C		Volume resistivity (ohm-cm)			Dielectric strength at 25° C
	1 kHz	1 MHz	1 Hz	1 MHz	25° C	65° C	100° C	(V/mil)
None	3.8	3.7	0.0035	0.015	8.7×10^{14}		5×10^{11}	400-500
Silica, 325 mesh, 100 phr	3.4	3.4	0.003	0.012	1.3×10^{14}	6.2×10^{13}		>330
Phenolic spheres, 15 phr	3.2	2.7	0.003	0.014	1.0×10^{14}	5.3×10^{13}		>330

[a] Reprinted from Ref. [40] by courtesy of the American Chemical Society.

[b] Quantities in parts per 100 parts resin (phr).

TABLE 10

Comparison of the Hot-Spot Temperatures of Unembedded Tubes with Those Embedded in Epoxy Formulations Containing Various Fillers[a]

Filler	Operating unembedded tubes at 23° C ambient	Operating embedded tubes at 23° C ambient	Operating embedded tubes at 82° C ambient
Silica	100	75	112
Kanamite	100	85	125
Colfoam	100	100	135
Phenolic microballoons	100	105	140
Glass microballoons	100	95	129

[a]Reprinted from Ref. [40] by courtesy of the American Chemical Society.

TABLE 11

Properties of Epoxylite #8982 High-Temperature Syntactic Foam[a,b]

Property	Value
Specific gravity	0.721
Compressive strength, psi	18,000
Service temperature, °C:	
Continuous	260
Short time	425
Lap shear (on unetched 24 ST aluminum), psi:	
25° C	840
120° C	750
180° C	640
260° C	350
315° C	300

[a]Data supplied by the Epoxylite Corp.

[b]Foam is resistant to moisture, solvents, hydraulic oils, and jet fuels.

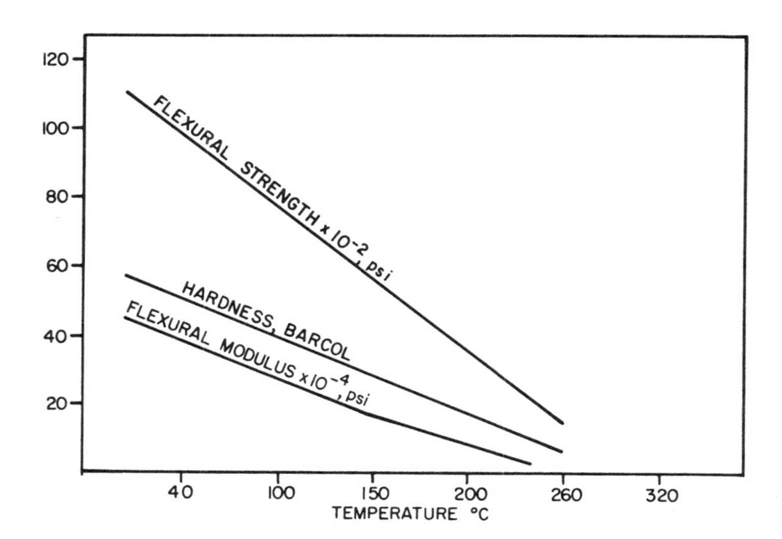

FIG. 8. Strength retention as a function of temperature for Epoxylite high-temperature formulation. Courtesy of the Epoxylite Corp.

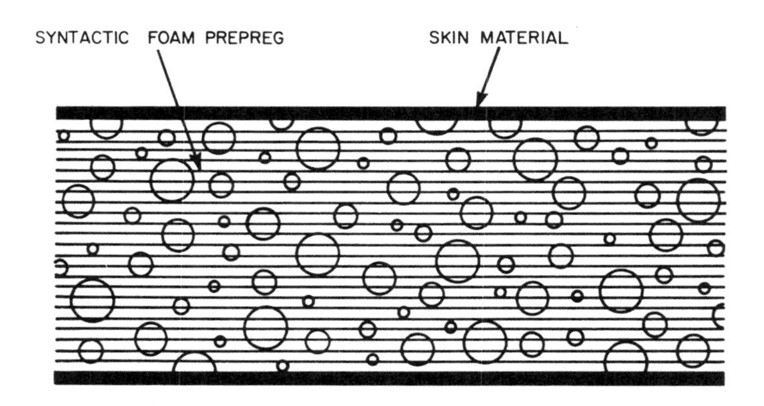

FIG. 9. Schematic construction of syntactic foam prepreg composite. Reprinted from Ref. [57] by courtesy of Narmco Materials Division, Whittaker Corp.

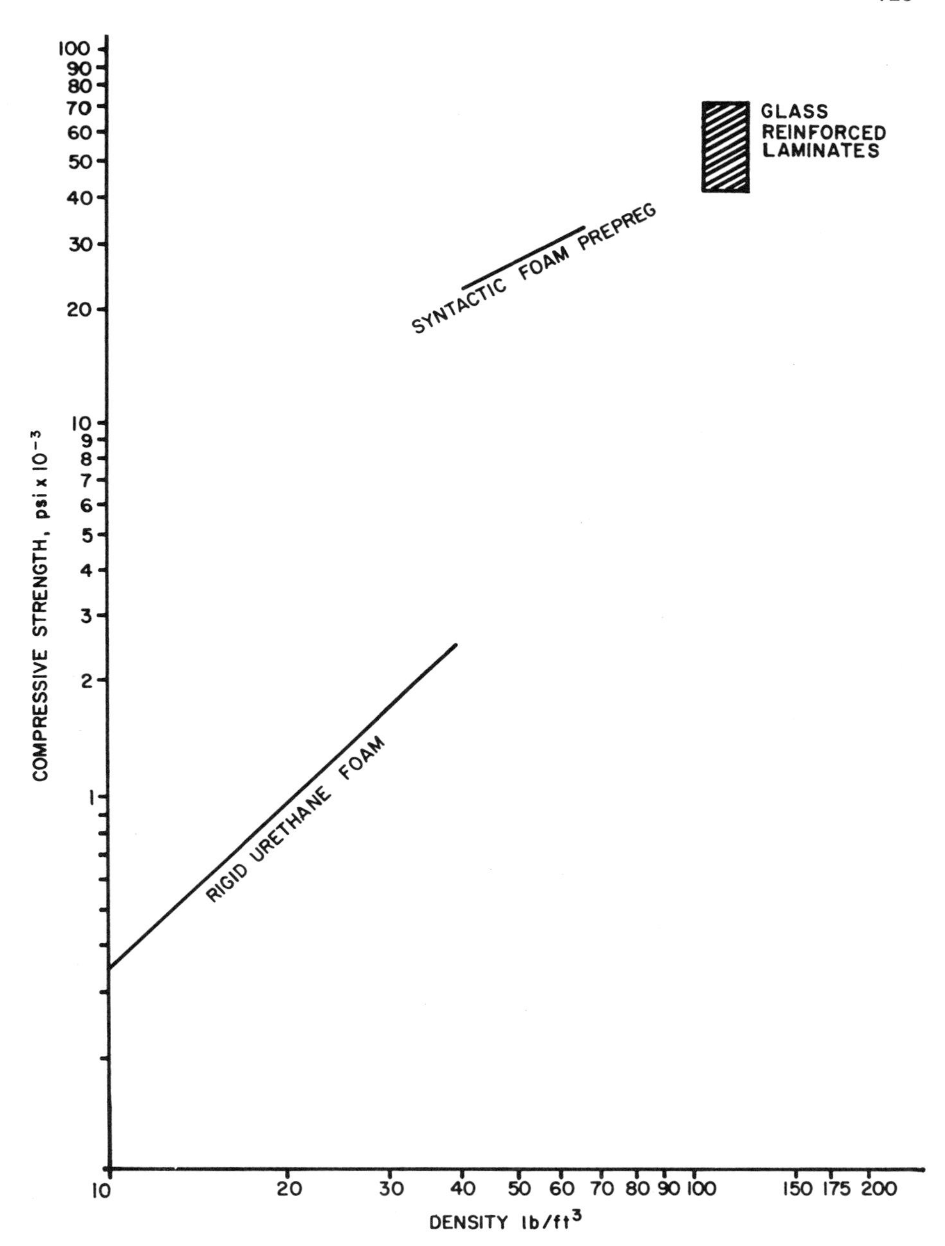

FIG. 10. Strength of typical composite materials as a function of density. Reprinted from Ref. [57] by courtesy of Narmco Materials Division, Whittaker Corp.

TABLE 12

Properties of Epoxy Laminates with Syntactic Foam Cores[a,b]

Property	Formulation 7801[c]	7802[d]	7803[e]	7804[f]	7801[g]
Specific gravity	0.7–0.9	0.7–0.9	0.7–0.9	1.7–1.9	0.7–0.9
Thickness, in.	0.365	0.400	0.420	0.100	0.400
Tensile strength, psi	13,000	13,000	13,000	60,000	16,000
Tensile modulus, psi x 10^{-6}	1.0	1.0	1.0	3.6	1.3
Flexural strength, psi	20,000	20,000	19,000	92,000	
Flexural modulus, psi x 10^{-6}	1.0	1.0	1.0	3.7	
Compressive strength, psi:					
Flatwise	19,000	17,000	15,000		6,000
Ultimate	22,000	20,000	16,000		
Edgewise	23,000	23,000	23,000	55,000	12,000
Compressive modulus, psi x 10^{-6}	1.0	1.0	1.0	3.8	1.1
Water absorption, wt%	<3[h]		<1[i]	0.5	
Izod impact, in.-lb/in.	7	7	6	16	6
Dielectric constant, X band	2.3	2.3	2.3	4.4	2.5
Loss tangent, X band	0.014	0.014	0.010	0.020	0.017
Thermal conductivity, (cal/(sec)(cm^2)(°C/cm)) x 10^4	2.6	2.6	2.6		2.6
Thermal expansion, (in./in.-°F) x 10^6	9.1	9.1	9.1	5	9.1

[a]Reprinted from Ref. [57] by courtesy of Narmco Materials Division, Whittaker Corp.

[b]Standard 10 x 10-in. test panels are fabricated from 10 layers of syntactic foam prepreg. Any surface skins are additional. The laminate is bagged with top and edge bleeder on caul sheets. The laminate is cured 9 min/ply at 150° C and 50 psi. The 7804 laminates are 12 ply. All specimens are tested according to FTMS 406.

[c]For use at ocean depths of up to 20,000 ft.

[d]For use at ocean depths of up to 10,000 ft.

[e]For use at ocean depths of up to 1000 ft.

[f]Compatible fabric prepregs for surface finishing (type 181).

[g]Fire-resistant syntactic prepreg primarily for surface use.

[h]At 13,500 psi, cycled 1500 times.

[i]At 300 psi, cycled 1500 times.

V. INFLUENCE OF STRUCTURE ON PROPERTIES

The formulations can be made less rigid by incorporating long-chain modifiers or long-chain curing agents. Conversely, they can be made more rigid by the use of short-chain elements, thereby creating tightly crosslinked networks. Depending on the specific chemical bonds involved, these provide increasingly higher thermal stability as the network becomes tighter.

In formulating for specific chemical resistance, ether, amine, or carboxyl groups may be selected to provide network bridging, with the former two offering improved alkali resistance and the latter, improved acid resistance. By designing systems with low oxygen- or nitrogen-to-carbon ratios, improvements in water resistance can be achieved, and further improvements can be obtained through the introduction of hydrophobic groups into the chain. Solvent resistance and water-vapor permeability are to a considerable extent functions of crosslinking density. A number of modifiers can be added to provide unusual characteristics.

These generalizations, often in the absence of specific foam-formulation data, are based on experience with nonfoamed epoxy resins. They follow the general industrial epoxy-resin formulating practice.

The foam's density, regularity, pore size, etc. affect the cured properties in the same fashion as with the other types of foams.

VI. APPLICATIONS

Perhaps the most significant application for epoxy-resin foams, entering the 1970s, is the use of sytactics for the construction of laminate cores for submergence applications (Fig. 11). This construction represents a new design concept, with properties between those of conventional glass-cloth laminates and lightweight foams. Such constructions have high impact strength and can withstand high local compressive forces. Resistance to microcracking ensures low water absorption. It has been stated that the syntactic foam prepreg is "an ideal choice where it is necessary to use relatively thin panels for stiffness and where relatively high local compressive forces may be encountered" [57]. Applications are expected to extend into the airframe industry as test results justify [58].

Porous epoxy-resin laminates, produced by a volatile solvent being released during cure (Table 13), are currently employed in upper-arm prostheses to alleviate the perspiration problems that have routinely been encountered with the older systems. Here the excellent chemical resistance of the epoxy formulations permits the prostheses to withstand attack from body fluids as well as from cleaning solutions. The net effect is lightweight structures that offer increased patient comfort.

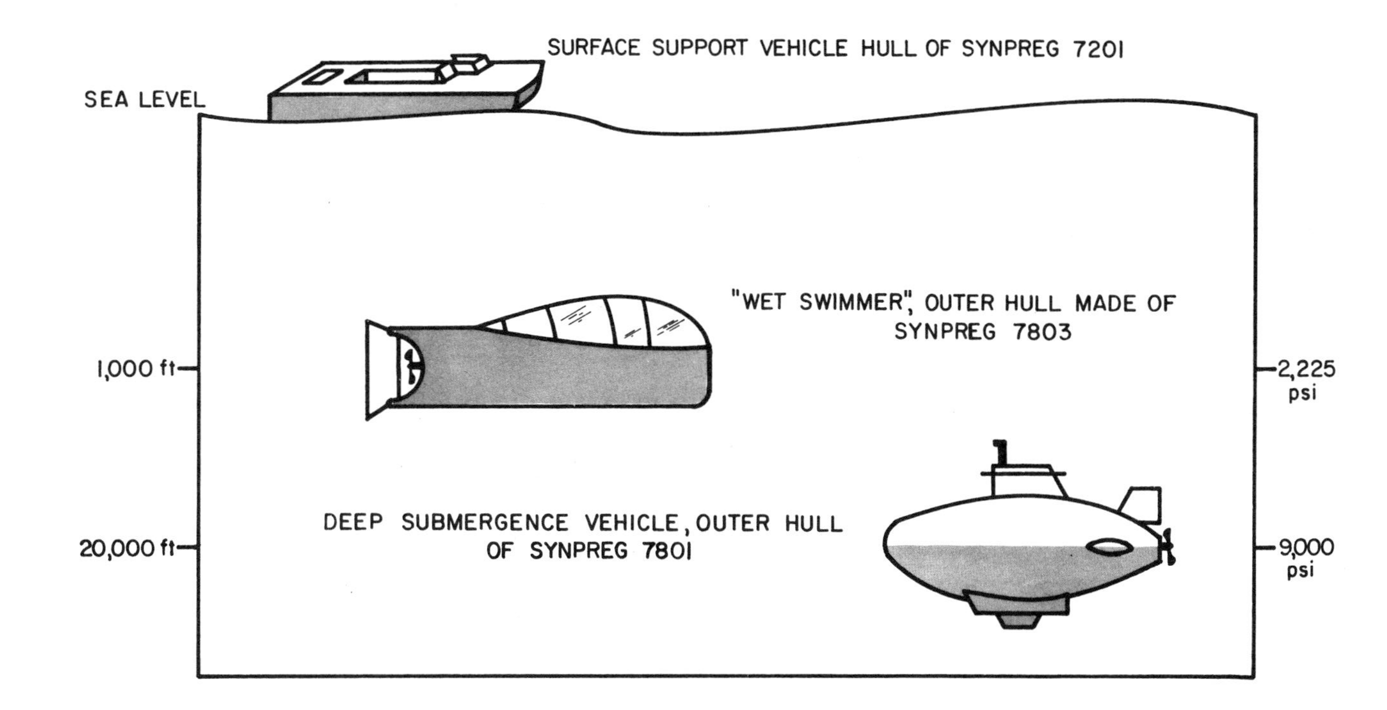

FIG. 11. Three marine applications of sytactic foam prepregs. Reprinted from Ref. [57] by courtesy of Narmco Materials Division, Whittaker Corp.

TABLE 13

Strength and Porosity of Epoxy Laminates as a Function of Resin Content[a]

Resin content (%)	Average compression load (lb)	Effective porosity (%)
68	2660	67
67	2645	58
66	2197	66
64	1847	71
63	1847	74

[a]Reprinted from Ref. [59] by courtesy of Interscience.

Epoxy-resin syntactic foams of intermediate densities (35-45 lb/ft^3) have been reported to be useful as pattern boards. Obtained as slabs, they can be worked like natural wood and offer the advantages of dimensional stability and lightweight [41]. The surfaces, after machining, are rubbed with plaster of paris and then sanded and shellacked [51]. Prefabricated chemical foam blocks have been used as lightweight bases for checking fixtures [2], the resistance of the epoxies to oils and solvents contributing to long-term stability.

High-temperature-resistant syntactic foams are used in a number of specialized applications in the aerospace industry, for edge-sealing honeycomb, and for rib reinforcement [16]. The high-temperature compounds are also used for potting attachments in honeycomb structures. Here they are preferable to the polyimide foams, which are cohesively weak.

Following the preference for epoxy-resin formulations for the protection of critical electronic components against severe environmental contaminants, the epoxy foams see limited service in airborne electronics, where weight saving is at a preminum and the thermal insulation qualities of the foam are not disabling.

One of the more widely reported applications in the early 1960s for the epoxy foams was as artificial dielectrics for radome sandwiches and as lens antennas for Doppler systems on aircraft. The epoxies are superior to the polyurethanes in these applications in terms of preserving stability under thermal and mechanical stresses [60]. These controlled dielectric materials [61] are obtained through the admixture of various metallic powders or flakes, and though their use dielectric constants can be raised to values of 7 or 8, or to intermediate values as required to match the dielectric constant of the adjacent structure. With dielectric foams, to the extent that the metal flakes or particles are not in contact (i. e., are electrically insulated from one another), the dissipation factor remains low. Relatively small amounts of metal have large effects on the dielectric constant.

REFERENCES

1. J. M. Buist, Soc. of Chem. Ind. (London), 50, 485 (1965).
2. H. S. Schnitzer and S. Richter, Mod. Plastics, October 1959, p. 99.
3. W. R. Cuming and P. M. Andress, Elec. Mfgr. (May 1958).
4. H. Lee and K. Neville, Handbook of Epoxy Resins, McGraw-Hill, New York, 1967.
5. H. Lee and K. Neville, Epoxy Resins, Their Applications and Technology, McGraw-Hill, New York, 1957.
6. K. Schrade, The Epoxy Resins, Dunod, Paris, 1957.
7. I. Skeist, Epoxy Resins, Reinhold, New York, 1958.
8. A. Paquin, Epoxyverbindung und Epoxyharz, Springer-Verlag, Berlin, 1958.
9. P. F. Bruins, ed., Epoxy Resin Technology, New York, 1968.
10. N. B. Graham (to Canadian Industries), Canadian Pat. 761,048 (1967).
11. D. G. Gluck (to U.S. Stoneware), British Pat. 1,082,305 (1967).
12. British Pat. 1,102,189 (to Fabwerke Hoeschst), (1968).
13. R. T. Dowd (to Shell Oil), U. S. Pat. 3,322,700 (1967).
14. Preliminary information sheet, Epon foam H-10.1, revised Feb. 28, 1964, Shell Chemical Co., 1964.
15. V. D. Valgin et al., Plast. Massy., No. 2, 34 (1967).
16. M. R. Pollock and H. K. Zahn, Mater. Des. Eng. (Aug. 1961).
17. W. Fisch et al. (to Ciba Ltd.), U.S. Pat. 2,965,586 (1960).
18. L. Bolstad and A. Stenerson, paper presented at the 13th Annual Technical Conference of the Society of Plastic Engineers, 1957.
19. K. Sekmakas (to Minnesota Mining and Manufacturing), U.S. Pat. 2,906,717 (1959).
20. V. D. Valgin et al., Plast. Massy., No. 3, 23 (1967).
21. J. E. Koroly (to Rohm & Haas), U.S. Pat. 2,623,023 (1952).
22. H. Lee and K. Neville, SPE Journal, 16, 315 (1960).
23. J. W. Shepherd (to Callery Chemical), U.S. Pat. 3,310,507 (1967).
24. T. Reetz and W. D. Dixon (to Monsanto), U.S. Pat. 3,119,853 (1964).
25. Netherlands Pat. Appl. 6,611,127 (to Fabwerke Hoeschst) (1967).

26. Belgian Pat. 667,085 (to Fabwerke Hoeschst) (1966).
27. H. Puchala and H. Anselm (to Consortium fur Elektrochemische Industrie), German Pat. 1,199,993 (1965).
28. R. C. Kohrn (to U.S. Rubber), U.S. Pat. 2,936,294 (1960).
29. H. Sander and F. Meyer (to Badische Anilin- and Soda-Fabrik), German Pat. 1,110,860 (1960).
30. British Pat. 967,259 (to De Bell & Richardson) (1964).
31. G. W. Schardt (to Wilson Products Mfgr.), U.S. Pat. 2,993,014 (1961).
32. British Pat. 919,779 (to Shell International Research Maatschappij N.V.) (1963).
33. A. S. Aase and L. L. Bolstad (to Minneapolis-Honeywell Regulator), U.S. Pat. 2,831,820 (1958).
34. H. L. Parry and B. O. Blackburn (to Shell Development), U.S. Pat. 2,739,134 (1956).
35. British Pat. 1,097,756 (to Ontario Research Foundation) (1968).
36. K. Andres, H. H. Steintrach, and K. Damm (to Farbenfabriken Bayer), Belgian Pat. 626,937 (1963).
37. V. D. Valgin et al., USSR Pat. 168,881 (1965).
38. W. R. Cuming, Electronic Design (Apr. 16, 1958).
39. R. Petri and K. Kochling (to Badische Anilin- and Soda-Fabrik AG), German Pat. 1,266,680 (1968).
40. F. T. Parr, ACS Symposium, Vol. 27, No. 2, 1959, pp. 265-275.
41. J. Delmonte, paper presented at the 15th Annual Meeting of the Reinforced Plastics Division of the Society of Plastic Industries, Inc., Chicago, Feb. 1960.
42. J. V. Milewski and E. G. Egbert (to Thiokol Chemical), U.S. Pat. 3,103,406 (1963).
43. K. D. Cressey, SPE Journal, May 1960, p. 557.
44. Syntactic foam, Laminating Tech. Release No. 4 (revised), Union Carbide Corp., March 1956.
45. Experimental Plastic Q-4124.2. Styrene-Epoxy Self-expanding Molding Material, the Dow Chemical Co. Bull. 171-81, June 1957.
46. Use of Epoxy Resins in Foams, Reinforced Styrene Foam, Reinforced Styrene Foam (Pre-expanded Beads); and Reinforced Styrene Foam (High Density), Jones-Dabney Co. Tech. Data Sheets, September 1957.
47. W. R. Cuming, H. E. Alford, and F. Veatch, paper presented at the 15th Ann. Tech. Conf. Society of Plastic Engineers, January 1959.
48. H. E. Alford and F. Veatch, Mod. Plastics, November 1961, pp. 141-150, 223.
49. H. E. Alford and F. Veatch (to the Standard Oil Co.), U.S. Pat. 3,316,139 (1967).
50. P. D. Jones, UTSL Rept. No. PD-21-8-1, Shell Chemical Co., April 1961.

51. W. H. Nickerson, paper presented at the 14th Ann. Tech. Conf. Society of Plastic Engineers, 1958.
52. R. P. Toohy, Chem. Eng. Progr., October 1961.
53. R. M. McClintok, SPE Journal, November 1958.
54. C. L. Segal, paper presented at the Epoxy Resin Symposium, Society of Plastic Engineers, Minneapolis, October 1958.
55. R. E. Burge, A. J. Landau, and R. P. Toohy, paper presented at the Regional Tech. Conf., Buffalo Section of the Society of Plastic Engineers, October 1961.
56. R. B. Feuchtbaum and R. A. Dunaetz, Insulation, January 1963, pp. 17-20.
57. F. E. Corse and R. C. Kausen, Syntactic Foam Prepregs, a New Composite Construction Material, Narmco Materials Division, Whittaker Corp., Costa Mesa, Calif., 1969.
58. W. R. Beck, D. L. O'Brien, and E. P. Davis, SPE Journal, 25, 83 (1969).
59. J. T. Kill et al., J. Biomed. Mater. Res., 1, 253 (1967).
60. H. B. Goldberg and H. S. Schnitzer, Elec. Des. News, February 1964.
61. M. H. Nickerson et al., (to De Bell & Richardson), U.S. Pat. 3,129,191 (1964).

Chapter 14

NEW HIGH-TEMPERATURE-RESISTANT PLASTIC FOAMS

Edgar E. Hardy

Monsanto Research Corporation
Dayton, Ohio

and

J. H. Saunders

Monsanto Company
Pensacola, Florida

I. INTRODUCTION

Though foamed materials have found very wide use in applications in which high-temperature resistance is not a critical factor, solid materials and honeycomb structures have been preferred when high-temperature performance was needed. Most conventional foamed plastics have been limited to operating environments below about 200° C. Recent research in high-temperature-resistant polymers has resulted in several structures that can be obtained in foamed as well as solid form. This chapter reviews three such new systems. Older foam types that may also be suitable for high-temperature applications are described in the chapters on phenolic, silicone, and inorganic foams.

For optimal high-temperature properties polymer systems should contain a maximum of aromatic units, a minimum of labile hydrogen or other atoms, no possible mechanism for easy depolymerization or chain scission, and a high molecular weight in the cured state. The synthesis of thermally stable polymers in the solid state has been reviewed recently by Jones [1]. Research on such materials has provided a very useful guide to the development of improved foams.

Though phenolic and silicone foams have in specific instances been employed for higher temperature applications, three heterocyclic ring systems have seen the widest application in the high-temperature field to date and hold the most promise for the future. These are the polyisocyanurate (1), polybenzimidazole (2), and polyimide (3) foams.

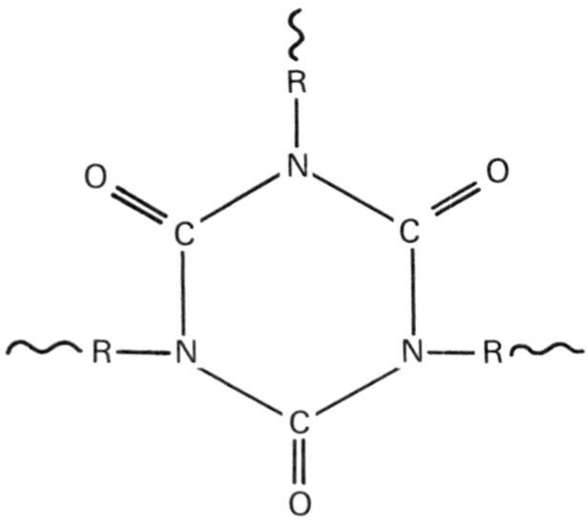

(1)

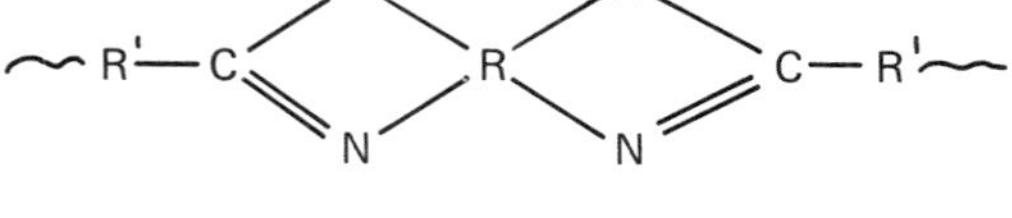

(2)

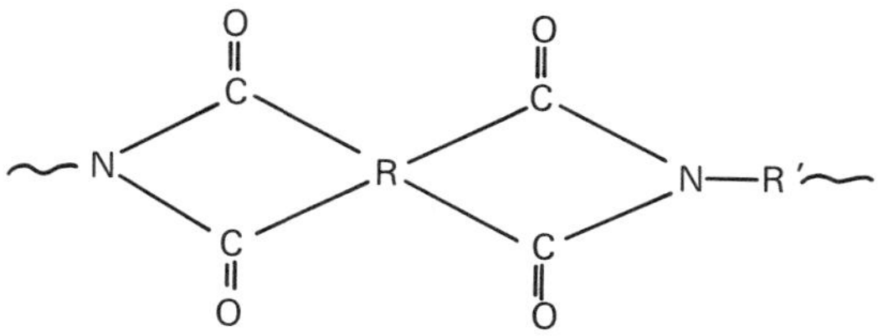

(3)

The literature on these foam systems is still somewhat limited, but considerable experimental work has been done and enough data are available to discuss these systems fairly widely. In each case we shall discuss the chemistry of the foam system, the temperature use-range of the foam, method of fabrication, and a variety of properties, especially structural, electrical, and thermal behavior.

As far as high-temperature resistance is concerned, we are interested not only in the prolonged resistance to high temperatures but also in the behavior on short-time exposure to high temperatures, especially in the presence of oxygen. In addition, fire resistance is becoming of ever greater significance, as is ablation behavior. Applications of these foams are outlined only briefly, since they are too new to have become firmly established in commercial use. The indicated applications do suggest reasonable expectations, based on available information.

II. POLYISOCYANURATE FOAMS

Polyurethane foams have been produced in large volumes for many applications, but the urethane group itself does not have good thermal stability above about 200° C (see Chapter 3). Decomposition occurs at about 200 to 250° C, regenerating isocyanate and polyol, and also producing olefins, amines, carbon dioxide, and other products. However, the isocyanates, key raw materials for polyurethanes, can be polymerized to form isocyanurates, which have much better heat resistance. The chemistry of these reactions of isocyanates, and the decomposition of urethanes, has been reviewed in detail [2].

One of the first successful steps in raising the temperature resistance of isocyanate-based foams, utilizing isocyanurate formation, was reported by Nicholas and Gmitter [3], who used prepolymers from 80:20 tolylene diisocyanate (TDI; 80:20 mixture of 2,4- and 2,6-isomers) and a hydroxyl-terminated polyester as the reactant:

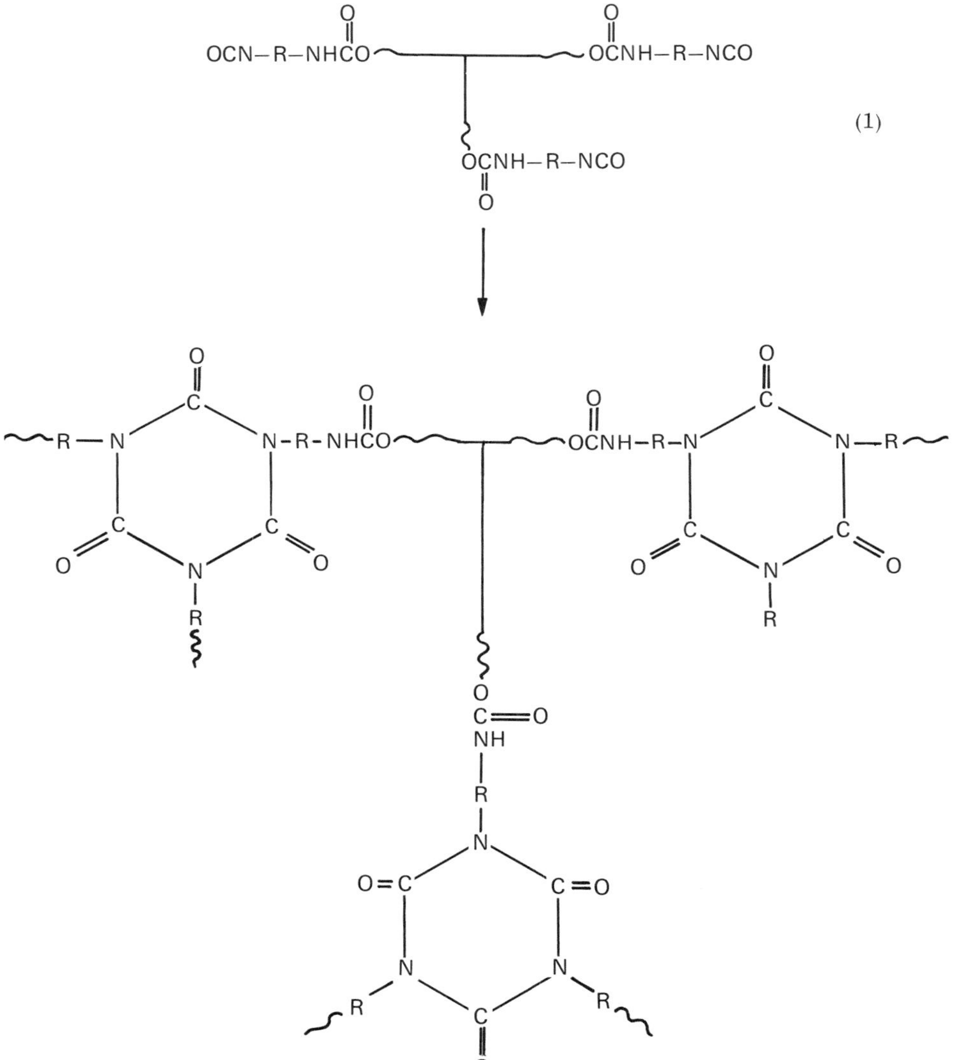

(1)

(4)

The resulting polymer (4) contained ester, urethane, and isocyanurate structures, and hence would likely be somewhat limited in heat resistance by the urethane and ester groups.

Many catalysts have been reported to be effective for promoting trimerization of isocyanates, or isocyanurate formation. Nicholas and Gmitter [3] reported the evaluation of numerous catalysts and chose a combination of a tertiary amine and an epoxide as being most suitable for their system. A typical foam recipe is shown in Table 1.

TABLE 1

Typical Formulation for the Preparation of a Rigid Isocyanurate Foam[a]

Composition	Parts by weight
Component A:	
Fomrez R-420 (32.4% NCO)	100.0
Trichlorofluoromethane	8.0
Silicone surfactant (alkyl siloxane-polyoxyalkylene copolymer)[b]	1.0
Component B;	
N, N', N''-Tris(dimethylaminopropyl) sym-hexahydrotriazine	0.8
Diglycidyl ether of bisphenol A[c]	0.8

[a]Data from Ref. [3].

[b]Silicone L-520, Silicones Division, Union Carbide Corp.

[c]DER-332, Dow Chemical Co.

With the indicated catalysts, the isocyanurate formation was sufficiently exothermic to volatilize the fluorocarbon blowing agent for foaming and also provide adequate cure for foam stabilization. An additional 24-h cure at 110° C was used before testing. Typical foam properties are shown in Table 2.

TABLE 2

Physical Properties of a Typical Rigid Isocyanurate Foam[a]

Property	Value
Density, lb/ft^3	4.5
Thermal conductivity (k factor), $Btu/(h)(ft^2)(°F/in.)$	0.14
Compressive strength, psi:	
Before aging, at	
23° C	81
150° C	84
175° C	68
205° C	56
230° C	48
After heat-aging 24 h at	
150° C	88
205° C	82
230° C	54
After humid-aging 7 days at 70° C and 95 to 100% R. H.	81
Dimensional stability (volume change, %):	
After heat-aging 24 h at	
150° C	0
175° C	0
205° C	-2
230° C	-5
After humid-aging 7 days at 70° C	0

[a]Data from Ref. [3].

The compressive strength of the isocyanurate foam compared with that of a polyurethane foam from the same polyester and of almost the same density is shown in Fig. 1, and weight-retention data for the isocyanurate foam at various temperatures are shown in Fig. 2.

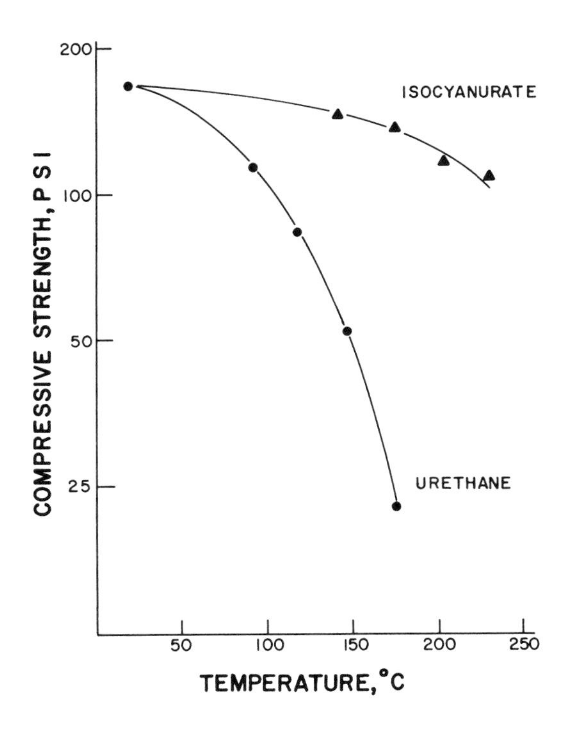

FIG. 1. Compressive strength versus temperature for polyisocyanurate (6.6-lb/ft^3 density) and polyurethane (6.5-lb/ft^3 density) foams. Data from Ref. [3].

It was to be expected that further improvements could be made by eliminating as many urethane groups as possible, reducing the amount of polyol component or eliminating it completely, and replacing the tolylene

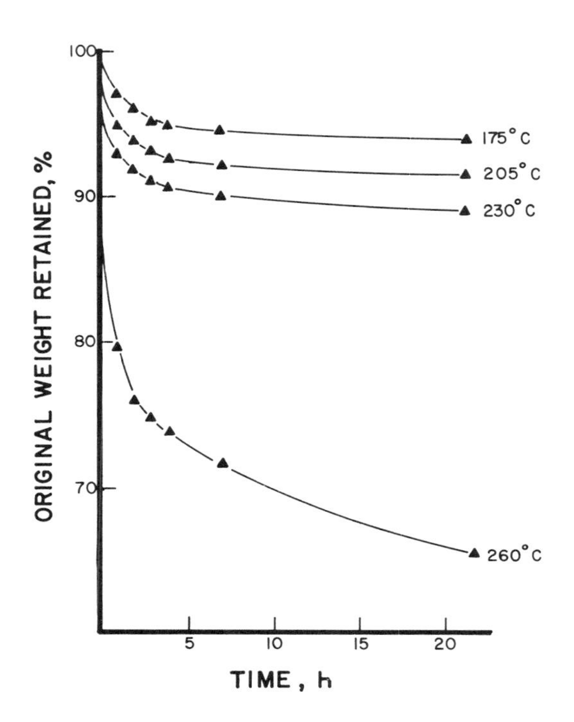

FIG. 2. Weight retention of a typical polyisocyanurate foam at elevated temperatures. Data from Ref. [3].

diisocyanate by an aromatic polyisocyanate (5) derived from an aniline-formaldehyde condensation products.

NCO NCO NCO

$$C_6H_4(NCO)-[CH_2-C_6H_3(NCO)]_n-CH_2-C_6H_4(NCO)$$

(5)

Such a polyisocyanate ("MDI") was shown to be an excellent char former itself [4], so a foam derived from a maximum quantity of this reactant would be expected to have good thermal stability and flame resistance. Such an approach was followed by both Imperial Chemical Industries Ltd [5] and the Mobay Chemical Co. [6].

Ball and co-workers [5] found that the isocyanurate foam from the polymeric isocyanate alone was too friable to be of practical value. They were able to overcome this deficiency by including enough polyol to convert 10 to 30% of the isocyanate groups to urethane and trimerizing the rest. Basic catalysts, fluorocarbon blowing agents, and polyoxyalkylene siloxane surfactants were used in foaming.

Thermogravimetric analyses of two conventional urethane and two modified isocyanurate foams of different urethane content are shown in Fig. 3. The excellent weight retention of the sample with a low urethane content is particularly noteworthy. Typical properties of a conventional surcose-based polyether-polymeric isocyanate foam and of the polyisocyanurate foams are shown in Table 3.

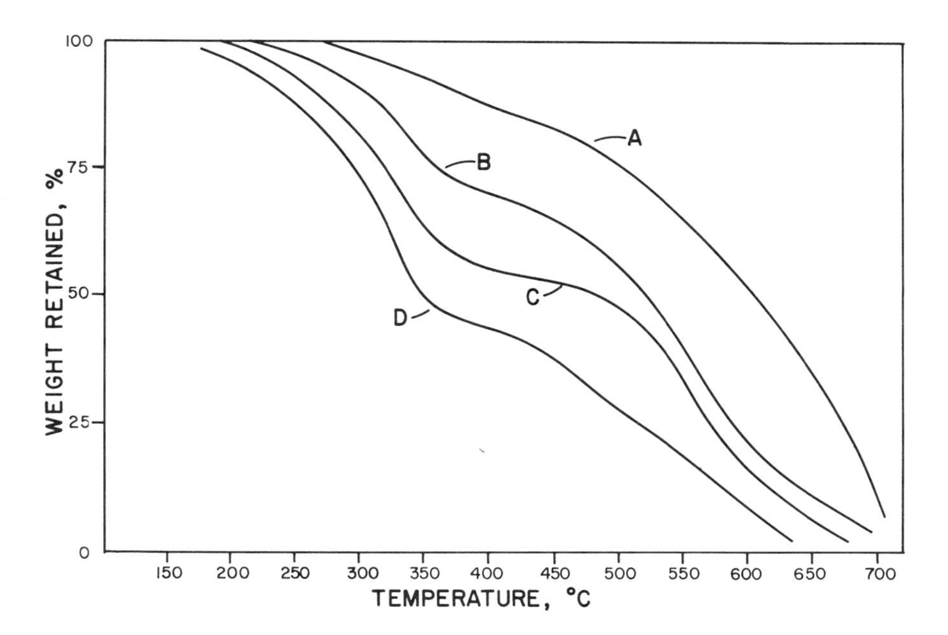

FIG. 3. Thermogravimetric curves obtained in air, at a heating rate of 5° C/min. Curve A, isocyanurate foam with low urethane content; curve B, isocyanurate foam, 20% urethane; curve C, sucrose-based polyol-MDI-based urethane foam; curve D, sucrose-based polyol-TDI-based urethane foam. Data from Ref. [5].

TABLE 3

Physical Properties of Free-Rise Foams[a]

Property	Sucrose/MDI	Isocyanurate
Density (core material), lb/ft^3	2.2	2.3
Closed cells, %	92	95
Compressive strength at 10% strain, psi:		
Parallel to foam rise	32	30
Perpendicular to foam rise	9	11
Tensile strength parallel to foam rise, psi	48	42
Shear data parallel to foam rise:		
Strength, psi	18	17
Modulus, psi	260	240
Water vapor transmission rate, perm.-in.	2	4
Thermal conductivity at 0° C, initial, $Btu/(h)(ft^2)(°F/in.)$	--	0.108

[a]Data from Ref. [5].

Both the flame resistance and dimensional stability at 150° C were shown to be much better for this isocyanurate foam than for the polyurethane. Its application in the building trade was discussed at some length, with preparation by molding, spraying, and panel-forming techniques [5].

A closely related polymeric isocyanate, designated Mondur MRS, was developed by the Mobay Chemical Co. for use in isocyanurate foams [6]. Typical formulations for its foaming are listed in Table 4, and its properties are shown in Table 5. Foams were prepared by using conventional polyurethane-foam machinery, including continuous slab formation, with slabs 3 ft or more wide, 1 to 2 ft high, and many feet in length.

Much of the weight loss at 200° C was doubtless blowing agent, with the subsequent shrinkage indicated. Open-cell foams would have been more stable, but these data are excellent for 1.6- to 1.8-lb/ft^3 closed-cell foams.

TABLE 4

Formulations of Polyisocyanurate Rigid Foam[a]

	Formulation		
	1	2	3
Raw materials, parts by weight:			
Polymeric isocyanate	100	100	100
Trichlorofluoromethane	18	22	23
Surfactant, DC-193[b]	2.0	2.0	2.0
Catalyst, DMP-30[c]	11.0	10.0	10.0
Polyether polyol[d]	0	7.5	15.0
Foaming characteristics:			
Cream time, sec	21	15	19
Rise time, min	2.5	2.3	2.0
Tackfree time, min	2.0	2.3	1.8
Firm time, min	5.0	3.2	4.0

[a]Data from Ref. [6].

[b]Silicone foam stabilizer, Dow Corning.

[c]2,4,6-Tri(dimethylaminomethyl)phenol, Rohm and Haas.

[d]Hydroxyl number 400 to 560, functionality 3 to 4.

TABLE 5

Physical Properties of Isocyanurate Foams[a]

	Formulation[b]		
Property	1	2	3
Density, lb/ft^3	1.6	1.8	1.6
Compressive strength at yield, psi:			
Parallel to rise	21.2	22.8	19.8
Perpendicular to rise	18.1	17.4	12.1
Closed cells, %	87.1	87.8	90.3

TABLE 5 (Continued)

Physical Properties of Isocyanurate Foams[a]

Property	Formulation[b]		
	1	2	3
Moisture vapor permeability, perm.-in.:			
Wet cup	8.1	13.0	12.4
Dry cup	3.0	5.2	4.6
Tabor abrasion resistance, cycles/in. foam (H-18 wheel)	63.3	75.9	108.0
k Factor, Btu/(h)(ft^2)(° F/in.)	0.152	0.126	0.124
Dimensional stability (volume change, %):			
At -40° C, ambient R.H.:			
1 day	None	None	None
1 week	None	None	None
2 weeks	None	None	None
At 70° C, 100% R.H.:			
1 day	+0.6	+1.0	+1.2
1 week	+5.2	+5.1	+6.2
2 weeks	+6.3	+7.5	+10.4
At 200° C, ambient R.H.:			
1 day	-3.7	-1.0	4.0
1 week	-10.7	-10.5	-9.5
2 weeks	-12.2	-13.1	-12.6
Weight loss at 200° C, ambient R.H., %:			
1 day	14.8	15.6	18.4
1 week	13.5	18.3	24.1
2 weeks	17.2	20.0	26.6

[a]Data from Ref. [6].

[b]See Table 4.

The pure isocyanurate foam made from formulation 1 yielded the most flame-resistant product (see Table 6), although friability and a tendency to crack at flame temperatures were problems. Foams prepared from formulations 2 and 3 in Table 6 exhibited reduced friability without seriously increasing flammability. Flame-spread ratings of these polyether foams were slightly higher in the tunnel test. The U. S. Bureau of Mines torch penetration test (Table 6) showed the polyether-modified foam to be superior because of its ability to form a cohesive char more readily. In addition, fuel-contribution and smoke-density factors were exceptionally low. The unmodified polyisocyanurate foam will be useful where maximum flame resistance is required and friability is not a serious problem.

TABLE 6

Flammability of Isocyanurate Foams[a]

Test method	Formulation[b] 1	2	3
ASTM D1692	Nonburning	Nonburning	Nonburning
Vertical bar test, flame height, in.	5.0	4.7	7.1
Bureau of Mines torch penetration test, min/in. foam	22.6	37.0	25.0
Tunnel test, ASTM E84:			
Flame-spread rate	25	35	35
Fuel-contribution factor	15	20	10
Smoke-density factor	25	50	100

[a]Data from Ref. [6].

[b]See Table 4.

Polyisocyanurate foams have also been described in one additional patent, but few details were given [7].

The polyisocyanurate foams, though not outstanding in retention of properties above 200° C, do offer a remarkable combination of ease of processing on commercial polyurethane-foam machinery, the use of low-cost raw materials, and outstanding flame resistance, low fuel contribution, and low smoke generation at low density. They would appear to offer a major advance for the building industry. The resistance of these foams to temperatures above 200° C would doubtless be improved by preparing them at higher densities and with an open-cell structure, as is frequently done for the polybenzimidazole and polyimide foams discussed in Sections III and IV.

III. POLYBENZIMIDAZOLE FOAMS

Polybenzimidazoles (PBI) have been shown to have outstanding thermal resistance, and the polymerization reaction is such that foaming can be achieved, along with the increase in molecular weight. Crosslinking can also be introduced. For convenience in foaming a prepolymer approach is generally used. This may be illustrated by the reaction of diphenyl isophthalate (6) with 3,3'-diaminobenzidine (7)

3 $C_6H_4(COOC_6H_5)_2$ (6) + 2 $(H_2N)_2C_6H_3{-}C_6H_3(NH_2)_2$ (7) →

$C_6H_5OOC-[C_6H_4-C(=N)(NH)-C_6H_3-C_6H_3-(N)(NH)C-]_2-C_6H_4-COOC_6H_5$ + $C_6H_5OH + H_2O$

Such a prepolymer is a solid that can be handled as a powder. Heating under appropriate conditions will cause foaming, due to further evolution of phenol and water as blowing agents. In addition the powder can be combined with reinforcing fibers, such as graphite, and microballoons for further density control. Crosslinking can be obtained thermally or by introducing higher functionality in the reactants.

Reed and Feher described polybenzimidazole foams of 12- to 80-lb/ft^3 density in 1968 [8]. These were prepared by blending the prepolymer powder with carbon fiber and silica microballoons in a mold, which was then heated in a press at 120° C and 15 psi for 30 min. Foaming occurred, but curing was not complete. An additional heating at about 0.5° C/min to 315° C and holding at that temperature for 2 h gave a moderate degree of cure. Postcuring was accomplished by raising the temperature at 0.5 to 1° C/min to 450° C in an inert atmosphere and holding at that temperature for 2 h. The foam was then cooled to 260° C or less in an inert atmosphere. Density was presumably controlled largely by the volume of microballoons used.

The mechanical properties could be varied by the density as well as the type and amount of fibrous filler used. Properties obtained for a 31-lb/ft^3 foam are shown in Table 7. Unfortunately no exact statement of the fiber or microballoon content was given.

TABLE 7

Mechanical Properties of a Molded Polybenzimidazole Foam[a, b]

Test temp. (° C)[c]	Compressive strength (psi)	Compressive modulus (psi x 10^5)	Tensile strength (psi)	Tensile modulus (psi x 10^5)
23	3000	2.41	1300	1.09
315	2900	0.36	1278	1.06
535	776	--	284	--
950	300	--	--	--

[a]Imidite SA foam, 31-lb/ft^3 density.

[b]Data from Ref. [8] .

[c]Tested after 5-min soak at temperature.

The isothermal aging of a similar foam of 25-lb/ft^3 density at 315° C is shown in Fig. 4. The excellent weight stability was also confirmed by differential thermal analysis, which shows weight losses for polybenzimidazole foams beginning at 510 to 570° C, with weight retentions of 26 to 36% at 1000° C (6° C/min heating rate, in helium). This was significantly better than the performance of the polyimide foams that were used in the comparison.

The electrical properties of these foams were also found to be good. The X-band dielectric constant of a 35-lb/ft^3 foam increased slowly from about 1.8 at room temperature to about 2.2 at 1100° C. In the same range the loss tangent (X band) increased from about 0.004 to about 0.04 [8].

Additional data on similar polybenzimidazole foams were reported by Dickey and co-workers [9]. In particular the effects of crosslinking on ablative properties (See Section V) were investigated. The best ablative performance was found when crosslinking was induced thermally by post-curing at a heating rate of 3° C/min, to 800° C, in an inert atmosphere. In this process the hydrogen content was reduced, giving a polymer with

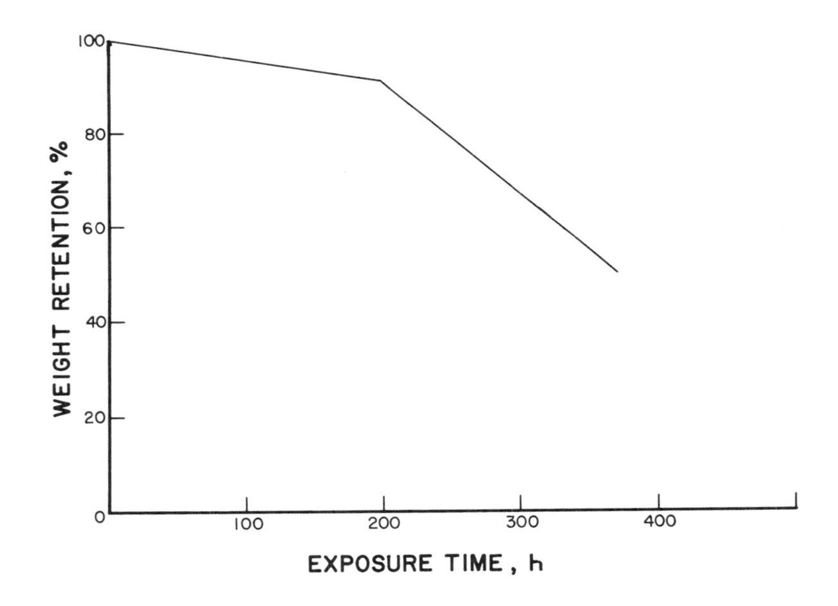

FIG. 4. Isothermal aging curve for Imidite SA polybenzimidazole foam, density 25 lb/ft^3, exposed at 315° C in static air at atmospheric pressure. Data from Ref. [8].

the empirical formula $(C_{47}H_{14}N_8)_n$. Oxidation of the diaminobenzidine to a higher functional polyamine before its reaction with diphenyl isophthalate resulted in crosslinking that was almost as effective. Triphenyl trimesate was not a highly effective crosslinking agent.

The polybenzimidazole foams have the best thermal stability of the foams described in this chapter but generally are less resistant to oxidation at elevated temperatures than are the polyimide foams. The polybenzimidazole systems are the most expensive of the three types considered here. So far their fabrication technology appears to be limited to the molding of small to moderate-sized items.

IV. POLYIMIDE FOAMS

Because of their excellent thermal and flame resistance polyimides have been used for films, coatings, adhesives, molding compounds, and most recently as rigid foams. These polymers have been prepared and foamed by two general methods, the first being the more conventional reaction of aromatic dianhydrides (8) with aromatic diamines (9) [10]:

$$n\ \mathrm{O(CO)_2R(CO)_2O}\ (8) + n\ H_2N-R'-NH_2\ (9) \longrightarrow \left[\begin{matrix} HOOC & & COOH \\ & R & \\ -NHC(=O) & & C(=O)NH-R'- \end{matrix} \right]_n \quad (3)$$

$$\xrightarrow{\text{heat}} \left[-N(CO)_2R(CO)_2N-R'- \right]_n + 2nH_2O$$

A typical dianhydride would be that from benzophenonetetracarboxylic acid (10), and such aromatic diamines as benzidine, diaminodiphenyl ether, and diaminodiphenylmethane may be used.

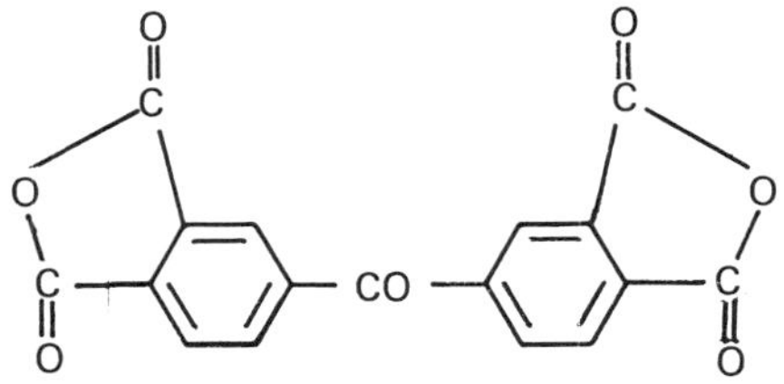

(10)

The second method utilizes the reaction between dianhydrides (8) and polyisocyanates: [11]

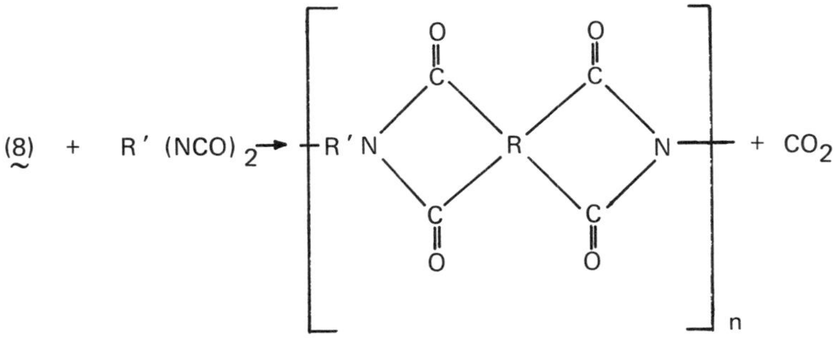

$$(8) + R'(NCO)_2 \rightarrow \left[-R'N(CO)_2R(CO)_2N- \right]_n + CO_2 \qquad (4)$$

Hurd and Prapas [12] have suggested that imide formation occurs by an initial addition of the isocyanate to the anhydride, as in the following case:

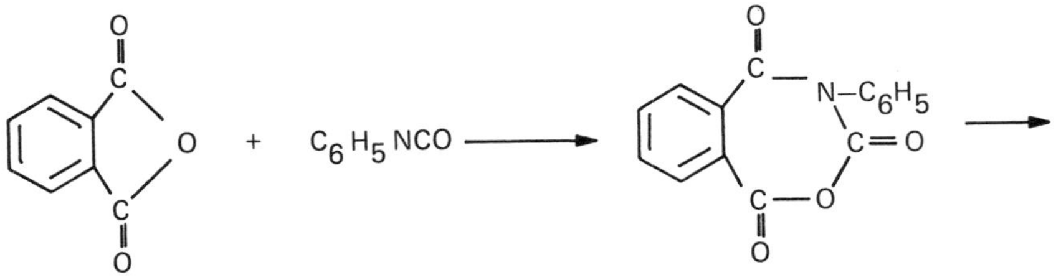

$$C_6H_5NCO \longrightarrow \quad N{-}C_6H_5 \quad C{=}O \longrightarrow$$

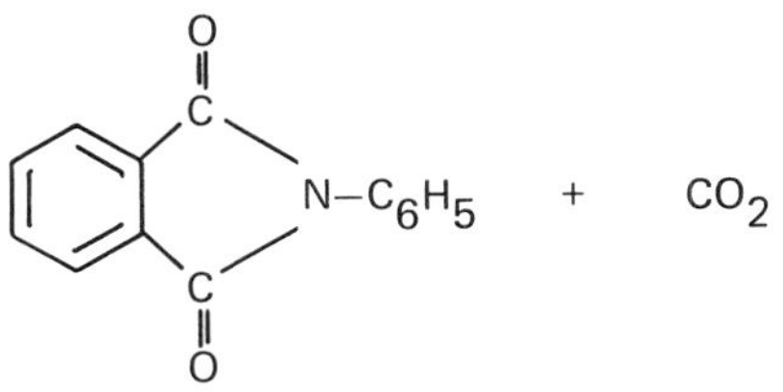

$$C_6H_4(CO)_2N{-}C_6H_5 + CO_2 \quad (5)$$

Fincke and Wilson [10] have described the preparation of polyimide foams from proprietary liquid dianhydride-diamine components to give densities in the range of 6 to 20 lb/ft^3. These contained both open and closed cells, and did not have as good thermal stability as foams prepared from solid polyamic-acid prepolymers. The compressive-strength properties of the foams from the liquid reactants are shown in Table 8.

The solid prepolymer was converted to foams by powdering, placing in a mold, and heating to foam and cure. Densities of 23 lb/ft^3 and higher were prepared in this way. The mechanical properties of these foams are shown in Tables 9 and 10.

TABLE 8

Compressive Strength of Polyimide Foams from Liquid Reactants[a]

	Compressive strength (psi) Density (lb/ft^3)			
Temperature	5	10	15	20
Room temperature	46	219	550	1000
315° C	14	17	20	21

[a]Data from Ref. [10].

TABLE 9

Compressive Strength, Modulus, and Porosity of Polyimide Foams from Solid Prepolymers[a]

Property	Foam density (lb/ft^3)				
	23.6	31.1	41.5	59.7	71.3
Compressive strength[b], psi:					
Room temperature	565	1200	3880	7880	9330[c]
93° C	479	992	3830	6210	10000[d]
205° C	346	746	2120	4890	9000
260° C	260	--	1625	--	--
315° C	148	--	668	1740	4270
Compressive modulus, psi x 10^3:					
Room temperature	--	18.3	73.8	130.0	271.6[c]
93° C	--	15.3	70.2	99.3	208.0[d]
205° C	--	13.0	44.1	76.8	175.5
260° C	--	--	40.6	--	--
315° C	--	--	21.7	--	118.0
Open cells, %[e]	--	66.3	54.5	34.0	22.0
Closed cells, %[e]	--	0	0	0	0

[a]Data from Ref. [10].

[b]Compressed to 10% deflection parallel to direction of pressure applied in molding.

[c]Compressed to 4% deflection.

[d]Compressed to 6% deflection.

[e]Measurements were made with a Beckman Model 930 Air Comparison Pycnometer.

TABLE 10

Tensile Strength and Modulus of Polyimide Foams from Solid Prepolymers[a]

Property	Foam density (lb/ft^3)			
	34.5	43.6	59.1	68.4
Tensile strength, psi x 10^3:				
Room temperature	0.90	2.10	1.01	3.00
205° C	0.76	1.76	0.70	1.48
315° C	1.05	0.76	0.76	0.90
Tensile modulus, psi x 10^3:				
Room temperature	37.7	72.2	74.2	134.0
205° C	24.2	54.7	53.2	83.7
315°C	17.8	29.1	46.0	63.2

[a]Data from Ref. [10].

The dielectric constants of these foams at room temperature were about 2 to 3, and tan δ was about 0.002 to 0.02 for the 34.5- to 69.3-lb/ft^3 density range and 0.1- to 1000-kHz frequency range. Water-vapor permeability was low, and the foams did not ignite when tested according to ASTM D-1692-59T.

Reed and Feher [8] also reported some properties of polyimide foams in the 12- to 20-lb/ft^3 density range. The compressive strength at room temperature and at 260° C was reported to be about the same as that shown in Table 8 at room temperature. Their technique, which was not disclosed, provided better temperature resistance than that used by Fincke and Wilson.

Farrissey and co-workers [11] utilized the same type of aromatic polyisocyanate in making polyimide foams as was described in Section II for polyisocyanurate foams. Crosslinking was increased by adding a highly functional polyol to react with part of the isocyanate, so that the foam contained some urethane structure. A solvent and surfactant were also included.

In order to foam, a premix was prepared by vigorous mixing of benzophenonetetracarboxylic acid dianhydride (BTDA) and the polyisocyanate; the amount of isocyanate was equivalent to the anhydride plus polyol. A second component was composed of a solution in dimethyl sulfoxide of a basic polyol and surfactant. The two components were rapidly mixed and allowed to foam. Toward the end of the strongly exothermic foaming process, sulfurous fumes were detectable. At the end of the rise, the foam was fairly soft, but it cured to a rigid foam in 1 to 2 h. Curing was completed, and the solvent was removed from the foams, which were of the open-cell structure, by heating in a vacuum oven at 100 to 110° C at 1 to 10 mm Hg, and finally at 200 to 230° C in a nitrogen-purged oven for 1 to 2 h. The properties of this foam are shown in Table 11.

TABLE 11

Properties of Polyimide Foam from Polyisocyanate[a]

Property	Value
Density, lb/ft^3	3.97
k-Factor (open cell), Btu/(h)(ft^2)(° F/in.)	0.26
Compressive strength, psi:	
Parallel to rise	34.9
Perpendicular to rise	45.6
Aging, volume change, %:	
Humid, 70° C, 28 days	+1.2
Dry, 95° C, 28 days	-0.6
Dry, 232° C, 14 days	-6.1, 2% weight loss
Compressive strength, psi;	
Parallel to rise	41.4
Perpendicular to rise	53.3
ASTM E84, tunnel test, flame-spread rating	10

[a]Data from Ref. [11].

The excellent thermal stability and fire-resistance properties of this foam are apparent. Samples held at 232° C in air for 14 days suffered weight losses of less than 2%, with no loss of compressive strength. A 6% weight loss was measured in 10 min at 450° C. Thermogravimetric analysis data indicate maximum weight loss at 640° C. The foam was also prepared in a range of densities (2.5-18.5 lb/ft^3) and compressive strengths (25-1340 psi).

Polyimide foams have also been described in patents [13-15], and the chemistry of solid polyimides has been reviewed recently by Dine-Hart and Wright [16].

V. APPLICATIONS

The excellent flame resistance of the foams described in this chapter should lead to extensive use in the building industry when costs and production technology are appropriate. At this time the polyisocyanurate foams appear to be readily adaptable to conventional polyurethane production techniques, and raw-material costs are competitive with those of many rigid polyurethane foams. Ball and co-workers, in particular, have placed emphasis on building applications [5], and very low flame spread, fuel contribution, and smoke generation have been demonstrated, even in the ASTM E-84 tunnel test [6].

Polyimide and polybenzimidazole foams so far appear to be limited to uses in smaller items of critical importance, so that the higher cost of raw materials and fabrication can be tolerated. Dickey and co-workers have shown the applicability of polybenzimidazole foams as ablative heat shields [9], as have Reed and Feher [8]. The dielectric properties of both polyimide and polybenzimidazole foams make them suitable for radome construction [8].

The insulating properties of the fluorocarbon-blown, closed-cell polyisocyanurate foams are excellent up to about 200° C. The thermal conductivity of the polyimide and polybenzimidazole foams is about twice as high, but for temperatures above about 200° C they are preferred as insulators because of their better thermal resistance.

REFERENCES

1. J. I. Jones, J. Macromol. Sci.-Rev. Macromol. Chem., C2 (2), 303 (1968).
2. J. H. Saunders and K. C. Frisch, Polyurethanes, Chemistry and Technology, I. Chemistry, Wiley-Interscience, New York, 1962, Chapter 3.
3. L. Nicholas and G. T. Gmitter, J. Cell. Plastics, 1, 85 (1965).
4. J. K. Backus, W. C. Darr, P. G. Gemeinhardt, and J. H. Saunders, J. Cell. Plastics, 1, 178 (1965).
5. G. W. Ball, G. A. Haggis, R. Hurd, and J. F. Wood, J. Cell. Plastics, 4, 248 (1968); (to ICI) Belgian Pat. 680,380 (1966).
6. Mobay Chemical Co., Technical Information Bulletin No. 101-F36, A Progress Report on Rigid Foam Displaying Low Flame Spread, Smoke Generation and Fuel Contribution.

7. J. Burkus (to U.S. Rubber), U.S. Pat. 2,993,870 (1961).
8. R. Reed and S. Feher, paper presented at the AIAA/ASME Ninth Structures, Structural Dynamics, and Materials Conference, Palm Springs, Calif., April 1-3, 1968, AIAA paper No. 68-303.
9. R. B. Dickey, J. H. Lundell, and J. A. Parker, J. Macromol. Sci. Chem., A3, 573 (1969).
10. J. K. Fincke and G. R. Wilson, Mod. Plastics, 46 (4), 108 (1969); see also E. Lavin and I. Serlin (to Monsanto), U.S. Pat. 3,483,144 (1969).
11. W. J. Farrissey, J. S. Rose, and P. S. Carleton, Polymer Preprints, 9 (2), 1581 (1968).
12. C. D. Hurd and A. G. Prapas, J. Org. Chem., 24, 388 (1959).
13. W. R. Hendrix (to DuPont), U.S. Pat. 3,249,561 (1966).
14. L. E. Amborski and W. P. Weisenberger (to DuPont), U.S. Pat. 3,310,506 (1967).
15. H. E. Frey (to Standard Oil), U.S. Pat. 3,300,420 (1967).
16. R. A. Dine-Hart and W. W. Wright, J. Appl. Polymer Sci., 11, 609 (1967).

Chapter 15

MISCELLANEOUS FOAMS

K. C. Frisch

Polymer Institute
University of Detroit
Detroit, Michigan

I. POLYVINYL CARBAZOLE FOAMS

A. Introduction

Polyvinyl carbazole has found commercial application in the fields of electrical insulation and electronics. This is due to its good electrical properties at radar frequencies, in which it resembles polystyrene, but it possesses much better heat resistance and a softening temperature of about 200°C. Although polyvinyl carbazole is a fairly polar polymer, the dielectric losses are low because of the relative immobility of the bulky carbazole side groups. The polymer is manufactured in the form of molding powders in the United States (Polectron, General Aniline and Film Corp.) and also in Germany (Luvican M-150, Badische Anilin- und Soda-Fabrik (BASF)).

The preparation of foams from polyvinyl carbazole has been reported in a number of patents, notably those of Stastny et al. [1-4], Dumont and Reinhardt [5], and Ellinger [6, 7].

B. Monomer Preparation

N-Vinyl carbazole (1) is prepared commercially from carbazole and acetylene as follows:

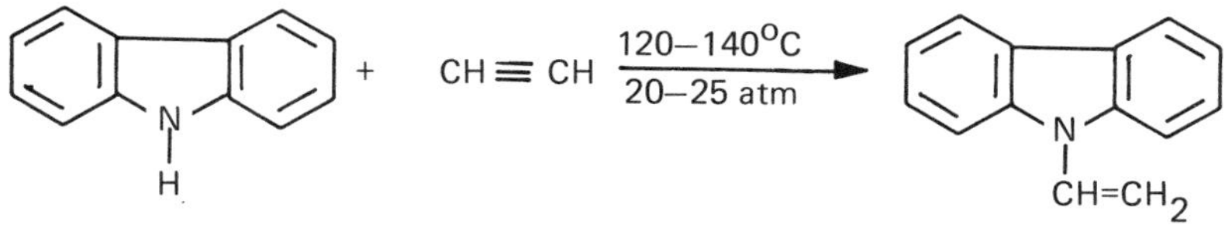

$$\text{carbazole (N-H)} + CH\equiv CH \xrightarrow[\text{20–25 atm}]{120\text{–}140^{\circ}C} \text{N-vinyl carbazole } (N\text{–}CH{=}CH_2) \tag{1}$$

Alkali hydroxides or alcoholates and other strongly alkaline substances are employed as catalysts in the reaction. Metallic zinc and certain zinc compounds are also used as secondary catalysts. Zinc dust or zinc oxide acts as a promoter and prevents the formation of cuprenelike explosive compounds. The acetylene is diluted with an inert gas, such as nitrogen, or ammonia. The latter may serve both as a diluent and as a secondary catalyst. Other routes of preparation of N-vinyl carbazole include the reaction of potassium carbazole with ethylene oxide, followed by dehydration of the resulting N-ethylolcarbazole. N-Vinyl carbazole is a white solid, melting at about 65°C. The monomer should be free of impurities, especially of sulfur compounds and anthracene (usually introduced with the carbazole), which are known to have a deleterious effect on the free-radical polymerization of N-vinyl carbazole [8, 9].

C. Polymer Preparation

The polymerization of N-vinyl carbazole can be carried out in a number of ways. Bulk polymerization is carried out by using azobisisobutyronitrile alone [9, 10] or in combination with di-tert-butyl peroxide [8, 11, 12]:

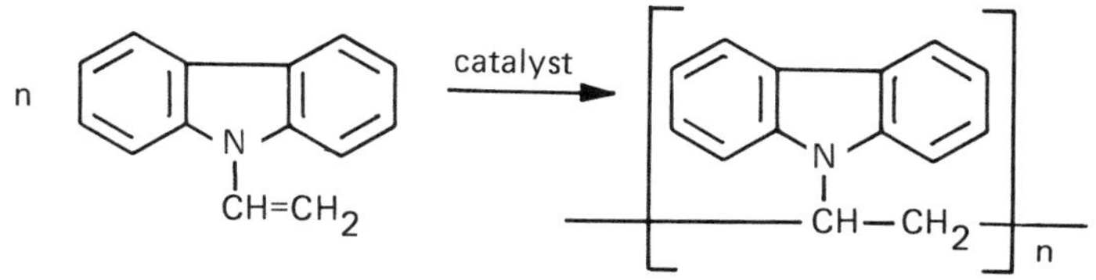

$$n \; \text{N-vinyl carbazole } (N\text{–}CH{=}CH_2) \xrightarrow{\text{catalyst}} \left[-CH-CH_2- \right]_n \text{ (N-carbazolyl on CH)} \tag{2}$$

If azobisisobutyronitrile is used as the sole initiator, bulk polymer of higher molecular weight and improved appearance is obtained.

Commercial molding powders are prepared by polymerizing vinyl carbazole in hot aqueous dispersion, employing alkaline dichromate as an initiator [13, 14].

Porous polymer in the form of beads or granules is obtained in aqueous dispersion [8] by using azobisisobutyronitrile and di-tert-butyl peroxide as initiators. It could be molded under low pressure to densities as low as 6.2 to 12.4 lb/ft^3.

The size of the expanded polymer granules depends on the conditions of polymerization. No expansion occurs with very small granules. Beads 0.5 to 1.5 mm in diameter expand on molding to yield low-density foams.

1. Preparation of Polymer Beads and Granules by Dispersion Polymerization

a. Beads. Beads 0.5 to 1 mm in diameter have been prepared by aqueous dispersion polymerization employing polyvinyl alcohol as a protective colloid and azobisisobutyronitrile as initiator. The concentration of the polyvinyl alcohol used in the polymerization depends on the purity of the monomer: more is required for once-recrystallized monomer than for the twice-recrystallized monomer. The use of the latter leads to polymers with higher molecular weight than when the less purified grade of monomer is used. The reaction conditions and the resulting dispersion polymers are listed in Table 1.

TABLE 1

Aqueous Dispersion Polymerization of Vinyl Carbazole to Foamed Beads[a]

Monomer[d]	PVA[b] (ml)	Weight of AIBN[c] (g)	Temp. of AIBN addition (°C)	Conversion (%)	Polyvinyl carbazole: Relative viscosity	Polyvinyl carbazole: Appearance
A	1.2	10	70	90	--	Fine beads
A	2.5	10	70	89	--	Fine beads
A	3	0.5	80	--	3.65	Small beads
A	3	0.5	85	--	4.03	Beads, 1.0-1.5 mm
A	3	0.5	90	--	5.0	Lump
A	6	0.5	87	--	5.5	Beads, 0.5-1 mm
B	3	0.5	80	--	2.22	Beads
B	5	1.0	85	--	2.99	Beads
B	5	0.5	90	--	2.95	Beads

[a]Reprinted from Ref. [6] by courtesy of Interscience.
[b]Polyvinyl alcohol solution (Elvanol, DuPont).
[c]Azobisisobutyronitrile.
[d]Quantity: 500 g. Monomer code: A, recrystallized once from methanol; B, as received.

b. Granules. Granules 2 to 5 mm in diameter have been prepared from twice-recrystallized monomer, using certain half-esters of polyethylene glycol (Nonex, Union Carbide) as protective colloids rather than polyvinyl alcohol, which proved not as satisfactory as the former.

The reaction conditions for the preparation of foamed polyvinyl carbazole granules are shown in Table 2. The polymerization is best carried out in air. In order to test the quality of the polymer produced, its expansion and cohesion on heating were determined, and whenever possible, a small disk was molded.

The dried foamed polymer granules or beads were extracted with methanol and with boiling acetone, and were subsequently air-dried. The extraction process was necessary in order to obtain satisfactory molding properties.

D. Impregnation and Molding of Polymer Granules and Beads

Although the foamed methanol- and acetone-extracted polymer granules expanded quite readily when heated above 150°C, the cohesion between the expending granules was found to be insufficient to yield molded articles with satisfactory mechanical strength. Impregnation of the granules or beads with solutions containing 3% of azobisisobutyronitrile in 25% solutions of dioxane-acetone or benzene-acetone followed by drying and subsequent molding at 200 to 250°C resulted in satisfactory molded panels and other shapes with good appearance in the density range 3.1 to 12.4 lb/ft^3 [6]. The panels had a beige or buff color and consisted of a high-density skin and a core of much lower density. The interior structure of the granule-based panels consisted of a network of strands and pores, with the strands tending to be directed perpendicular to the main plane. In general greater uniformity in cell structure was encountered at lower densities.

E. Properties

1. Properties of Base Polymer

Polyvinyl carbazole bulk polymer has a remarkably high softening range for a vinyl polymer. The Martens number has been reported to be 150°C [15], the Vicat softening point as 190°C [16], the heat-distortion temperature as 100 to 150°C [16], the flow temperature as 270°C [16], and the highest continuous temperature to which the polymer may be exposed to as 120°C [15].

2. Foam Properties

a. Heat Stability. The Vicat softening of foamed polyvinyl carbazole granules (relative viscosity 2.69) is 200°C, compared with 173.5°C for bulk polymer of higher molecular weight (relative viscosity 4.62) [7]. The heat stability of foamed panels of polyvinyl carbazole, previously

TABLE 2

Aqueous Dispersion Polymerization of Vinyl Carbazole in the Presence of Polyethylene Glycol Monoester-Type Protective Colloids[a]

Protective colloid[b]		Stirrer			Polymer				
Type	Vol. (ml)	Type[c]	Speed (rpm)	Conv. (%)	Rel. visc.	Appearance	Expansion[d]	Cohesion[e]	Disk[f]
N-64	0.5	A	600	94.7	2.07	Light, rocklike mass	+	++	+
N-64	1	A	600	96.5	2.07	Light, rocklike mass	+	++	+
N-64	5	A	600	98.5	1.77	Large, regular, rough granules	++	++	+++
N-64	5	B	600	86	2.80	Large, uniform granules	++	+	-
N-31	10	B	600	87.6	2.65	Large, expanded mass	+/++	+	+
N-52	10	B	600	88.2	2.30	Small lumps	++	++	+
N-139	10	B	600	85.4	2.65	Large expanded mass, lighter than water	+	+	+

[a]Reprinted from Ref. [6] by courtesy of Interscience.
[b]Code: N-64, polyethylene glycol 1000 monooleate; N-31, polyethylene glycol 200 monolaurate; N-52, polyethylene glycol 600 monooleate; N-139, polyethlene glycol 400 monolaurate.
[c]Stirrers: (A) glass link, diameter 9 cm; (B) stainless-steel wing stirrer, diameter 7.7 cm.

[d]Code: +++, mold filled; ++, moderate, some cavities; +, slight expansion; -, no expansion.
[e]Code: ++, moderate, granules fused, boundaries clearly visible; +, granules badly fused and easily broken; -, no fusion.
[f]Code: +++, smooth surfaces, well-formed edges; ++, fairly good surfaces, irregular edges; +, disk incompletely formed; -, no disk formed.

oven-dried at 80°C, was determined by subjecting the panels to a gradual heat cycle up to 170°C (24 h at 130°C + 24 h at 150°C + 30 h at 170°C) [7]. The total weight loss of these panels of 6.2- to 6.7-lb/ft^3 density ranged from 1.7 to 4.0%. Although some darkening was observed, little distortion or other deterioration was noted.

Other granule-based, foamed polyvinyl carbazole panels were exposed to 160°C for 85 h and exhibited weight losses of only 4.35 to 4.75% [7]. Additional heating of these panels for 21 h at 220 to 230°C showed a loss of only 0.51%. The foam, when heated at 220 to 230°C, contracted and became somewhat distorted. The volatile materials obtained during the heating cycle were condensed and were found to contain dioxane, tetramethylsuccinodinitrile, and carbazole.

The dimensional stability of foam panels of 3.1- to 4.4-lb/ft^3 density was determined [7]. Heating to 180°C for a period ranging from 4 to 240 h or at 200°C for 1 to 4 h showed a distortion of less than 1%. The dimensional stability appeared to decrease with increasing density. Foams of 10- to 11.2-lb/ft^3 density exhibited a distortion of 5% when exposed to temperatures of 190°C for 1 to 2 h.

b. Compressive Strength and Modulus. Ellinger's [7] compressive-strength data on granule-based foam of different densities, demonstrating the good high-temperature properties of these foams, are shown in Fig. 1, 2, and 3. Measurements could be made up to 180°C on the lightest foam (3.1- to 4.4-lb/ft^3), but the foam samples in the two higher density ranges deformed too much at this temperature without the application of a load, so that the maximum temperature for these foams was limited to 160°C.

The compressive modulus as a function of foam density at various temperatures is shown in Fig. 4 [17].

c. Tensile Strength. The tensile strength of samples cut from foam panels and bonded individually between the ground end surfaces of 1-in.-diameter steel cylinders with an adhesive was determined at a crosshead speed of 0.25 in./min [17]. The tensile strength as a function of density and temperature is shown in Fig. 5 [17].

d. Thermal Conductivity. The thermal conductivity data for polyvinyl carbazole foam are shown in Table 3 [7].

e. Electrical Properties. Polyvinyl carbazole has excellent dielectric properties at radar frequency, even under conditions of high humidity. The dielectric constant ϵ and loss factor tan δ of polyvinyl carbazole foam at various densities and two different temperatures (20 and 180°C) are presented in Table 4 [7]. Very little difference was observed in the dielectric constant measured at these temperatures, whereas the loss factor increased somewhat at the higher temperature. A linear relationship was found between the dielectric constant and the density of the foam [7].

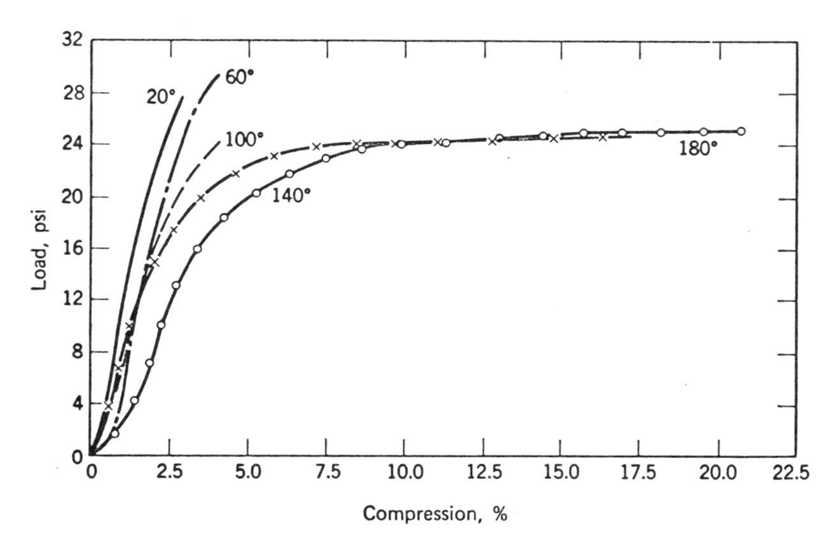

FIG. 1. Compressive strength of polyvinyl carbazole foam with densities of 3.1 to 4 lb/ft^3. Reprinted from Ref. [7] by courtesy of Interscience.

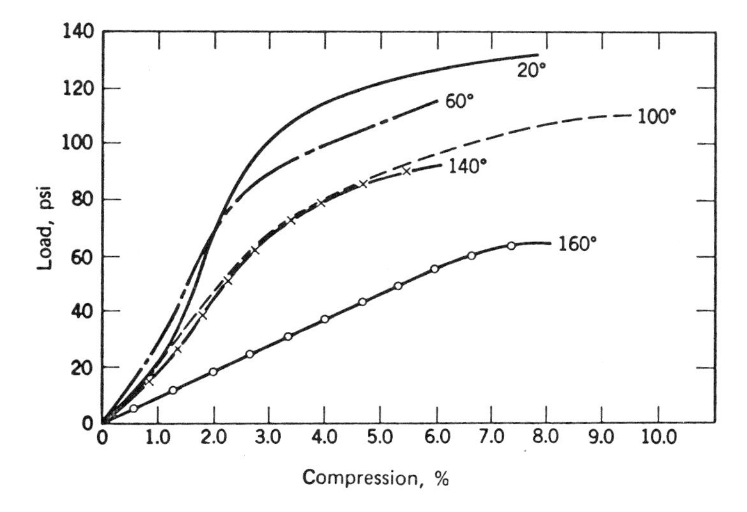

FIG. 2. Compressive strength of polyvinyl carbazole foam with densities of 6.2 to 7.5 lb/ft^3. Reprinted from Ref. [7] by courtesy of Interscience.

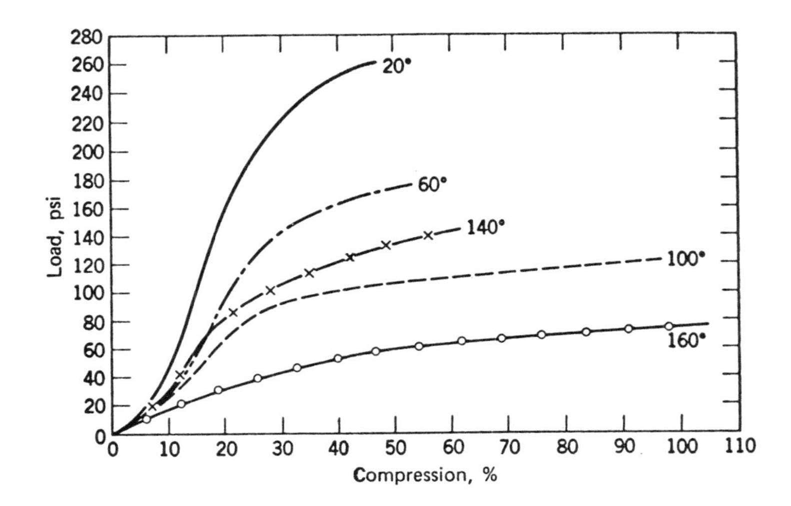

FIG. 3. Compressive strength of polyvinyl carbazole foam with densities of 10 to 11.2 lb/ft^3. Reprinted from Ref. [7] by courtesy of Interscience.

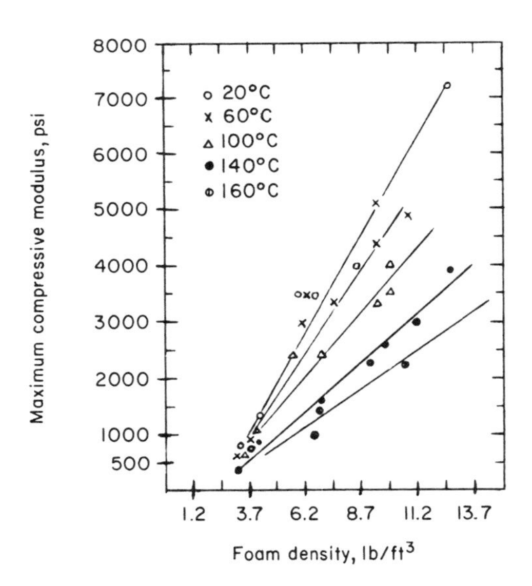

FIG. 4. Relationship between compressive modulus and foam density at various elevated temperatures. Reprinted from Ref. [7] by courtesy of Interscience.

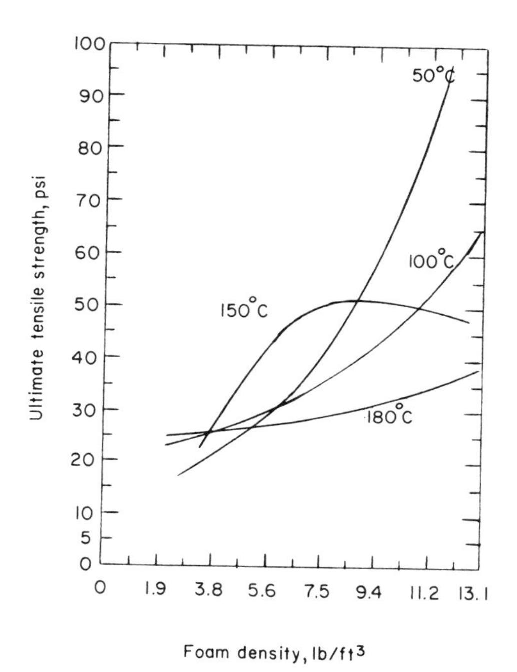

FIG. 5. Relationship between foam strength and density at various temperatures. Reprinted from Ref. [7] by courtesy of Interscience.

TABLE 3

Thermal Conductivity of Granule-Based Foamed Polyvinyl Carbazole[a]

Density (lb/ft^3)	Mean temperature (^oC)	Thermal conductivity	
		kcal/(sec)(cm^2)(^oC/cm)	Btu/(h)(ft^2)(^oF/in.)
3.3	83	0.000082	0.24
6.9	84	0.000077	0.22
11.2	92	0.000075	0.22

[a]Reprinted from Ref. [7] by courtesy of Interscience.

TABLE 4

Dielectric Constant ϵ and tan δ of Samples Cut from Granule-Based Foamed Polyvinyl Carbazole Panel[a]

Density (lb/ft^3)	Temperature (°C)	Dielectric constant ϵ	tan $\delta \times 10^4$
4.9	20	1.101	9
	180	1.100	18
6.0	20	1.149	9
	180	1.141	15
6.2	20	1.114	10
	180	1.110	16
6.2	20	1.107	12
	180	1.103	19
8.2	20	1.198	11
	180	1.190	18

[a]Reprinted from Ref. [7] by courtesy of Interscience.

Conditions of high humidity were found to have no noticeable effect on the dielectric constant and tan δ. It was also noticed that bead-based foam panels were more homogenous with regard to the dielectric constant and tan δ than the granule-based panels [7].

f. Effect of Chemicals and Solvents. The effects of selected liquids on polyvinyl carbazole foam during a 65-h exposure at ambient temperature are shown in Table 5 [7].

II. PYRANYL FOAMS

A. Introduction

In 1961 the Central Research Laboratory of Canadian Industries Ltd. developed a new foamed-in-place thermosetting cellular plastic, which was commercially introduced as Kayfax pyranyl foam [18-21]. ICI America, Inc., a subsidiary of Imperial Chemical Industries Ltd., first marketed this product in 1966 with the hope that its relatively high thermal stability and the ease of handling the basic foaming ingredients, due to their low viscosities, (less than 100 cP at 23°C) would provide an edge over rigid polyurethane foams [22]. Late in 1967, ICI America removed the product from the market due to its relatively high cost and short supply of raw materials.

TABLE 5

Effects of Selected Liquids on Granule-Based Foamed Polyvinyl Carbazole 65 h at Ambient Temperature[a]

Liquid	Effect
Dilute hydrochloric acid	None
Dilute sulfuric acid	None
Dilute nitric acid	Yellow coloration
n-Hexane	None after 16 h
Diethyl ether	Softened, a little polymer dissolved
Ethyl acetate	None
Methylene dichloride	Completely soluble in a few minutes
Chloroform	Completely soluble in a few minutes
Carbon tetrachloride	Almost no effect
Dimethylformamide	Completely soluble in a few minutes
Cyclohexane	None
Benzene	Completely soluble in 1 h
Toluene	Softened, some swelling, about 50% soluble
Xylene	Softened without swelling, slightly soluble
Dioxane	Slowly soluble
Tetrahydrofuran	Slowly and incompletely soluble
Pyridine	Soluble

[a]Reprinted from Ref. [7] by courtesy of Interscience.

B. Chemistry of Pyranyl Intermediates

The pyranyl monomer is obtained from propylene according to the following scheme; by which the acrolein dimer (2) is obtained:

$$CH_3CH{=}CH_2 \xrightarrow{O_2} CH_2{=}CHCHO \xrightarrow[\text{pressure}]{\text{heat}} \text{(dihydropyran ring)}{-}CHO \qquad (3)$$

The acrolein dimer (2-formyl dihydropyran) (3) is converted to the acrolein tetramer (pyranyl monomer) via either the Tishchenko reaction, which gives the "ester tetramer" (4), or aldol condensation, which yields the "aldol tetramer" (5).

2 (2) CHO $\xrightarrow[\text{Tishchenko reaction}]{Al(OR)_3}$ (3) $-CH_2OOC-$ (4)

2 (2) CHO $\xrightarrow[\text{aldol condensation}]{NaOH,\ 30^{o}C}$ (4) CH, OH, CHO (5)

It is necessary that the tetramers be heat- and vacuum-treated to remove such by-products as acrolein dimers and alkyl esters [21], the presence of which makes the foams likely to scorch. Another procedure for reducing the amount of by-products consists of not carrying the Tishchenko reaction to completion and thus avoiding the formation of polycondensation products [21].

C. Preparation of Pyranyl Foams

The foams are formed by exothermic reaction caused by the crosslinking of the pyranyl double bonds in the presence of an acidic catalyst:

[ ]$_2$ R

monomer stream — cationic polymerization → —R— (6)

catalyst + blowing agent

The heat of the reaction causes the blowing agent (methylene chloride, chloroform, or dichlorodifluoromethane or trichlorofluoromethane) to vaporize, yielding low-density cellular plastics.

The concentration of the catalyst is normally from 0.1 to 1.0 wt%, based on polymer weight. The catalysts are Lewis-type acids, which include boron trifluoride etherate, fluosilicic acid, and fluoboric acid. A combination of the latter two acids provides increased time of rise and foam with low friability [23].

Surfactants can be incorporated to control cell structure and foam properties. Siloxane-ethylene oxide and siloxane-propylene oxide block copolymers produce excellent foam structures, exhibiting a high degree of isotropicity and low densities [21]. Incorporation of organic phosphates (e.g., tris(2-chloro-3,4-epoxybutyl) phosphate and tris(2-bromo-3,4-epoxybutyl) phosphate) imparts flame retardance to the pyranyl foams [24].

According to another patent issued to Canadian Industries Ltd., a mixture of "ester tetramer" (4) and "aldol tetramer" (5) produces foams whose properties are superior to those of foams derived from either tetramer alone [25].

D. Pyranyl-Foam Systems and Equipment

The pyranyl-foam system can be applied as either a two- or three-component system, although a two-component one is more suitable for applications in the appliance industry. The metering ratio is 10 parts of the resin to 1 part of the catalyst stream. In the case of flame-retardant pyranyl foams the mixing ratio is 6:1.

The catalyst concentration can be varied to control the cream time without influencing the foam properties. Conventional foam-processing equipment can be used for metering the two components since they are readily miscible and possess low viscosities. A methylene chloride solvent flush is used after pouring the foam.

The reaction conditions for appliance filling with pyranyl foam can be varied. Typical conditions are as follows:

Resin and catalyst temperature	$25 \pm 2^{\circ}C$
Jig temperature	$41 \pm 3^{\circ}C$

From the start of dispensing the foam the reaction times are as follows:

Cream time	18 sec
Rise time	35 sec
Tack-free time	38 sec
Jig dwell time	8 min

Total pressures developed are similar to those obtained for urethane foams, but the rates of pressure buildup and dissipation are faster with pyranyl systems. Hence shorter jig dwell times are achieved with the latter.

E. Properties

1. Mechanical Properties

Typical values for compressive and tensile strengths of 2.2-lb/ft^3 overall density (core density 1.8-lb/ft^3) pyranyl foams are shown in Table 6 [18]. The relatively close values between the strength properties parallel and perpendicular to the direction of foam rise indicate a high degree of isotropicity for pyranyl foam. Pyranyl foam exhibits good adhesion to degreased metal surfaces, such as aluminum, steel, phosphated steel, and stainless steel, as well as to such other substrates as wood and paper [18].

2. Thermal Conductivity

The thermal conductivity (k factor) of cut samples of Kayfax pyranyl foam of 2.2-lb/ft^3 density as a function of time is shown in Fig. 6 [18]. The initial value for the K factor was 0.117 Btu/(h)(ft^2)(^{o}F/in.), and after 28 days at 70°C the samples had a value of 0.153 [18]. The aged value after 6 months at 70°C was still under 0.16. In the case of poured 48 x 24 x 2-in. pyranyl-foam panels initial and aged values of 0.104 and 0.144, respectively, were obtained.

The effect of surfactant level on the k factor of cut foams is seen in Fig. 7 [18].

3. Cell Structure and Permeability

Table 7 presents water-absorption and water-vapor-transmission data for pyranyl foam [18]. The closed-cell content and the cell size of this foam are also listed. The data indicate that this foam possesses a very high content of fine, closed cells.

4. Dimensional Stability

The dimensional changes of 2.2-lb/ft^3-density foams were determined by exposing foam samples to various environmental conditions. The results shown in Table 8 [18] indicate good dimensional stability.

TABLE 6

Mechanical Properties of Pyranyl Foams[a,b]

Property	Value
Compressive strength (10%), psi:	
Perpendicular to direction of flow	18
Parallel to direction of flow	14
Yield (6%) perpendicular to direction of flow	22
Yield (6%) parallel to direction of flow	16
Tensile strength, psi:	
Perpendicular to direction of flow	20
Parallel to direction of flow	19
Elongation at break, %	4

[a]Data from Ref. [1].
[b]Overall foam density 2.2-lb/ft^3.

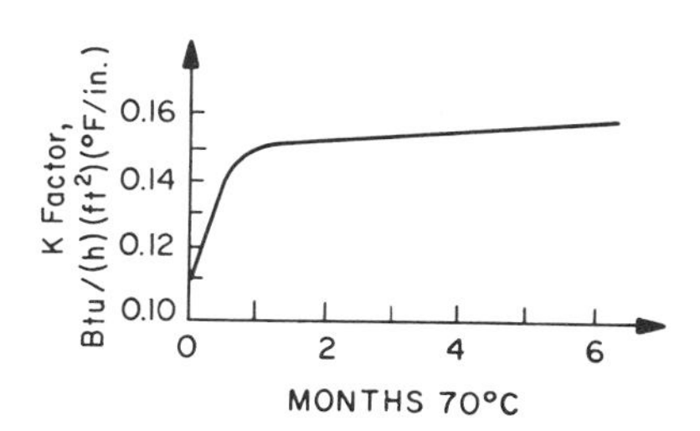

FIG. 6. Thermal conductivity (k factor) of trichlorofluoromethane-blown 2.2-lb/ft^3 pyranyl foam as a function of time. Reprinted from Ref. [18] by courtesy of Technomic Publishing Co.

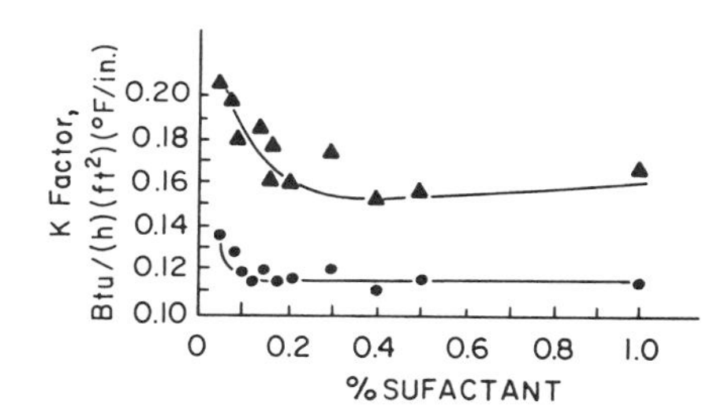

FIG. 7 Effect of surfactant level on the thermal conductivity (k factor) of trichlorofluoromethane-blown 2.2-lb/ft^3 cut pyranyl foam. Key: ▲, 28 days at 70°C; ●, 24 h after preparation. Reprinted from Ref. [18] by courtesy of Technomic Publishing Co.

TABLE 7

Cell Structure and Permeability of Pyranyl Foam[a]

Property	Value
Closed-cell content (corrected), %	98
Cell size (freezer-wall sample), cm^3	2×10^{-4}
Water absorption (24 h):	
g/1000 cm^2	14.4
g/1000 cm^3	16.2
Water-vapor transmission, perms	2.6

[a] Data from Ref. [18].

TABLE 8

Dimensional Stability of Pyranyl Foam[a]

Temperature (°C)	Change after 28 days (%)		
	Dimension	Freezer mockup	Panel[b]
-25	Length	+0.2	0.0
	Volume	+0.4	+0.1
+70	Length	+2.2	+1.7
	Volume	+5.3	+3.9 to 5.1
+70, 100% R.H.	Length	+2.3	+0.9 to 1.9
	Volume	+5.0	+2.7 to 4.9

[a]Adapted from Ref. [18]. Overall foam density 2.2-lb/ft^3.
[b]Size: 48 x 24 x 2 in.

5. Thermal Stability

Pyranyl foams withstand immersion in molten asphalt at 205°C for 30 sec without undergoing delamination of facings or distortion [18]. Foams that had been heated in metal molds at 165°C for 20 min did not exhibit any decrease in properties. The k factor remained constant, and the compressive strength increased on occasion by as much as 20%.

Three insulated cabinets were subjected to a paint-baking cycle of 20 min at 165°C and a total residence time of 30 min in the oven. An average wall-thickness increase of 6, 7, and 9%, respectively, for foams of 2.4-, 2.2-, and 2.0-lb/ft^3 overall density [18] was observed.

The maximum service temperature for low-density pyranyl foams is 135°C, but higher temperatures are possible for short periods. When exposed to 78°C for 24 h, pyranyl foams showed a 20% reduction in tensile strength, but the elongation at break remained unchanged [18].

6. Flame Resistance

Incorporation of flame retardants can render pyranyl foams self-extinguishing or nonburning according to ASTM D1692. The dimensional stability of the foams under humid aging conditions is only slightly affected, and a combination of good fire resistance and good physical properties can thus be obtained.

7. Chemical Resistance

The resistance of pyranyl foams to many chemicals and solvents is very good, as indicated in Table 9 [18]. Although they absorb many solvents, the foams retain their original shape after drying.

TABLE 9

Chemical Resistance of Pyranyl Foam[a,b]

Chemical	Effect
Sulfuric acid (5%) Hydrochloric acid (10%) Acetic acid (10%) Sodium hydroxide (40%) Gasoline Cyclohexane	Foam stable for >1 month in each medium
Benzene Isopropanol	Foam stable for >1 month in each medium; although solvent-logged, foam retains shape on drying

[a]Reprinted from Ref. [18] by courtesy of Technomic Publishing Co.
[b]Foam density 2.2-lb/ft^3.

F. Applications

Pyranyl foams were developed for use in appliance insulation, buoyancy applications, and construction applications, including sandwich panels, continuous slab, sprayed foam, foamed-in-place foam, and pipe insulation [18]. Other intended applications are for insulation in the automotive and railroad industries.

III. POLYESTER FOAMS

A. Introduction

Polyester foams are generally based on unsaturated polyester resins crosslinked with vinyl monomers in the presence of a blowing agent and a catalyst.

Thus, if G is a glycol, and A is an unsaturated acid (maleic or fumaric), the esterification reaction will produce a linear unsaturated polyester of the following structure:

$$—G—A—G—A—G—A—$$

If another unsaturated monomer (e.g., styrene [26]) should be introduced, with a suitable initiator like a peroxide and a blowing agent, the result will be a complex three-dimensional, cellular polymer that can be represented schematically as follows:

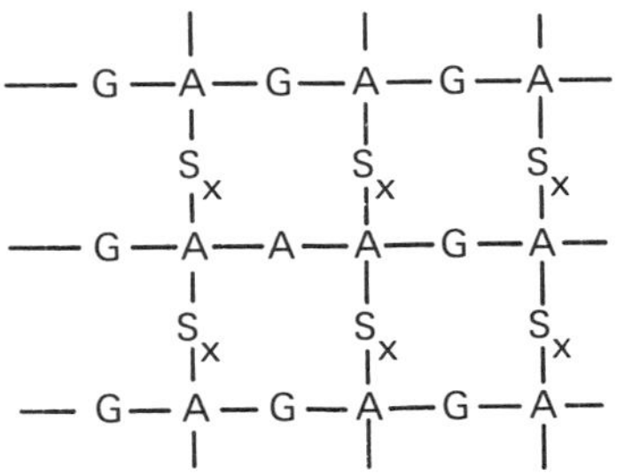

```
         |       |       |
  — G — A — G — A — G — A —
         |       |       |
         Sx      Sx      Sx
         |       |       |
  — G — A — A — A — G — A —
         |       |       |
         Sx      Sx      Sx
         |       |       |
  — G — A — G — A — G — A —
         |       |       |
```

The polyester resins can be made flame retardant by the use of halogenated acids or acid anhydrides (e.g., chlorendic anhydride) either alone or in combination with antimony oxide.

B. Preparation of Foams

In a procedure described by Chemische Werke Huels [27] unsaturated polyesters (e.g., those based on fumaric acid, phthalic anhydride, ethylene glycol, diethylene glycol, and a small amount of hydroquinone) are first mixed with styrene, preferably in a polyester-monomer ratio of 2.1:1 to 2.5:1. To this polyester-monomer blend are added water, benzoyl peroxide, and some zinc stearate. Sodium bicarbonate is slowly added along with a solution of maleic anhydride and a small amount of dimethylaniline in styrene. The addition of an acid or acid anhydride to the bicarbonate generates the carbon dioxide blowing agent. The mixture is agitated for 15 min and then poured into a mold and kept at a temperature of up to 75°C to yield a foam of about 3.4-lb/ft^3 density with a pore diameter of 3 to 7 mm.

Sasse [28] has described a process for the formation of molded polyester-foam products in which air bubbles are introduced by vigorous mechanical stirring into a polyester molding compound consisting of an unsaturated polyester, polymerizable monomers, catalyst, and accelerator. The mixture is poured into a mold and placed under vacuum immediately before gelation. The vacuum is controlled so as to yield a fivefold to eightfold expansion.

In another method [29] the resin mixture under pressure is passed through a pressure-reduction zone with an orifice of less than 1/8-in. diameter into a foaming zone, where proper pressure and temperature are maintained to promote foaming and curing. The resulting polyester foams of about 2 to 4 lb/ft^3 density exhibited good strength and rigidity.

In a process described by Suter and Ammon [30] a partially cross-linked polyester resin (obtained by heating an approximately equimolar mixture of a dicarboxylic acid and a triol or a triol-containing mixture of glycols) is expanded by evaporation of water that forms in the final stage of the condensation. A significant improvement of this method was obtained by performing the reaction at 180 to 220°C in two stages: 5 to 15 min at 600 to 800 mm Hg and 35 to 60 min at 80 to 120 mm Hg. For best results a cylindrical reactor with a free-moving piston to eject the expanded resin is recommended.

Parker and Rood [26] have described the preparation of polyester foams by treating a polyester-styrene mixture with a blowing agent (consisting of 95 parts of dichlorodifluoromethane and 5 parts of trichlorofluoromethane) in one zone, and a curing agent in a second zone. The foam was made flame resistant by the incorporation of chlorinated paraffin wax and antimony oxide.

The use of trichlorofluoromethane as a blowing agent for the production of medium-high-density (7-16 lb/ft^3) polyester foams has also been disclosed by Salgado and Berlinger [31].

A polyester foam, designated Estafoam, was announced in 1957 by Vanguard Products [32] but has not found any significant commercial acceptance. This product is based on the incorporation of a gas, such as air, into an unsaturated polyester by employing mechanical agitation. After release through a nozzle the fully expanded catalyzed foam has the appearance of whipped cream. The stable froth is poured into a mold or onto a moving belt where it is leveled by a doctor blade. The time of foam solidification depends on the amount of catalyst in the foam.

Aminoalkylation of polyesters with polyamine and aldehyde has been shown to produce foams of high strength at a density of about 4.4-lb/ft^3 [33]. Addition of up to 60% alumina gave foams with superior toughness, compressive strength, and good refractory properties, with a charring temperature of up to 1650°C [34]. Powdered aluminum, sodium silicate, dolomite, and fluosilicate have also been used in admixture with polyester-styrene blends containing benzoyl peroxide to yield good foam insulators on evaporation of the water [35]. Addition of 0.01 to 5 wt% of hexyl, lauryl, or cetyl β-resorcylate, β-resorcylanilide, or β-resorcyl-p-laurylanilide has been shown to increase the heat stability of polyester foams [36].

Since polyesters generally contain residual hydroxyl and carboxyl end groups, diisocyanates can be used to provide some crosslinking and to generate carbon dioxide. The addition of tolylene diisocyanate (5-20% based on the polyester), a surfactant, water, and a peroxide to unsaturated polyesters has been reported to result in foams possessing high compressive strengths [37]. The use of a flexible urethane foam to provide a form for a rigid foam has been reported [31]. After the urethane foaming resin, saturated with styrene, has been allowed to expand, it is impregnated with a polyester-resin solution in styrene, containing added peroxide. The product after hardening consists of an open-celled, rigid foam.

Martin and Schumann [38] described a process for the production of polyester foams employing an N-vinylpyrrolidone-vinyl propionate copolymer foam stabilizer. In a typical procedure a mixture of 8.5 parts of formic acid, 7.5 parts of water, and 0.35 part of N,N-bis(α-methyl-β-hydroxyethyl)-p-toluidine is homogenized with another mixture consisting of 65 parts of a polyester, made from maleic anhydride, phthalic anhydride, and 1,2-propanediol; 35 parts of styrene; 0.01 part of hydroquinone; 15 parts of sodium carbonate; 2 parts of 50% benzoyl peroxide; and 7.5 parts of the N-vinylpyrrolidone-vinyl propionate copolymer. Foaming takes place in a few seconds and is complete in 3 min. The foam sets in 10 min and is hard and colorless, with uniform pore structure and a density of about 3.7-lb/ft^3. These foams can be used as structural elements where heat and solvent resistance are desired.

Chemische Werke Huels AG [39] has disclosed a process for the preparation of foams from mixtures of esters of saturated glycols and α,β-unsaturated acids and anhydrides admixed with copolymerizable monomers (e.g., styrene or an ester of acrylic or fumaric acid) using an azo or hydrazo compound simultaneously as a foaming agent and a polymerization initiator.

Water-filled polyester foams are being produced by Reichhold Chemicals and Ashland Chemicals [40]. A process described by Leitheiser and Coderre [41] consists of mixing an unsaturated polyester with water to form an emulsion that cures spontaneously on addition of catalyst. The resulting product is a hard, white, fine-grained material consisting of a plastic matrix containing microscopic particles of water distributed uniformly throughout the matrix. The water droplets are between 2 and 5 microns in diameter. The resin can be extended by as much as 90% water, although the material develops optimum strength at 50 to 60%. At this water concentration the cell walls are strong enough to withstand freeze-thaw cycling to temperatures as low as -34°C. This material handles and fabricates similarly to wood.

C. Applications

The development of polyester foams has been relatively slow despite greatly increased efforts in recent years. Polyester foams have been recommended as a core material for sandwich panels [42, 43], as thermal and acoustical insulation [26, 28, 35, 44], as cushioning materials [26], and as building materials [28].

Uses for water-filled polyester foam have included such molded items as lamps, plaques, and furniture parts [40]. Other applications for this foam as a replacement for concrete in paving and construction have been described [45].

IV. CELLULOSE ACETATE FOAMS

A. Introduction

Cellulose acetate was one of the first rigid foams finding commercial use in aircraft manufacture during World War II and in specialty construction uses [46]. Developed by E.I. du Pont de Nemours & Co. [47], the foam has been produced for over 20 years by the Strux Corp. [48].

B. Manufacture

The production of cellulose acetate foam involves extrusion of flaked cellulose acetate, employing a mixture of acetone and ethyl alcohol as blowing agent and comminuted mineral of 100 to 325 mesh as nucleating agent. The solvent mixture serves also as a solvent for the cellulose acetate above 70°C as well as a plasticizer during the hot extrusion. The extrusion process leads to the formation of unicellular foam by the sudden release of pressure of the superheated, pressurized molten polymer mass as it emerges from the extrusion die. The extruded foam has a denser outer skin, which may be removed by cutting off the outer edges. Cellulose acetate foam is extruded in the form of boards and rods.

C. Properties

Cellulose acetate foam is produced at relatively high densities, 6 to 8 lb/ft^3. It has excellent strength, low water absorption (after 14-days immersion it absorbs only 3 lb of water per cubic foot), and resistance to attack by vermin and fungus. The foam has a broad service-temperature range (from about -57 to +177°C) and can withstand temperatures as high as 194°C for short periods. It has good electrical properties, which makes it useful for electronics and X-ray equipment. A summary of its properties is presented in Table 10.

D. Applications

The most widely used current applications for cellulose acetate foam include rib structures in the fabrication of lightweight reinforced plastic parts and as a core material in sandwich construction, bonded to metal, wood, or glass. Other applications are as reinforcement for aircraft control surfaces, radome housings, filler blocks under fuel cells, tank floats for indicating devices, and ribs, posts, and framing in houses and shelters. Due to its buoyancy characteristics, cellulose acetate foam has been used in lifeboats, buoys, and other flotation devices.

TABLE 10

Typical Properties of Cellular Cellulose Acetate[a]

Property	Value
Density, lb/ft^3	6-8
Tensile strength, psi	170
Compressive strength, ultimate, psi	125-150
Flexural strength, psi	147
Flexural modulus, psi	5500
Shear strength, psi:	
In tension	140
In compression	125
Impact strength, Izod, unnotched, ft-lb	0.1
Coefficient of linear expansion per ^oC	4.5×10^{-5}
Thermal conductivity, k factor, $Btu/(h)(ft^2)(^oF/in.)$	0.31
Water absorption, lb/ft^3:	
At 50% R.H.	0.15
At 100% R.H.	1.05
Electrical properties:	
Dielectric constant	1.12
Loss tangent	0.002-0.003
Burning rate, in./min	4.9

[a]Data from Ref. [49].

V. POLYVINYL ALCOHOL-FORMALDEHYDE FOAMS

Polyvinyl alcohol-formaldehyde (polyvinyl-formal) foams were first reported in 1945 by Revertex Ltd. [50]. These foams have high water absorption and good strength at low densities. Due to their good abrasion resistance and resiliency when wet, they are very suitable as synthetic sponges.

Polyvinyl alcohol (PVA) resin is crosslinked with formaldehyde in the presence of an acid catalyst:

$$-CH_2-\underset{OH}{CH}-\left[CH_2-\underset{OH}{CH}\right]_n-CH_2- + nCH_2O \xrightarrow{H^+}$$

$$-CH_2-\underset{OH}{CH}-\left[CH_2-\underset{|}{CH}\right]-CH_2- \quad (O-CH_2-O \text{ bridge}) \quad -CH_2-\underset{OH}{CH}-\left[CH_2-CH\right]-CH_2- \qquad (7)$$

Benning [51] believes that another competing reaction also contributes, to some extent, to the final properties of the product, although he considers reaction (7) to be the dominating reaction.

$$-CH_2-\underset{OH}{CH}-CH_2-\underset{OH}{CH}- + -CH_2O \rightarrow -CH(O)-CH_2-CH(O)- \; (O-CH_2-O \text{ ring}) \qquad (8)$$

Polyvinyl alcohols containing less than 10% and preferably less than 1% of residual acetate groups are most suitable for the foaming reaction. In order to obtain stable polyvinyl-formal foams it is preferable to have 35 to 80% of the hydroxyl groups of polyvinyl alcohol react with formaldehyde [52]. Although other aldehydes or dialdehydes can be employed, formaldehyde is preferred because of its low cost and the short curing time of the resulting foams. Besides an acid catalyst (with a dissociation constant of at least 10^{-2}), a variety of anionic or nonionic surfactants can be used to obtain a stable froth. Both the chemical nature and the concentration of the surfactant are important factors in determining the froth volume.

Benning [51] has given a detailed description of the chemistry and physics of the polyvinyl-formal foam formation.

Foaming can be achieved either by incorporating gas-releasing agents or, more economically, by frothing the reaction mixture by beating air into it with a high-speed stirrer or propeller [50, 52]. In the former method several variations have been employed--for example, evolution of carbon dioxide or hydrogen by action of the acid on carbonates or metal hydrides [53] and decomposition of diazo [54], N-nitroso [55], or borohydride [56] compounds.

Polyvinyl alcohol can be modified for improved end properties. A patent issued to the Borden Co. discloses the use of polyvinyl alcohol, alkoxylated with ethylene-, propylene-, butylene-, or styrene oxide [57]. It can also be modified by introduction of acid or ester groups (e.g., carboxylic or sulfonic acid), in which case sponges with improved softness and resistance to detergents are obtained [58].

An increase in the abrasive character of the foams is desirable for certain applications. Use of granular paraformaldehyde produces localized higher crosslinking and hence hardened surface areas [59]. Polyvinyl-formal resins, precured and finely divided, can be added to the initial solution before the frothing stage [60]. Polyvinyl-formal foams suitable for grinding and polishing contain silicon carbide as the abrasive filler [61].

If the foamed mixture of the polyvinyl-formal resin containing abrasive grains is treated with gelatin, before washing with water, a foam with staggered porosity and improved abrasive properties can be obtained [62].

Polyvinyl-formal foams are also used to obtain sponge-coated textile sheets [63] or as construction materials in the form of rods and bars, when reinforced with fillers, such as polyester fibers, clay, polyvinyl chloride, melamine resins, and calcium sulfate [64].

VI. POLYAMIDE FOAMS

Polyamides, more commonly known as nylons, are usually prepared by a condensation polymerization of a diamine with a dibasic acid (or their respective derivatives) (reaction (9)) or by a ring-opening polymerization of a cyclic lactam (reaction (10)):

$$H_2N(CH_2)_xNH_2 + HOOC(CH_2)_yCOOH \xrightarrow[-H_2O]{\text{heat}} \left[NH(CH_2)_xNHCO(CH_2)_yCO\right]_n \tag{9}$$

$$n\left((CH_2)_x\begin{matrix}NH\\ |\\ CO\end{matrix}\right) \longrightarrow \left[NH(CH_2)_xCO\right]_n \tag{10}$$

Polyamide foams have been produced by the anionic polymerization of lactams with seven-membered or larger rings by mixing a blowing agent with the monomer [65].

In a typical formulation 20 parts of benzene is added to 400 parts of caprolactam, which is melted at 80°C. This mixture is raised to 130°C, and 1.5 parts of sodium methylate catalyst is added. Separately a mixture of 100 parts of caprolactam and 15 parts of bis(capryllactam-N-carboxylic acid) hexamethylenediamide are heated to 130°C. The two melts are then mixed together and transferred to a mold at 150°C. A foam of about 11-lb/ft^3 density is obtained [65].

Low-density polyamide foams have been described by Beyerlein and Wilhelm [66]. In this procedure sodium caprolactam is added to molten caprolactam that contains gypsum and ligroin. Hexamethylene bisureido-caprolactam is then added at 130°C, producing foams with uniform pore structure and densities of 3 to 4-lb/ft^3.

Various anionic catalysts have been used, including sodium methylate [65, 67], alkali or alkaline-earth formates [68], sodium hydroxide-in-oil dispersion [69], and alkali-metal lactam [66, 70]. Isocyanates [68], diisocyanate [67], and an organic carbonate [69] have also been used as cocatalysts or promoters.

Chemical blowing agents have been utilized for the expansion of poly-lactams. These include azodicarboxamide or trihydrazino triazine [71], α^1, α^4-diazidohexamethylbenzene [70]. As already mentioned, such solvents as benzene, heptane, octane, or cyclohexane can be used as blowing agents [65, 66]. Decomposition of carboxylic acid ester to liberate low-boiling alcohol has also been employed for the foaming of polyamides [72]. In this process the cellular polyamides are prepared by the condensation of amino acids, dicarboxylic salts of diamines, and oxamic acid or its esters at 180 to 240°C [72]. Very tough and heat-resistant foams are obtained by this procedure.

The use of 1 to 5% of potassium stearate as a surfactant resulted in a uniform cell structure in foamed ϵ-caprolactam [67]. Surfactants of the type $RN(XCO_2M)_2$ (where R = C_{10-20} aliphatic hydrocarbon, X = alkylene group, and M = salt-forming cation) have been successfully employed [73].

Reinforced objects from cellular polylactams are obtained by using 30% by volume of such fibrous fillers as glass fibers, fibrous potassium titanates, aluminum, copper, or steel [74]. These may also be blended with such nonfibrous fillers as chalk, alumina, titanium dioxide, calcium silicate, aluminum silicate, and portland cement.

A patent granted to Badische Anilin- und Soda-Fabrik [75] discloses that addition of 1 to 10 wt% of an epoxy compound to the monomer results in substantially stressfree products or tough, elastic foams.

Polyamide foam coated with solid polyamide resin has been made into bowling pins. These exhibit crack resistance and resilience and will not delaminate on impact [76]. Polyamide foam is also used for artificial limbs [77] as well as insulation and elastic sealer [78]. These foams have also been shown to be suitable for the manufacture of oil-impregnated bearings and bushings [71].

A new foam, called ECN foamed nylon, of undisclosed composition, has been developed by DuPont [79]. It is said to possess unusual toughness, abrasion resistance, tensile strength, and ability to absorb energy. At densities of 2 to 4 lb/ft^3 the material is a semirigid foam, and at densities of about 8 lb/ft^3 it resembles balsa wood in rigidity and load-bearing characteristics. In the density range 1.5 to 3 lb/ft^3 ECN has a k factor of 0.22 to 0.27 $Btu/(h)(ft^2)(^oF/in.)$. This foam can withstand ambient temperatures as high as 120^oC. Though polyamide is not noted for unusual outdoor durability, uncoated ECN foam, when exposed for 14 months in Florida, did not exhibit any visible signs of degradation or discoloration [79]. Painting, coating, and pigmenting are obvious methods for improving the weatherability of this foam.

VII. IRRADIATED ACRYLIC FOAMS

A. Preparation

Irradiated, rigid acrylic foam was developed by the Sekisui Chemical Co. in Japan. It is produced by forming first a prepolymer of acrylonitrile, acrylamide, and acrylic acid. The prepolymer is irradiated with β or α rays [80, 81] and is then cured in a mold at elevated temperatures. The reaction mixture expands to about 40 times its original volume and, when completely cured, yields a tough and rigid foam with a woodlike appearance. The details of this process have as yet not been published.

B. Properties

1. Mechanical Properties

The tensile strength of acrylic foam as a function of foam density is shown in Fig. 8. The corresponding curves for a urethane foam and a polystyrene foam are included for comparison. It can be seen that at a density of 3.12 lb/ft^3, the tensile strength of the acrylic foam is about 185 psi, and at a density of 6.24 lb/ft^3, about 255 psi. These values are considerably greater than those shown for the urethane and polystyrene foams. The flexural and compressive strengths of these acrylic foams are seen in Figs. 9 and 10 [80], indicating likewise higher values for the acrylic foams when compared with both urethane and polystyrene foams.

2. Thermal Properties

a. Thermal Resistance. Table 11 [80] lists the thermal resistance of irradiated acrylic foam in comparison with other plastic foams. The

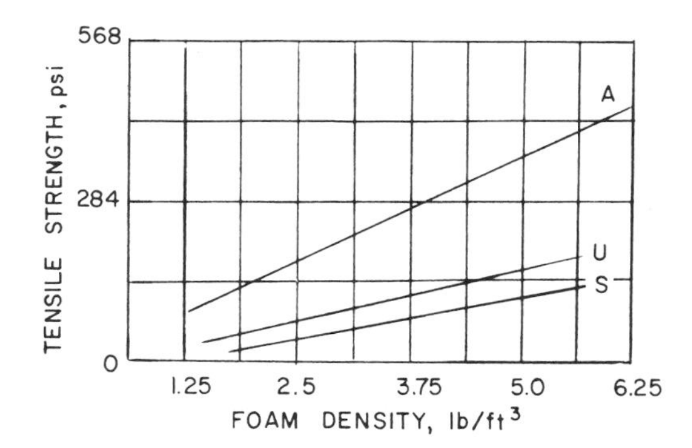

FIG. 8. Tensile strength of acrylic (A), urethane (U), and polystyrene (S) foams as a function of density at $20^{o}C$, 65% relative humidity. Reprinted from Ref. [80] by courtesy of Technomic Publishing Co.

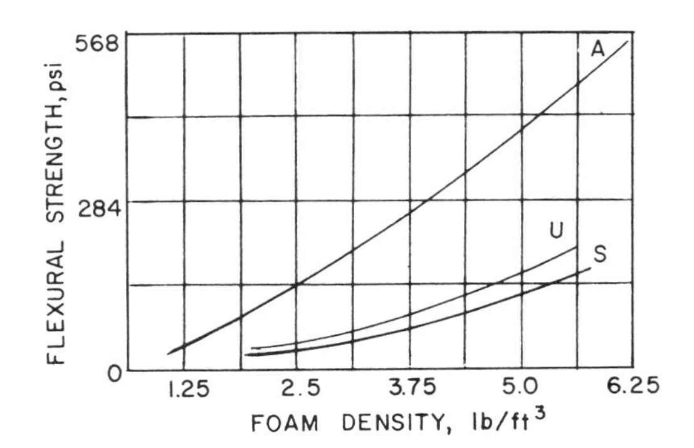

FIG. 9. Flexural strength of acrylic (A), urethane (U), and polystyrene (S) foams as a function of density at $20^{o}C$, 65% relative humidity. Reprinted from Ref. [80] by courtesy of Technomic Publishing Co.

maximum service temperature, used as a measure of thermal resistance, was taken as the maximum temperature at which there was no noticeable change in dimensions or appearance over an extended period at that temperature. The maximum service temperature for the acrylic foam was $140^{o}C$ for a foam of 2.06 lb/ft^3 density, which is higher than that of the urethane or polystyrene foam. However, it should be pointed out that certain types of urethane foams (e.g., isocyanurate-modified urethane foams) possess higher service temperatures than those listed in this table.

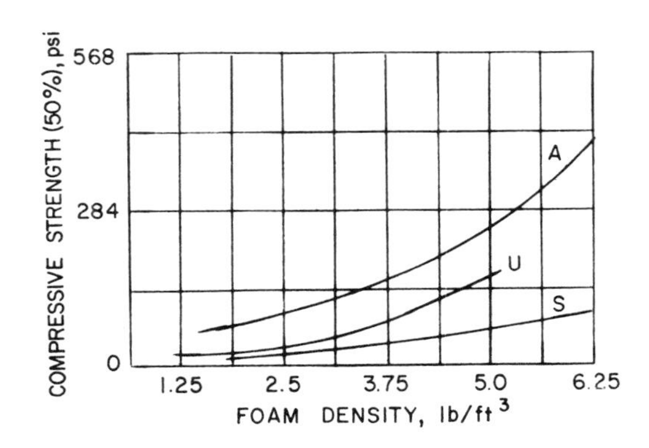

FIG. 10. Compressive strength (50%) of acrylic (A), urethane (U), and polystyrene (S) foams as a function of density at 20°C, 65% relative humidity. Reprinted from Ref. [80] by courtesy of Technomic Publishing Co.

TABLE 11

Thermal Resistance of Some Plastic Foams[a]

Foam	Maximum service temperature (°C)
Acrylic	140
Polystyrene	70
Urethane	120
Polyvinyl chloride	80
Polystyrene	120
Urea	120

[a]Reprinted from Ref. [80] by courtesy of Technomic Publishing Co.

b. Thermal Conductivity. The thermal conductivity of acrylic foam is reported by Kuroiwa [80] to be about 0.141 Btu/(h)(ft^2)(°F/in.). As can be seen from Table 12, the thermal conductivity of acrylic foam, though higher than that of fluorocarbon-blown rigid urethane foam, is lower than that of a carbon dioxide-blown urethane foam or a polystyrene foam.

3. Sound Insulation

Acrylic foam possesses outstanding sound-absorbing qualities. The sound-insulation values for acrylic foam are shown in Table 13 [80]. Acrylic foam may also be used in combination with other materials to form soundproof composite structures.

TABLE 12

Thermal Conductivity of Some Plastic Foams[a]

Foam	Density (lb/ft^3)	Thermal conductivity ($Btu/(h)(ft^2)(^{o}F/in.)$)
Acrylic	4.6	0.14
Rigid urethane[b]	2.1	0.10
Rigid urethane[c]	2.0	0.20
Polystyrene[d]	2.2	0.18
Polyvinyl chloride	6.4	0.21

[a]Adapted from Ref. [80].
[b]Blown with fluorocarbon.
[c]Blown with carbon dioxide.
[d]Beads.

TABLE 13

Sound Absorption of Acrylic Foam[a]

Frequency, Hz	5.3×10^2	6.2×10^2	3.2×10^3	2.0×10^4	6.0×10^4
Transmission loss, dB	23	20	5	15	16

[a]Reprinted from Ref. [80] by courtesy of Technomic Publishing Co.

4. Solvent Resistance

Acrylic foam exhibits good solvent resistance [80]. Aliphatic and aromatic hydrocarbons, ethyl alcohol, chloroform, and carbon disulfide have no effect. Sodium hydroxide, in 50% solution, will induce a color change without causing any other effect, but acetic acid and 10% nitric acid bring about a fair degree of swelling.

C. Applications

Applications for acrylic foam have been developing in such markets as building construction, structural applications, and thermal insulation, where the properties of this foam are best suited (see Table 14).

TABLE 14

Application of Acrylic Foams[a]

Property	Application
High mechanical strength	Panels, partition board
High thermal resistance	Heat insulation, roofing material
Solvent resistance	Core material for chemical storage tank walls
Machinability and workability	Casting patterns; vacuum-forming patterns
Others	Ceiling, wall, and other interior finish

[a]Reprinted from Ref. [80] by courtesy of Technomic Publishing Co.

VIII. FLUOROCARBON FOAMS

Polymers based on fluorocarbon resins are well known for their heat stability and chemical inertness. However, these very properties make it necessary to use special techniques of foaming.

In the case of polytetrafluoroethylene the finely powdered polymer is mixed with a finely powdered foaming agent. The latter must be either water soluble, to be leached out by boiling water, or must decompose at higher temperature to expand the composite.

A process used to produce low-density, spongy polytetrafluoroethylene [82] employs 1,1-difluoro-1,2,2,2-tetrachloroethane as the foaming agent. The polymer and the foaming agent, 20 g of each in the form of fine powders, are mixed and molded to shape at 20°C. The foaming agent is removed by sublimation under vacuum at 20 to 20°C for 4 h. This intermediate gives a porous material with an apparent density of approximately 6 lb/ft^3 when sintered at 370°C for 3 h.

Foaming of the polytetrafluoroethylene can occur by sublimation of temporary fillers like camphor or naphthalene [83] or by generating gas within the polymer [84]. In the latter case a gas-generating agent, such as aluminum carbide, barium peroxide, or aluminum powder, is mixed with the polymer, and the molded piece, while heating, is exposed to a diffusing agent, such as hydrogen chloride, steam, or carbon dioxide. Inclusion of polymethylmethacrylate and naphtha in the extruded polytetrafluoroethylene causes foaming on sintering [85].

A closed-cell foam can be produced by subjecting a mixture of shredded polymer and tetrafluoromethane to a pressure of 7 psi for 5 days to imbibe the halocarbon and then extruding the cellular mass [86].

Due to its high-temperature stability and chemical inertness foamed polytetrafluoroethylene is useful as a catalyst support, filler in distillation and filtration columns, and especially as a support in gas-liquid and liquid-liquid chromatography [87].

Foamed fluoroethylene-propylene (FEP) resin is used for wire insulation. Randa [88] has designed an extruder with a wire crosshead for a smooth, symmetrical resin flow.

A 70:30 copolymer of vinylidene fluoride-perfluoropropene (Viton A) has been foamed by using dinitrosopentamethylenetetramine as the blowing agent [89]. At lower densities (7 lb/ft^3) the Viton A foam is soft and flexible. A solid rocket propellant containing Viton A foam has been described [90]. A typical formulation is shown in Table 15.

TABLE 15

Typical Formulation for Viton A Foam

Component	Parts by weight
Vinylidene fluoride-hexafluoropropylene	100
Powdered fuel	40
Hexamethylenediamine dicarbamate	1.5
Dibasic lead phosphite	10
Zinc oxide	10
Tetrachlorohexafluorobutane	15

The components listed in Table 15 are compounded on a cold mill and cured for 30 min at 155°C in a closed mold. The resulting foam has a density of 28 lb/ft^3 and a tensile strength of 50 psi.

Foamed fluorocarbon resins like FEP are being used to insulate coaxial cable, principally for electronic applications [91]. Other uses are for the production of electronic wire of 0.05-in. diameter [91]. The dielectric constant of the foamed FEP insulation is about 1.6, as compared with 2.0 or 2.2 for the solid fluorocarbon resin.

IX. POLYSULFONE FOAMS

A. Introduction

Polysulfone resins were introduced by the Union Carbide Corp. in 1965 as a stable and self-extinguishing thermoplastic [92]. The polysulfones contain the sulfur atom in its highest oxidation state

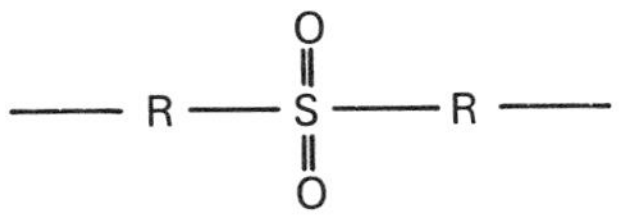

$$—R—\overset{\overset{\displaystyle O}{\|}}{\underset{\underset{\displaystyle O}{\|}}{S}}—R—$$

and hence are resistant to oxidation.

However, relatively little information has been published on polysulfone foams. Perhaps the most interesting type of polysulfone foam is that made from expandable beads, developed by research workers in France, which resemble polystyrene beads with regard to their processing.

B. Preparation of Foams

In a typical formulation 1-butene and sulfur dioxide are copolymerized in such a way that the copolymer contains excess unpolymerized 1-butene, which volatilizes to cause foaming:

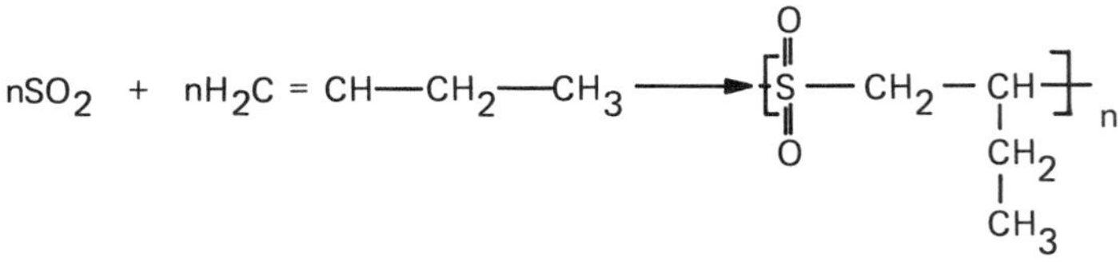

$$nSO_2 + nH_2C=CH—CH_2—CH_3 \longrightarrow \left[\overset{\overset{\displaystyle O}{\|}}{\underset{\underset{\displaystyle O}{\|}}{S}} —CH_2—\underset{\underset{\displaystyle CH_3}{|}}{\underset{\displaystyle CH_2}{\underset{|}{CH}}} \right]_n \quad (11)$$

Thus 400 g of water, 0.2 g of ethyl hydroxyethyl cellulose, 59 g of 1-butene, 128 g of sulfur dioxide, and 0.74 ml of isopropyl peroxydicarbonate solution in ethyl maleate are stirred at 250 rpm for 6 h at 48°C and heated at 75°C for 1 h to give pearls, 0.5 to 2 mm in diameter, which are washed and dried at 25°C [93]. These pearls, which contain 9.5% unpolymerized 1-butene, are foamed at 100°C to give cellular polysulfone of about 3.1 lb/ft^3 density [93].

A cellular polysulfone based on divinyl sulfone with trichlorofluoromethane as the blowing agent has been shown to possess low density and fine structure [94].

C. Applications of Polysulfone Foams

Molded articles from cellular polysulfones, coated with polystyrene, polyesters, epoxy resin, or vinyl resin, are reported to have improved hardness and surface rigidity [95]. Polysulfone foams are also useful as insulating or packaging materials [96] or, when modified with urea-formaldehyde resin, as flowerpots that are permeable to air and water [97].

Due to their excellent self-extinguishing and low smoke-generating properties these foams could be of interest in a number of other rigid-foam applications. The economics of these foam systems and the possible elimination of sulfur dioxide could be major factors affecting the future of these foams.

X. IONOMER FOAMS

The term "ionomer" has been used to denote carboxyl-containing polymers or copolymers of olefins, associated with monovalent or divalent cations. Although these materials are not crosslinked in the conventional sense, since they are thermoplastic and can be processed in standard thermoplastic processing equipment, they do exhibit characteristics resembling those of partially crosslinked plastics.

The most commonly used ionomers consist of copolymers of ethylene and methacrylic acid, which can be represented schematically as follows:

$$—(CH_2—CH_2)_n—(CH_2—\overset{\displaystyle CH_3}{\underset{\displaystyle COOH}{\overset{|}{\underset{|}{C}}}}—)_m$$

The major distinction between ionomers and polyelectrolytes is that the ionic comonomer units in the ionomers are only a minor portion of the entire polymer chain (ranging on the average between 3 and 5 mole %), whereas polyelectrolytes consist mainly of ionic monomer units. The introduction of ionic bonds between the pendant carboxyl-containing polymer chains has been claimed to lead to reduced crystallinity of the semicrystalline polymers (which is evident by a high degree of transparency) while increasing at the same time the modulus, yield point, and oil resistance [98].

Both DuPont and Union Carbide introduced ionomers commercially. DuPont introduced its product under the trade name of Surlyn A [98]; Union Carbide has described extensively the properties of the ionomer HXQD-2137 [99].

Ionomer foams were introduced commercially in 1968 by the Gilman Brothers Co., based on DuPont's ionomer resin [100, 101]. They are closed-cell foams, produced by either extrusion or rotational or injection molding. The relatively low melting point and high melt strength of the ionomer resins add in broadening the processing range. Extrusion is carried out through an annular die, and foaming of the resinous mass can be accomplished either by directly injecting a low-boiling-point solvent, such as trichlorofluoromethane, or by employing chemical blowing agents [102].

Ionomer foam can be readily vacuum-formed without the use of a mold [103]. Application of heat for about 8 sec enables the foam to assume the configuration of the product. The foam can also be easily heat-sealed.

Foamed ionomer systems range in density from 3 to 30 lb/ft^3 and exhibit the typical toughness and solvent-resistance properties that are characteristic of ionomer resins. A summary of the properties of foamed sheets, made from Surlyn A-1800 ionomer resin has been given by Whitfield [101] (see Tables 16 and 17).

The low-temperature properties of these foams are outstanding, but the upper use temperature of about 70°C must be considered a serious limitation in some applications. However, current work is directed toward raising the upper use temperature to 100°C or higher [103].

Ionomer foams find applications in the construction and transportation industries because of the combination of low-temperature flex strength and good thermal and moisture-barrier properties [104].

Ionomer foams, modified with flame-retardant additives, have been used in the aircraft and packaging industries [104]. Due to the adhesive character of the molten ionomer, which acts as a binder and allows for direct thermal lamination of the foamed sheet to various substrates, these foams can be used for the production of a variety of laminates and filled systems. Because of their excellent energy-absorbing characteristics and high tear strength, these foams have found use in athletics for protective padding and in reusable packaging systems.

The outer skin of these foams is smooth and accepts printing and embossing [105].

In a recent application ionomer foam was used as protective insulation for concrete pours in the construction of the Libby Dam near Libby, Montana [106]. At the dam site the foam sheets were wired to the inside of the steel concrete forms. After the pour, the foam remained on the concrete surface for over 2 weeks, allowing the concrete to cure evenly, without cracking by limiting the heat transmittance from the material.

TABLE 16

Typical Properties of Cellular Sheet from Surlyn A-1800 Ionomer Resin[a]

Property	ASTM test method	Typical value: Density (lb/ft^3) 2	10	20
Mechanical:				
Compressive strength (10%) psi	D1621	0.3	19	110
Compressive modulus, psi	D1621	70	1150	4000
Compressive load deflection (25%), psi	D1565	2.5	65	270
Tensile strength, psi	D1623	140	600	1250
Elmendorf tear strength, g/mil	D1922	1.8	24	65
Bashore resilience, % rebound	D1564R	35	37	41
Compression at 30% deflection, %	C1621	39	43	--
Thermal:				
Maximum continuous-service temperature, °C		71	71	71
Low-temperature brittleness, 180-degree bend at -40°C			No failures	
Thermal conductivity, Btu/(h)(ft^2)(°F/in.)	C177	0.3	0.7	--
Dimensional stability, linear change, %:	D2126			
At 60°C, 1 week		0	0	
At 70°C, 24 h		--	1.5	--
At 104°C, 24 h		--	66	--
At 104°C, 24 h, after curing		--	19	--

Table 16 (cont'd.)

Property	ASTM test method	Typical value: Density (lb/ft^3) 2	10	20
Compression set at 25% strain, 24 h, % not recovered:	D1564			
At 70°C		80	70	
At 104°C		99	99	
Flammability:				
Burning rate, in./min	D1692	2	2	
Burning rate (stabilized), in./min	D1692	0.5	0.5	
Chemical and physical:				
Water-vapor transmission:				
perm.-in.	E96	0.1	0.02	--
g/(mil)(24 h)(100 $in.^2$)		40	10	
Oxygen permeability, cm^3/(mil)(24 h)(100 $in.^2$)(atm)		8×10^3	2×10^3	9×10^2
Dissipation factor:				
At 10 kHz		0.001	--	--
At 1 MHz		0.003	--	--
Dielectric constant at 1 MHz		1.5	--	--

[a] Adapted from Ref. [101].

TABLE 17

Solvent Resistance of Ionomer Foam[a,b]

Solvent	Weight change (%)[c]	Appearance[c]
Water	4	No change
Methanol	9	No change
Ethylene glycol	7	No change
Acetone	4	No change
Benzene	7	Clear, limp
Ethyl acetate	3	No change
HC1, dilute	9	No change
NaOH, dilute	20	No change
SAE motor oil	1.5	No change

[a]Data from Ref. [101] .
[b]Foam made from Surlyn A-1800 ionomer resin; density 8 lb/ft^3.
[c]After 30-day immersion in solvent.

XI. TEMPERATURE-ADAPTABLE FABRICS AND TEMPERATURE-REVERSIBLE FOAMS

R. H. Hansen*
Bell Laboratories
Murray Hill, New Jersey

A novel concept recently announced by J. R. Stevens and Co. has made it possible to produce fabrics that adapt to temperature changes [107]. A reversible process that causes materials to expand when cooled and contract when heated results in a fabric that (a) automatically increases in thickness and decreases in its rate of air and moisture transmission in a cold atmosphere and (b) spontaneously decreases in thickness and increases in permeability to air and moisture when placed in a warmer invironment.

It would seem that this behavior would only be possible by cheating on the laws of nature. Actually it is based on a very simple concept: a gas that is soluble in a liquid solvent is usually expelled when the solvent is frozen. If the solution of gas in solvent is confined within an inflatable component, solidification of the solvent causes expulsion of the dissolved gas and inflation of the inflatable component. The volume of the expelled gas can be varied from a fraction of the volume of the original solution to an amount orders of magnitude greater than the volume of the original solution by proper choice of gas, solvent, and the pressure at which the solution

*Work done at J. P. Stevens and Co. Technical Center, Garfield, N. J.

is prepared. The process is reversible because the expelled gas redissolves when the solvent melts.

A temperature-adaptable fabric made of hollow fibers with sealed ends is illustrated in Fig. 11 (uninflated) and Fig. 12 (inflated). In this example all of the hollow fibers are filled with the same solution of a gas in a liquid, but it is obvious that different solvents could be employed in the same fabric (e.g., one in the warp and the other in the fill) to obtain a stepwise increase of thermal insulation properties as environmental temperature decreases.

The insulating efficiencies of (a) a simple hollow-fiber fabric, (b) the same hollow-fiber fabric filled with solvent only, and (c) with the same solvent containing an expellable gas are shown in Fig. 13. In these experiments the fabrics were placed between an aluminum plate embedded in crushed Dry Ice and an insulated copper block containing a thermocouple. A control experiment in which the fabric was omitted was also performed. It is seen from Fig. 13 that when there was no interposed fabric, the temperature of the copper block in direct contact with the chilled aluminum plate reached -10°C in less than 5 min. There was little difference in the insulating effectiveness of an empty (air-filled) hollow-fiber fabric and the same fabric filled with acetophenone (mp 19°C), although placing these fabrics between the aluminum and copper blocks did increase the time required for the temperature of the copper block to decrease from room temperature to -10°C to about 25 min. However, when the same fabric was filled with a solution of monochlorotrifluoromethane in acetophenone (prepared by saturating liquid acetophenone with the gas at room temperature and at 1-atm pressure), inflation occurred as the solvent froze, and a marked change in insulating efficiency was observed. In this case the temperature of the copper block did not reach -10°C for over 2 h. It is also seen by comparing Figs. 11 and 12 that the air and moisture permeabilities of the inflated fabric are substantially lower than those of the uninflated fabric because interstitial space between fibers is greatly reduced.

FIG. 11. Uninflated temperature-adaptable fabric made of hollow fibers with sealed ends.

FIG. 12. Inflated temperature-adaptable fabric made of hollow fibers with sealed ends.

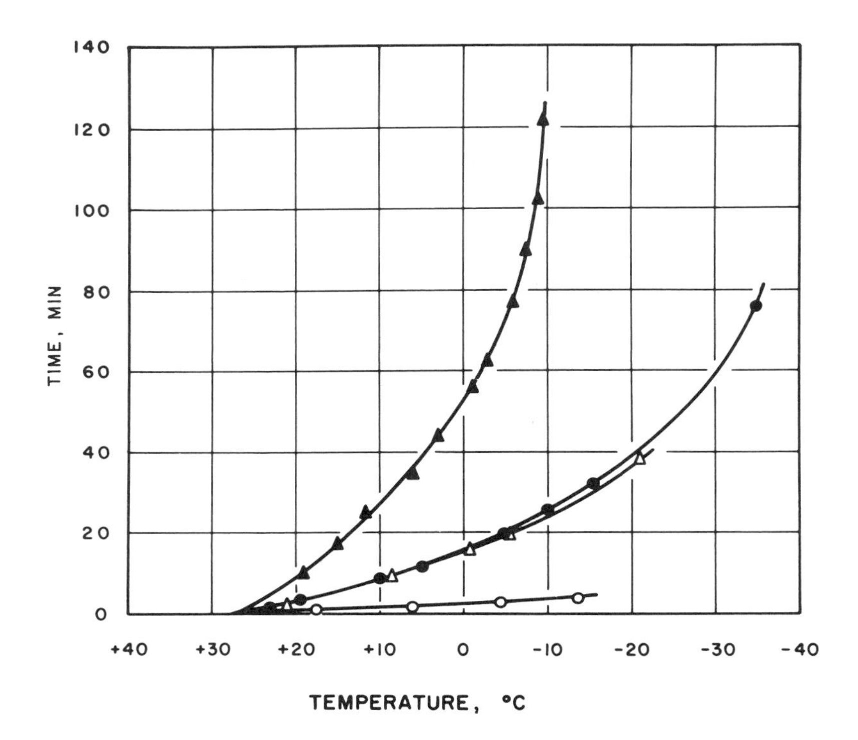

FIG. 13. Insulating efficiencies of hollow-fiber fabrics: △, empty fibers; ●, fibers filled with acetophenone; ▲, fibers filled with acetophenone and monochlorotrifluoromethane; ○, copper block in direct contact with aluminum plate.

A variety of structures can be used to prepare temperature-adaptable fabrics. Instead of hollow fibers, capsules, sealed envelopes, materials that have a curved or zigzag shape uninflated and assume a straighter configuration when gas is expelled, and foamed fibers (which differ from hollow fibers only in that they have a multitude of small inflatable elements rather than a single one) can be used. The fabric itself can be knit, woven, or nonwoven.

Virtually any gas-solvent system can be used, provided that they are compatible with, and are substantially retained by, the inflatable element. The only other criteria are that the gas have the desired solubility in the liquid solvent and that the system convert on freezing from a homogeneous liquid phase comprised of a solution of at least one gas in at least one solvent to two phases, gaseous and solid. Almost all solutions of gases exhibit this behavior. Two notable exceptions are concentrated aqueous hydrochloric acid and concentrated aqueous ammonium hydroxide. Both freeze, but neither evolves a gas. Aqueous hydrochloric acid forms solid crystalline hydrates of hydrogen chloride, and aqueous ammonium hydroxide forms a polymer that can be drawn into a fiber. Water is useful in other systems (e.g., CO_2 dissolved in water), and ammonia can be expelled from acetophenone on freezing. Presumably there are solvents that will expel hydrogen chloride on freezing.

There are wide limits on the temperatures at which fabrics can be caused to expand because many materials that are not ordinarily considered solvents melt at temperatures even above 200°C (at which point they can be used as solvents), and most polymers used for fabrics would not be employed at such high temperatures. Lower limits for gas expulsion would be temperatures below that at which the gas liquefies. A two-phase system with little or no change in volume would probably result if expulsion occurred below the boiling point of the gas.

One might consider a sheet of plastic as a fabric (e.g., plastic tarpaulins). If the sheet contained a multitude of capsules containing a gas-solvent system, the sheet would be converted into a closed-cell foam at temperatures below the freezing point of the solvent and would revert to an essentially noncellular material at temperatures above the melting point of the solvent. The solvent could be selected so that freezing (and foam formation) would occur at the desired temperature, just as in the case of a fabric.

Closed-cell reversible foams will probably have more varied applications than temperature-adaptable fabrics, but both materials excite one's imagination.

REFERENCES

1. F. Stastny and H. Gerlich (to BASF), German Pat. 934,692 (1955).
2. F. Stastny and K. Buchholz (to BASF), German Pat. 951,299 (1956).
3. F. Stastny (to BASF), German Pat. 956,808 (1957).
4. H. E. Kobloch, F. Meyer, and F. Stastny (to BASF), British Pat. 933,621 (1963).
5. E. Dumont and H. Reinhardt, German Pat. 1,001,488 (1957).
6. L. P. Ellinger, J. Appl. Polymer Sci, 10, 551 (1966).
7. L. P. Ellinger, J. Appl. Polymer Sci., 10, 576 (1966).
8. H. Fikentscher and R. Fricker (to BASF), German Pat. 931,731 (1955).
9. L. P. Ellinger, J. Appl. Polymer Sci., 9, 3939 (1965).
10. L. P. Ellinger (to British Oxygen Co.), British Pat. Appl. 25, 171 (1964).
11. Badische Anilin- und Soda-Fabrik, British Pat. 739,438 (1955).
12. H. Davidge, J. Appl. Chem., 9, 553 (1959).
13. W. Reppe, Neue Entwicklungen auf dem Gebiet der Chemie des Azetylens und Kohlenoxyds, Springer, Berlin, 1949, p. 18.
14. J. W. Copenhaver and H. M. Bigelow, Acetylene and Carbon Monoxide Chemistry, Reinhold, New York, 1949, p. 63.
15. J. M. DeBell, W. C. Goggin, and W. E. Gloor, German Plastics Practice, DeBell and Richardson, Springfield, Mass., 1946.
16. E. H. Cornish, Plastics, 28 (3), 61 (1963).
17. C. J. Benning, Plastic Foams, Vol. I, Chemistry and Physics of Foam Formation, Wiley-Interscience, New York 1969.
18. I. G. Morrison, J. Cell. Plastics, 3, 364 (1967).
19. N. B. Graham and G. D. Murdock (to Canadian Industries), U. S. Pat. 3,311,573 (1967).
20. W. D. Bowering and N. B. Graham (to Canadian Industries), U. S. Pat. 3,311,574 (1967).
21. N. B. Graham (to Canadian Industries), U.S. Pat. 3,311,575 (1967); U.S. Pat. 3,318,824 (1967).
22. Anon., Plastics Technol., 13, 15 (1967).
23. G. E. Bernier, T. Gilchrist, and R. H. Pollen (to Canadian Industries), U.S. Pat. 3,386,928 (1968).
24. G. J. Trudel (to Canadian Industries), U.S. Pat. 3,433,809 (1969).
25. T. Gilchrist and M. Ternbah (to Canadian Industries), U.S. Pat. 3,364,154 (1968).
26. E. G. Parker and L. D. Rood (to Pittsburgh Plate Glass), British Pat. 1,053,234 (1966).
27. Chemische Werke Huels AG, French Pat. 1,440,512 (1966).
28. H. Sasse (to H. Sasse and Firma Wilhelm Petry), German Pat. 1,247,648 (1967).

29. L. D. Rood (to Pittsburgh Plate Glass), U.S. Pat. 3,362,919 (1968).
30. H. Suter and H. V. Ammon (to BASF), Belgium Pat 667,076 (1966).
31. A. Salgado and I. Berlinger (to Reichhold Chemicals), U.S. Pat. 3,232,893 (1966).
32. Anon., Chem. Eng., 64 (12), 174 (1957).
33. Farbwerke Hoechst AG, Netherlands Pat. 6,505,968 (1965).
34. M. Wismer and G. F. Bosso, Natl. Acad. Sci.-Natl. Res. Council Publ. No. 1462, 1967, p. 153.
35. Y. du Tertre, French addition 81,527 (1963) to French Pat. 1,195,734.
36. Mitsubishi Rayon Co. Ltd., Japanese Pat. Appl. 629 (1963).
37. A. Cooper and L. B. McQueen (to Expanded Rubber Co.), U.S. Pat. 2,888,407 (1959).
38. K. Martin and K. Schumann (to BASF), German Pat. 1,230,561 (1966).
39. Chemische Werke Huels AG, British Pat. 1,044,611 (1966).
40. R. J. Fabian, Mater. Eng., 70 (1), 18 (1969).
41. R. H. Leitheiser and R. Coderre, Chem. Eng. News, 46 (2), 34 (1968).
42. Patronato de Investigacion Cientifica y Technica "Juan de la Cierra," Spanish Pat. 294,999 (1964).
43. J. F. Yanes, French Pat. 1,468,237 (1967).
44. G. Will, German Pat. 1,150,524 (1963).
45. R. H. Leitheiser, R. J. Hellmer, and E. T. Clocker, in. Proc. 23rd Conf. SPI Reinforced Plastics/Composites Div., Washington D.C., 1968.
46. R. A. Hoffer, Mats. and Methods, 25, 58 (1947)
47. R. E. Jauer (to DuPont), U.S. Pat. 2,312,463 (1950).
48. Strux Corp. Tech. Bull. Cellular Cellulose Acetate.
49. Modern Plastics Encyclopedia, McGraw-Hill, New York, 1970.
50. Revertex Ltd., British Pat. 573,966 (1945).
51. C. E. Benning, Plastic Foams, Vol. I, Wiley-Interscience, New York, 1969, p. 493.
52. C. L. Wilson, U.S. Pat. 2,609,347 (1952).
53. G. Schultz (to Farbwerke Hoechst), German Pat. 1,141,147 (1962).
54. A. D. Rogers and M. A. Stevens (to DuPont) U.S. Pat. 2,825,747 (1958).
55. R. L. Frank (to Ringwood Chemical), U.S. Pat. 2,708,661 (1955).
56. R. C. Wade (to Metal Hydrides), U.S. Pat. 2,930,770 (1960).
57. B. D. Halpern and B. O. Krueger (to Borden Co.), U.S. Pat. 3,052,652 (1962).
58. Kalle AG, British Pat. 973,951 (1964).
59. H. G. Hammon, U.S. Pat. 2,668,153 (1954).
60. C. L. Wilson, U.S. Pat. 2,636,013 (1953).
61. Kalle AG, British Pat. 942,756 (1963).

62. E. E. Tarakanova (to Vladimir Scientific Research Institute of Synthetic Resins), USSR Pat. 195,095 (1967).
63. T. Ishizuka, Japanese Pat. 14,580 (1966).
64. Kurashiki Rayon Co. Ltd., French Pat. 1,490,844 (1967).
65. H. Brueggemann et al. (to BASF), German Pat. 1,159,643 (1963).
66. F. Beyerlein and H. Wilhelm (to BASF), German Pat. 1,177,340 (1964).
67. C. F. Fischer (to DuPont), U.S. Pat. 3,232,892 (1966).
68. Farbenfabriken Bayer AG, Netherlands Pat. Appl. 6,603,887. (1966).
69. T. G. Hyde (to DuPont), U.S. Pat. 3,207,713 (1965).
70. M. F. Fuller (to DuPont), Belgian Pat. 628,004 (1963).
71. D. J. Cram and J. G. Hawkins (to Whiffen and Sons), British Pat. 949,581 (1964).
72. F. Becke and K. Wick (to BASF), U.S. Pat. 3,060,135 (1962).
73. M. F. Fuller (to DuPont), French Pat. 1,477,567 (1967).
74. E. I. DuPont de Nemours & Co., Netherlands Pat. Appl. 6,512,554 (1966).
75. Badische Anilin- und Soda-Fabrik AG, Netherlands Pat. Appl. 6,513,025 (1966).
76. Polymer Corp., British Pat. 1,058,307 (1967).
77. K. Dachs et al. (to BASF), Belgian Pat. 669,248 (1966).
78. Allhag AG, German Pat. 1,245,522 (1967).
79. ECN Foamed Nylon, Product Development Bull., E. I. duPont de Nemours & Co.
80. K. Kuroiwa, J. Cell. Plastics, 3, 38 (1967).
81. Anon., Chem. Eng. News, 43 (47), 44 (1965).
82. Y. Kometani and S. Koizumi (to Daikin Kogyo Co), Japanese Pat. 4974 (1967).
83. R. T. Fields (to DuPont), U.S. Pat. 3,058,166 (1962).
84. C. L. Lightfood (to Hercules Powder), U.S. Pat. 3,137,745 (1964).
85. R. J. Moore and W. G. Atwell (to Raybestos-Manhattan), U.S. Pat. 3,054,761 (1962).
86. S. K. Randa (to DuPont), U.S. Pat 3,072,583 (1963).
87. Société d'Electro-Chimie, d'Eléctro Metallurgie et des Acieries Eléctriques d'Ugine, French Pat. 1,357,643 (1964).
88. S. K. Randa, Plast. Technol., 10, 42 (1964).
89. A. Cooper, Plastic Inst. Trans., 29, 39 (April 1961).
90. L. Spenadel, J. Homer, and I. Kirschenbaum (to Esso Research and Engineering), U.S. Pat. 3,398,215 (1968).
91. R. Bender, Handbook of Cellular Plastics, Lake Publishing, Libertyville, Ill., 1965, p. 331.
92. Anon., Mod. Plastics, 42 (9), 87 (1965).
93. Produits Chemiques Pechiney-Saint-Gobain, Netherlands Pat. Appl. 6,603,273 (1966).
94. T. G. Finnerty and V. Kerrigan (to ICI), British Pat. 995,935 (1965).

95. J. P. Tanneur (to Produits Chimiques Pechiney-Saint-Gobain), French Pat. 1,484,679 (1967).
96. Produits Chimiques Pechiney-Saint-Gobain, French Pat. 1,463,182 (1966).
97. Produits Chimiques Pechiney-Saint-Gobain, French Pat. 1,521,403 (1968).
98. R. W. Rees, Mod. Plastics, 42 (1), 98 (1969).
99. S. Bonotto and C. L. Purcell, Mod. Plastics, 42 (7), 135 (1965).
100. Anon., Chem. Week, 102 (11), 95 (1968).
101. R. L. Whitfield, in E. I. DuPont de Nemours & Co. Tech. Bull. Surlyn A-1800.
102. C. E. Benning, Plastic Foams, Vol. 1, Wiley-Interscience, New York, 1969, p. 611.
103. Modern Plastics Encyclopedia, 1968-1969, p. 387.
104. R. H. Kinsey, in Modern Plastics Encyclopedia, 1970-1971, p. 238.
105. Anon., Plastics Technol., 14 (7), 72 (1968).
106. Anon., Mod. Plastics, 47 (4), 68 (1970).
107. R. H. Hansen (to J. P. Stevens & Co.), U.S. Pat. 3,607,591 (1971).

Chapter 16

INORGANIC FOAMS*

Marco Wismer

PPG Industries, Inc.
Springdale, Pennsylvania

*This book deals mainly with organic foams, but a thorough coverage must include several types of inorganic foams, principally foams from glass, metals, refractories, concrete, and sulfur. Though little used as yet their unusual properties, and often low material costs, indicate significant growth possibilities.

I. CELLULAR GLASS

Cellular glass is a thermal insulation material composed of hermetically sealed glass cells. It is light, rigid, incombustible, moistureproof and waterproof, and is extensively used in the chemical and construction industries.

A. Methods of Manufacture

There are three general methods for the preparation of cellular glass. Since the fluid viscosity of molten glass is on the order of 10^5 and 10^7 P at 1050 1500°C, respectively [1], one method consists simply of whipping gas into the molten mass. Lytle [2] has described a process that forces molten glass under pressure through a passage with a restricted throat. Gases are introduced as bubbles into the glass adjacent to the throat, before the molten glass enters the reduced-pressure zone where the gases expand and form a cellulated glass.

Miller [3] forces a fluid into the molten glass through the pores of a permeable mold. Long [4] first described the preparation of cellular glass from pulverized glass mixed with coal and sodium sulfate.

The most frequently used method starts with ground glass, powdered carbon, and antimony oxide or sodium sulfate [5, 6]. The glass is usually a typical lime-soda glass, 70.7% silicon dioxide, 7.5% calcium and magnesium oxide combined, 12.4% sodium oxide, 3.8% boron trioxide, and 5.1% aluminum oxide and iron oxide. The glass is finely pulverized so that 90% will pass a 350-mesh screen. This mixture is introduced into a furnace with an entrance temperature of 760 to 788°C. In about 40 min the glass is completely sintered without much cellulation occurring. The melt containing the sintered mixture is then further heated to about 871°C in 30 min. During this period the carbon is oxidized by the sulfates and the metal oxides in the glass, and gases (sulfur dioxide, carbon dioxide, and, in some cases, hydrogen sulfide) are entrapped as bubbles and celluation occurs. A critical cooling step is conducted in the temperature range 649 to 704°C and occurs in 10 to 15 min. During this period generation of gases in the hot zone deep in the block approximately compensates for shrinkage in the surface due to the cooling of gases. During the end of the cooling period a semirigid shell forms around the glass block. Further cooling occurs to about 426°C, when a hard external shell is formed with a still soft interior. Another critical range of cooling is between 510 and 454°C, where annealing is conducted very slowly.

Most of the currently produced lime-soda-glass foam is dark, since powdered coal or carbon black in amounts of up to 1%, depending on the type of carbon used, is initially present in the process. Reduction of the amount of carbon to a level where all of it would be oxidized to a gas is not practical since cellulation would be seriously reduced. At the minimum carbon content necessary for satisfactory cellulation some carbon remains in the cellulated product. This minimum amount of carbon varies with the type of glass.

It is believed that each carbon particle is a nucleation site for gas formation. D'Eustachio [7] therefore, utilized, finely dispersed carbon and was able to reduce the carbon content without impairing cellulation. The finely divided carbon is prepared by adsorbing sugar solutions on diatomaceous earth or silica followed by heating to 300°C with a limited supply of air. The resulting char deposited on silica is ball-milled with a typical lime-soda glass containing 1.0% antimony oxide.

Weyl [8] has described the reaction of sodium sulfate with carbon:

$$Na_2SO_4 + 2C \longrightarrow Na_2S + 2CO_2 \quad (1)$$

Hyde and Stookey [9] produced white glass foams by replacing carbon with amino compounds such as urea.

The odor of such entrapped gases as hydrogen sulfide may be determined in some applications. Shoemaker [10] described the use of a sulfate-free glass for the preparation of a glass foam free of sulfur dioxide and hydrogen sulfide. Ferric oxide is used as the source of oxygen, and finely pulverized carbon is employed as the reducing agent. Carbon dioxide is formed by the reduction of ferric oxide by carbon. The ferric oxide reacts with the carbon to eliminate most of the free carbon, thus producing a lighter colored foam.

Ford [11] discovered that finely pulverized silica or complexes of alumina and silica, when mixed with glass powder, improved the moisture resistance of the resulting glass foam at elevated temperatures.

Udy [12] prepared silica foams by using carbon as the gas-producing agent in a two-stage process. This cellulation probably occurs according to the following equations:

$$C + SiO_2 \longrightarrow SiO + CO \quad (2)$$

$$SiC + 2SiO_2 \longrightarrow 3SiO + CO \quad (3)$$

In order to accelerate this foaming action, Ford [11] coated silica with silicon carbide.

High-purity silica (99.7%) may be mixed with 0.5% aluminum and foamed [13]. The cellulation reaction with aluminum may occur as follows:

$$3SiO_2 + 4Al \longrightarrow 3Si + 2Al_2O_3 \quad (4)$$

$$Si + SiO_2 \longrightarrow 2SiO \quad (5)$$

Silicon monoxide vaporizes at the fusion temperature of silica and acts as a foaming agent.

Several methods and apparatuses for the manufacture of cellular glass are suggested in the patent literature [14].

Schultz [15] determined by gas-chromatography measurements that the hydrogen sulfide content is as high as 11% by volume in the glass-foam cell. Since water is involved in the formation of hydrogen sulfide, Schultz accelerates the foaming process by using a high enough water-vapor pressure to prevent the loss of water from the glass surface.

A recently described new foamed-glass process [16] uses sodium silicate solution, rock wool, small amounts of chemicals to modify the properties, and foaming agents, such as sugar and polyols. Boiling in water yields a homogeneous slurry that is spray-dried to a powdery product. The residual water is eliminated at 299°C, and foaming occurs between 349 and 749°C. Cellulation is achieved at 610°C, whereupon the foam bun is cut to lengths and conveyed into a cooling oven. The cooling period is 4 h. Granules can be easily obtained by the same process.

The manufacture of cellular glass bodies with relatively small spherical shapes is described by Ford [17]. The spheres have closed cells and therefore are substantially impervious to moisture and vapors. The glass used may be a conventional lime-soda glass or borosilicate glass containing 5 to 15% boric acid. Such oxidizing agents as sulfates or antimony trioxide may be added to the pulverized glass-carbon mixture. No molds are required for the manufacture of nodules. There are several methods by which these glass nodules can be produced. In one, pettles or nodules may be formed in a conventional pettle or briquetting machine. Though many mixtures of glass and carbon black have a pronounced tendency to cohere, suitable binders, such as sucrose or dextrose in water or urea in water, may be added. After forming, the nodules must be heated for sintering and cellulating. For the purpose of sintering the nodules are gradually heated to a temperature of 760 to 788°C and are maintained at this temperature until the entire mass is vitreous. If heating is done in a partially oxidizing atmosphere, the carbon is burned out on the outer skin of the nodules, leaving only glass. The nodules are then further heated to 910°C for cellulation. The soft glass outer skin of the nodules will expand from the pressure of the gas within but does not cellulate since the carbon has burned out. After cellulation is completed the particles are cooled to room temperature and are ready for use without annealing or further processing.

Slayer and Soltis [18] use aluminum flake in combination with glass powder, a binder, barium sulfate, and calcium sulfate to obtain fine-celled, uniform foam pellets.

D'Eustachio [19] has described a process for the preparation of glass nodules that have lower density and a more chemical resistant outer skin. This is achieved by mixing particulate glass, a cellulating agent like carbon black, an oxidizing agent like sodium sulfate, and a liquid glass-forming binder like sodium silicate in water. The mixture is pelletized and dried so that at least part of the liquid binder migrates toward the surface of the pellet. The pellets are then coated with a parting agent, such as aluminum hydrate. The coated parts are heated in a rotary kiln to a temperature at which a portion of the parting agent and the glass form a skin on the pellet, which differs in chemical composition from the cellulated interior of the nodule. Cellulation of these nodules occurs at about 926°C.

Shannon [20] produced foam-glass nodules that are fused together and thus form a unitary structure with interconnecting voids throughout.

Burwell [21] has described the preparation of cellular volcanic ash. The ash is heated to 1100°C, which sinters the material to a reddish vitreous product. A rapid increase of the temperature to 1350°C bloats the volcanic ash. At these temperatures the vitreous mass is plastic and gas develops, but the viscosity is still high enough to prevent its escape. A cellular structure is formed and persists on cooling.

B. Properties

The typical physical properties of a commercial foamed glass are listed in Table 1 [22].

TABLE 1

Physical Properties of Cellular Glass[a]

Test	Result
Acid resistance	Impervious to common acids and acid fumes
Capillarity	0
Closed-cell content, %	99-100
Coefficient of expansion per °F	0.0000046

TABLE 1 (Cont.)

Physical Properties of Cellular Glass[a]

Test	Result
Combustibility	Incombustible
Composition	True glass-completely inorganic
Compressive strength (average, ultimate), psi	100
Density, lb/ft^3	9
Flexural strength, psi	75
Hygroscopicity	No increase in weight after 246 days in air at 90% R.H.
Modulus of elasticity, psi	180,000
Moisture absorption[b], vol %	0.2
Permeability, Water vapor	0.00 perm-in
Shear strength, psi	40
Specific heat, Btu/lb	0.20
Thermal conductivity, Btu/(h)(ft^2)(^oF/in.):	
25^oF	0.365
50^oF	0.38
75^oF	0.40
Thermal diffusivity, ft^2/day	0.42

[a]Data from Ref. [22].

[b]Tested according to ASTM Designation C260-61.

The pores in glass foam vary between 0.5 and 3.5 mm [23]. In general the wall thickness between the pores is relatively greater for the smaller pores than for the larger ones, and consequently very high porosities are more easily obtained in glasses containing large pores.

For a glass with a true density of 2.5 lb/ft^3 the percentage porosity P corresponding to an apparent density 0.5 is simply obtained as follows:

$$P = \frac{d_v - d_a}{d_v} \times 100 = \frac{2.5 - 0.5}{2.5} \times 100 = 80\%, \quad (6)$$

where d_v is the true density and d_a is the apparent density.

C. Applications

Foam glass is supplied in flat blocks, curved segments, as factory shaped fittings, and, for pipe insulation, in the form of hollow rods. It is applied in single layers or in multiple-layer form, using hot asphalt or cold-setting adhesives to provide bonding to the substrate [24].

Applications for cellular glass are numerous due to its good insulating properties and its high heat resistance. The most popular application is as insulation on monolithic concrete roof decks, precast concrete decks, steel decks, and wood roof decks. It provides good insulation as ceiling and wall lining. Long-life insulation performance on underground heating lines and piping has been provided by cellular glass. Ammonia tanks insulated with cellular glass have been in use for many years.

Cellular glass with open cells can be made as described by D'Eustachio [25] and Haux [26].

II. METAL FOAMS

A. Methods of Manufacture

One of the earliest references to metal foams describes the production of a spongelike metal foam by melting together two or more metals of different melting points. The melting occurs in a pressurized vessel, and the molten metal is subsequently released into a space of lower pressure. The lower melting metal vaporizes in the lower pressure chamber, thus producing a metal froth in which the voids are filled with the vaporized lower melting metal. On cooling the base metal solidifies into a foamed structure [27, 28]. The metal foam most frequently mentioned in the literature is aluminum foam. Since it has a low melting point, aluminum can be handled with relative ease in the liquid state. After introduction of the gas-producing materials the foaming reaction can proceed at atmospheric pressure. Hence most of the work has been concentrated on aluminum and aluminum alloys. This process has been described by Bjorksten [29] and in several patents [30].

A foaming agent, such as a metal hydride, is added to powdered aluminum or to a pool of molten aluminum with subsequent rapid cooling. The foaming agent has to possess chemical and thermal stability in order to allow complete dispersion in molten metal without undergoing much premature decomposition. The foamable composition is heated to a temperature capable of melting the metal to be foamed and simultaneously causing the decomposition of the foaming agent. Foams can be produced from pure aluminum, but for many operations aluminum-magnesium alloys in the 7 to 10% magnesium range are preferred. The favored foaming agents are the hydrides of zirconium or titanium. These are used in amounts of 0.4 to 0.5% by weight. The density of the foam is regulated by the amount of foaming agent used. The decomposition has to be complete. Fiedler [31] enhances the blowing action of the metal hydride by the injection of air, oxygen, or steam through a fusible tube. The foam can be poured from the mixing vessel into molds, where it is heated by a gas burner as it solidifies. The advantage of foaming an aluminum-magnesium alloy with oxygen is that the oxide formed on the inside of the bubbles keeps them from collapsing. Foam can be produced continuously by keeping the mixing vessels filled from a ladle and overflowing the mixture onto a moving belt where it can be kept hot until it has foamed sufficiently and then cooled by water sprays until it solidifies. Bjorksten [29] states that ceramic molds or other molds with low conductivity enhance the foaming action of molten metals containing foaming agents. Foaming action can be very vigorous, and 2-in. cross-sectional molds have been used to produce aluminum foams up to 8 ft high. The pressures developed in the molds are considerable.

An apparatus for the continuous production of aluminum, magnesium, zinc, lead, or copper foam consists of a covered melting chamber into which a foaming agent such as a hydride is separately fed. The mixture is stirred continuously, and the produced foam expands through an outlet from the chamber into the conveyor for cooling and removal [32].

A metal alloy containing aluminum, magnesium, and a blowing agent (e.g., cadmium and magnesium carbonate) has been extrusion-pressed at elevated temperatures. The carbonate blowing agents develop carbon dioxide at elevated temperatures but below the melting point of the alloy. The extruded material is then heated and foamed [33].

A method has been described for casting foam on the surface of a liquid supporting medium that is nonreactive with the metal. Foamed metal panels produced by this method have a smooth, even gloss on the surface that contacts the liquid supporting medium during casting. The liquid supporting medium can be molten lead [34].

A silver foam has been prepared from coal-saturated liquid silver. After the coal was burned out the pore volume of the silver was 92% [35].

Nickel foams have been prepared by polymerizing at an elevated temperature a mixture of nickel powder, a silicone resin, and a heterocyclic nitrogen foaming agent. Foaming occurs at about 204°C, and the foam is reduced in a hydrogen atmosphere. The remaining organic material is pyrolyzed and removed at about 649°C, whereupon the powdered form is fused at 1315°C. All these operations are conducted in a hydrogen atmosphere [36].

A magnesium foam has been produced by mixing a magensium-alloy powder containing 6% zinc and 0.6% zirconium with basic magnesium carbonate [37]. Porous metallic bodies of copper, silver, gold, or platinum are formed by utilizing an alloy with low-melting zinc or cadium. Subsequent evaporation of the low-melting zinc or cadmium by heating under a high vacuum produces the cellular structure [38].

Many porous metals can be produced by pressing and sintering metal powders. Zinc foams have been prepared by pressing zinc powder with ammonium chloride filler particles, which were then sublimed to form a foamed structure that was subsequently sintered [39]. Zirconium hydride and powdered wolfram were incorporated into cast iron. Gas evolved from the hydride alone does not produce a foam unless a suitable promoter, such as powdered wolfram, is present [40].

Most powder-based porous metals have pores of irregular shapes, which impairs uniformity of permeability [41, 42].

B. Properties

The workability of aluminum foam is good, and it can be foamed into 0.5-in. sections. The castings can be sawed and cut. Strength properties are described in Table 2 [29].

The k factor for foamed aluminum at 70°F as a function of density is given in Table 3.

As the material is compressed, it gains strength. A foam with a density of 17 lb/ft^3 and a compressive strength of 3000 psi will increase in strength to 6000 psi when compressed to 35 lb/ft^3. According to Bjorksten, compression of the foam does not destroy the cell structure but merely changes the cell shapes. Closed- or open-cell foam can be produced. It was found that ultrafine pores of submicro size, when closely spaced, behave like particles. They prevent grain growth and creep of materials at elevated temperatures, and it is speculated that it may be possible to produce porous metals whose strength will exceed that of dense metals [43].

TABLE 2

Strength of a Typical
Aluminum-Magnesium Alloy Containing 7% Magnesium[a]

Density (lb/ft^3)	Transverse rupture[b]
15	110
20	180
25	240
30	280

[a]Data from Ref. [29].

[b]Specimen 2 in. wide, 1 in. thick, 12-in span.

TABLE 3

Thermal Conductivity (k Factor) of Foamed Aluminum at 70°F

Foam density (lb/ft^3)	k Factor (Btu/(h)(ft^2)($^{\circ}$F/in.))
11.24	3.1
18.18	3.65
27.17	4.7

C. Applications

Many applications have been mentioned for foamed metals. Nonsinkable aluminum foams for boats and life preservers have been described by Elliott [44]. Bjorksten [29] has proposed the construction of fire-resistant wall panels. Built-up roofing, oil-storage-tank covers, and engine floats for aircraft are other applications listed [45]. Porous copper and nickel have been suggested for electrodes for secondary elements with alkali electrolytes [46]. Porous iron particles from sponge iron are recommended as oil-lubricated bearings [47]. These form the bulk of the open-cell porous-metal bodies used. The porosity of bearings is in the range 20 to 30%, whereas porous filters with porosities as high as 70% are used for the filtration of wanted or unwanted particles from fluids, for separating liquids, and for air cleaning. Porous beryllium is used extensively in the production of nuclear energy due to its low density (1.85 g/cm^3), high strength-to-weight ratio, and high heat capacity [41].

III. CELLULAR REFRACTORIES

The four methods for the production of low-density cellular refractories are (a) organic binder burnout, (b) sintering, (c) oxidation of cellular surface, and (d) blowing. Commercially the most prominent method is sintering.

A. Manufacture

1. Burnout Method

One approach is preparing high-purity alumina and zirconia cellular products centers on the use of additives which are mixed with slurries of the refractory materials. These additives may involve granulated cork, sawdust, coke, and tar [48].

An earlier process for manufacturing cellular refractory bodies describes the use of relatively large quantities of shredded or small fragments of waste paper intimately mixed dry with finely divided refractory materials, made into bricks, and fired [49].

By a novel method of producing low-density inorganic refractories, alumina and silica refractories with densities of less than 30 lb/ft^3 have been obtained by foaming the metal oxides in unsaturated polyester resins and firing the foamed objects up to 1900°C. The foamed objects undergo considerable shrinkage at such temperatures, but this method produces low-density refractories without cracks and without substantial change in the original foamed shapes. The fact that unsaturated aromatic- polyester foams perform better than other foamable systems has been explained by the kinetics of the thermal degradation of the unsaturated polyesters.

Differential thermal and thermogravimetric analyses demonstrate the rapid formation of solid high-melting-point residues in the thermal degradation of unsaturated polyesters. These residues prevent distortion and excessive flow of the foam structure during the burning process [50].

Porous ceramics of low thermal conductivity have been prepared by admixing refractory oxides with pure sucrose. The composites are ground together, molded, and heated in several steps [51].

Ispen, Inc. has prepared zirconia foams by impregnating a preformed organic sponge with a slurry of finely divided zirconia grains. The ceramic bond is obtained after burning out the organic structure and sintering the oxide. Since the structure of the original sponge can be closely controlled, the sintered product is capable of being made to a specified porosity and pore size. All pores are interconnecting [52].

Foamed, refractory alumina objects are obtained by using resole resins as the foamable binder, burning out the binder, and sintering the refractory materials [53].

Discrete particles made by firing approximately 1 volume of aluminum coated with about 1% sodium hydroxide and approximately 10 volumes of alumina (-100 mesh) are recovered on a 100-mesh screen. The particles contain about 29% aluminum and 71% alumina which represents about 56% conversion of the aluminum. The particles are mixed with 33% aqueous gum arabic as a temporary binder, molded into shape under 5000-psi pressure, dried, and fired [54].

Porous refractories are prepared with a homogeneous mixture of high-temperature-sinterable aggregate and a binder like tar. The mix is press-molded into shapes which are then heated to a temperature sufficient to produce substantial distillation of the tar. The composite is cooled and then further fired to sintering temperature in the presence of an oxidizing gas to burn the distillation residue of the carbon-containing constituent, thereby creating a multiplicity of point heat sources to ensure complete sintering of all sinterable particles [55].

2. Sintering Method

Refractories can be made from alumina bubbles. Fused alumina is poured from an electric furnace so that the molten stream flows past an air jet. The bubbles thus produced are bonded and pressed into lightweight bricks [56].

Ipsen Industries, Inc. has developed a process for manufacturing refractory foams which involves the entrainment of air into a slurry of refractory oxides, followed by drying and sintering. Firing of alumina and zirconia foams is conducted in an oxidizing atmosphere [57].

Lightweight, castable refractory compositions are prepared from vesicular granules with a bulk density of no more than 30 lb/ft^3 kyanite, and calcium aluminate. The composition is mixed with sufficient water to form a plastic mass that can be applied by pouring, tamping, or vibrating. Composites are dried at 110°C [58].

Refractory foams (especially alumina with densities of 30 lb/ft^3 or less are prepared by blending the refractory oxides with acid aluminum phosphate in the presence of water. Foaming is obtained by the evolution of hydrogen. Composites are then fired rapidly to 1315°C [59].

Zircon, mullite, and calcium aluminate foams with densities of not more than 60 lb/ft^3 have been prepared by mechanical whipping, with egg albumin as the foaming agent and a silicate and a phosphate as the binder system [60].

3. Oxidation Method

Alumina refractory bodies are prepared from aluminum or aluminum-based alloy particles that are first converted into irregular hollow shells of alumina with heat and a fluxing agent. The shells are then bonded with gum arabic and sodium hydroxide in the presence of water, pressed into shapes, and fired [54].

4. Blowing Method

Sodium silicate foam products of any thickness can be produced in commercial microwave ovens. When rapidly boiled, liquid sodium silicate forms a rigid, strong, white foam. The microwave oven allows the liquid to be heated uniformly throughout its thickness [61].

Hamilton Standard has developed a foamed-in-place ceramic composed of zirconium phosphate-bonded zirconia, lithium peroxide and potassium superoxide being used as the foaming agents [62]:

$$2Li_2O_2 + 2H_2O \longrightarrow 4LiOH + O_2 \tag{7}$$

$$4KO_2 + 2H_2O \longrightarrow 4KOH + 3O_2 \tag{8}$$

B. Properties

Low-density cellular refractories are used for thermal insulation at high temperatures. In order to be a suitable insulating material, they must possess good thermal shock resistance and dimensional stability over a wide temperature range. In some specialized applications resistance to a reducing atmosphere is also necessary.

The properties of some typical low-density cellular refractories are described in Table 4. Alumina does not undergo disruptive crystalline inversion as does zirconia, and therefore improvements in thermal shock resistance are obtained primarily by adjusting coarse-to-fine material ratios and variations in density [69]. Pure-alumina bricks are especially suited for hydrogen-atmosphere furnaces (Fig. 1) [70]. In order to impart

TABLE 4

Typical Properties of Low-Density Cellular Refractories

Property	Product						
	A[a]	B[b]	C[c]	D[d]	E[e]	F[f]	G[g]
Chemical analysis, %:							
Alumina	94.5	64.8	99.3	97.1	96.3	97.6	--
Silica	0.3	33.2	0.1	--	--	1.4-1.9	--
Zirconia	--	--	--	--	--	--	93.0[h]
Calcia	--	--	--	--	--	--	5.0
Density, lb/ft^3	49.5	55.5	30.0	29.5	32.0	28.0	37.5
Porosity, %	80	70	83	88	85	85-90	89
Cold crushing strength, psi	220	275	135	--	--	270	--
Modulus of rupture, psi	200	230	--	--	--	140	--
Reheat shrinkage in air:							
Linear shrinkage, %	0.30[i]	0.60[j]	2.00[k]	--	--	1.2[l]	--
Volume shrinkage, %	1.7[i]	--	--	--	--	4.8[l]	--
Thermal conductivity (air), $Btu/(h)(ft^2)(^{\circ}F/in.)$	2.93 at 2500°F[m]	4.14 at 2500°F[m]	3.64 at 2000°F[m]	3.0 at 2200°F[m]	3.0 at 2200°F[m]	3.7 at 2000°F,[m] 4.6 at 2400°F[m]	<1.0 at 2200°F[m]
Maximum use temperature, °C	--	1650	1480	1850	1650	1650-1750	2450

[a]Insulpure (Babcock & Wilcox); data from Ref. [63].

[b]Insulating firebrick K-3000 (Babcock & Wilcox); data from Ref. [64].

[c]Alundum TA-4062 (Norton); data from Ref. [65].

[d]Insulating ceramic brick 3400 (Ipsen Industries); data from Ref. [66].

[e]Insulating ceramic brick 3000 (Ipsen Industries); data from Ref. [67].

[f]Experimental product (PPG Industries); data from Ref. [50].

[g]Insulating ceramic brick 4200 (Ipsen Industries); data from Ref. [68].

[h]Also contains hafnia.

[i]Hold temperature 1510°C; hold time 24 h.

[j]Hold temperature 1620°C; hold time 24 h.

[k]Hold temperature 1455°C; hold time not specified.

[l]Hold temperature 1650°C; hold time 5 h.

[m]Mean temperature.

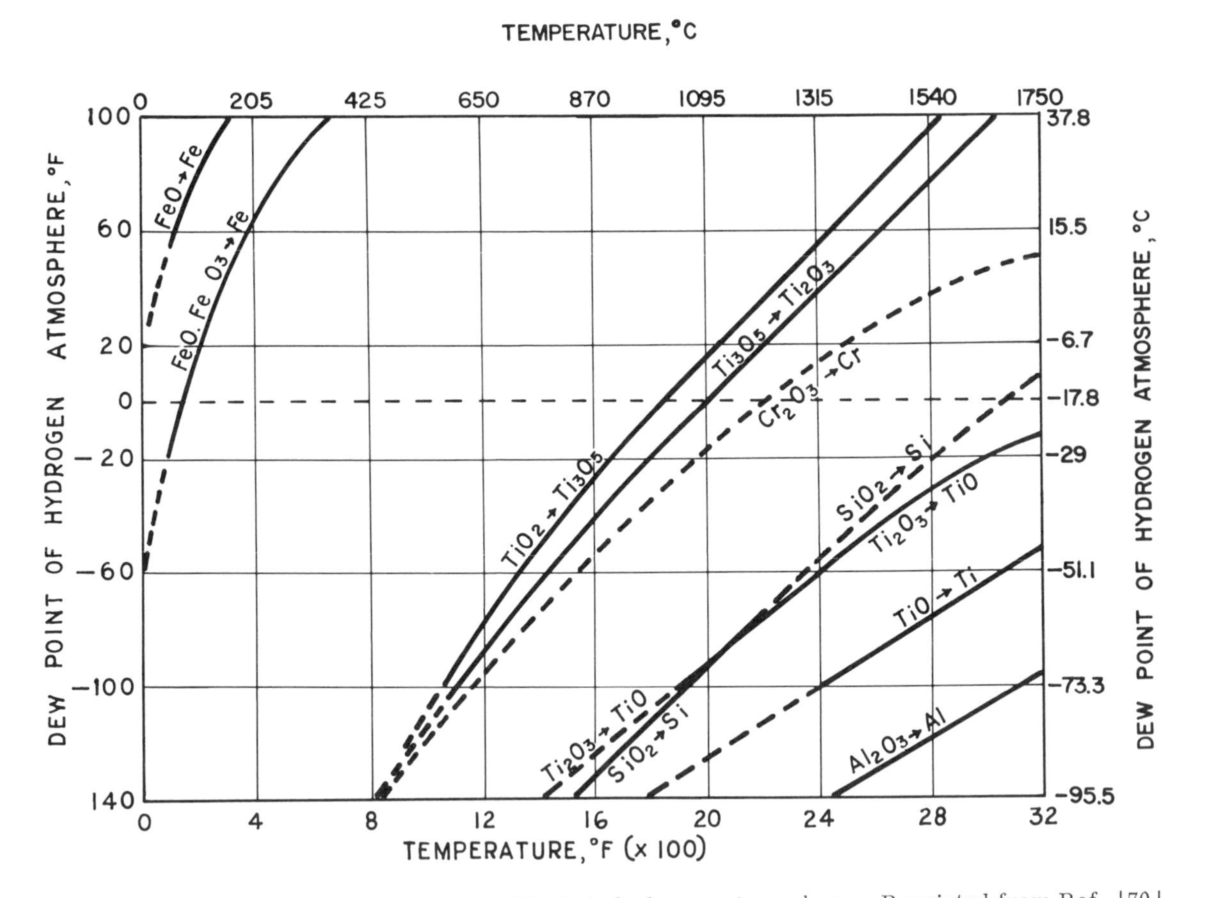

FIG. 1. Theoretical metal-metal oxide equilibria in hydrogen atmosphere. Reprinted from Ref. [70], by courtesy of the Babcock & Wilcox Co.

thermal shock resistance to zirconia refractories, zirconia has to be stabilized. The average thermal conductivity of calcia-stabilized zirconia is about one-third lower than that of magnesia-stabilized foamed zirconia. Calcia forms an intermediate compound with zirconia, namely, calcium zirconate, whereas magnesia does not [71]. Foamed zirconia at approximately 25% theoretical density is capable of withstanding a maximum temperature rise of 33°C/sec and can retain its integrity to well over 2200°C [72]. The incorporation of unstabilized zirconia into stabilized zirconia refractories lowers the thermal coefficient of linear expansion, which in turn improves the thermal shock resistance [73].

Resin-impregnated silica foams provide greater thermal insulation than do impregnated zirconia or alumina foams when compared on an equal-weight basis. Impregnation with phenolic resins increases the strength of the ceramic foams four to eight times, and the strengthening effect is still noted after the resin has been pyrolyzed to a carbon char [74]. Thoria, magnesia, zirconia, and beryllia have the highest temperature capabilities in that order; however, thoria is radioactive and has a high specific gravity. Beryllia as a foam could be hazardous because of dust formation, and magnesia is susceptible to hydration and is not stable in a reducing atmosphere [75].

C. Applications

Applications for high-temperature insulating bricks made of cellular refractories can be classified into two groups. The largest catagory includes such standard-atmosphere uses as insulation of crucible-type melting furnaces, reheat furnaces, aluminum-processing furnaces, structural clay and ceramic kilns, rotary driers, and combustion chambers. The other category covers insulation of controlled-atmosphere (hydrogen or vacuum) furnaces for brazing, sintering, bright annealing, and heat treating. Some specialized uses, such as foamed alumina and zirconia for nose cones and heat shields, have also been described.

IV. CELLULAR CONCRETE

Cellular concrete is used extensively in the building industry. The voids in this material are noninterconnecting, which is necessary to maintain good physical properties.

A. Methods of Manufacture

Two methods are commonly used to obtain a cellular structure. One is by chemical reaction in which aluminum powder and alkaline materials are added to the water-and-cement slurry [76-78]. A reaction takes place between the aluminum powder and added alkaline materials, such as caustic soda or potassium carbonate, as well as with the alkaline components of the cement. This results in the evolution of hydrogen gas, which acts as a blowing agent to provide the cellular structure.

The foaming occurs below 32°C in steel-reinforced molds, and the foam hardens at room temperature. At this stage the cellular concrete is shaped and machined to its desired form. The final curing process is carried out in an autoclave, where further chemical reaction strengthens the material. The densities vary from 20 to 50 lb/ft^3. Autoclaved cellular concrete has gained wide acceptance in Europe [79, 80].

The production of a factory-expanded cellular concrete by the addition of expandable-polystyrene beads to the cement has also been proposed. In this method portland cement, polystyrene-foam beads, natural fibers, and water were mixed with the pregenerated foam froth, and cellular foams with a density of about 25 lb/ft^3 were obtained [81-83].

The most common methods of producing cellular concrete are the entrainment of air by the foaming-in-place or the preformed-foam technique. In the foaming-in-place procedure a foam stabilizer is introduced directly into an aqueous slurry of portland cement and sand.

In the preformed-foam method a solution of emulsifiers and foam stabilizers is mixed with compressed air in a special foam generator to produce a stable foam. The resulting foam looks very much like shaving cream. The foam generator delivers foam at a given rate and hence the amount of foam added to the concrete mix can be gauged by the duration of its delivery. Water hardness, temperature changes, and other factors are not important. The only equipment required in addition to the foam-generating system is a standard concrete mixer and a source of compressed air. The pregenerated stable foam is added to a mixture of sand, cement, and water. For densities of 40 lb/ft^3 or less no sand is used [84].

The foaming agents most effectively used to foam concrete are mixtures of hydrolyzed protein and emulsifiers. Many emulsifiers are suitable, such as ammonium soaps of fatty acids, alkyl naphthalene sulfonates (e.g., sodium isobutylnaphthalenesulfonate), or nonionic emulsifiers [85-87]. The surface-active agents have a stiffening effect on the boundary surfaces of the foam lamellae. Semihydrophilic detergents have an optimum stiffening effect [88]. Additives like cellulose glycolates, starch, and methyl cellulose are suggested. Fibers, such as mineral fibers, are also recommended as reinforcements for cellular concrete [89].

B. Properties

The relationships between density, compressive strength, and thermal conductivity are described in Table 5.

TABLE 5

Properties of Elastizell-Type Insulating Concrete[a]

Dry density	Compressive strength (28-day)(psi)	Elastic modulus X 10^5 (psi)	Thermal conductivity (k factor) (Btu/(h)(ft^2)(^oF/in.))
20	100		.4
30	200		.7
40	300		1.1
50	400		1.4
60	550		1.8
70	700		2.3
80	900	6.60	2.9
85	1000	7.25	3.3
90	1250	7.90	3.7
95	1600	8.55	4.1
100	2000	11.40	4.5
105	2500	14.25	4.9
110	3000	17.10	5.5
115	3500	19.95	6.0

[a]Data from Ref. [90].

The National Bureau of Standards has studied the properties of cellular concretes, placing special emphasis on autoclave products [91]. McCormick [92] reports that mixes containing expanded shale aggregates produce higher strength values than those containing sand aggregate of the same density value. Kluge [93] has discussed the effect of modulus of elasticity and compressive strength, and shearing (diagnoal tension resistance of light-weight aggregate concrete) in relationship to ACI building codes.

C. Applications

Fine cellular concrete is quite a versatile building material. Density ranges from 25 to 40 lb/ft^3 for insulating roof fill placed over structural roof decks have been used successfully. This method has a compressive strength sufficient to withstand roof traffic and provides a good bond for built-up roofing. It can also be used as an insulation under slab on grade or basementless homes, or it can be precast into insulating panels. Cellular concrete in the density range 40 to 100 lb/ft^3 can be used as floor slabs as well as for lightweight floor fill on multistory buildings [90].

V. RIGID SULFUR FOAMS

Rigid sulfur foams were developed at the Southwest Research Institute, San Antonio, Texas [94]. A froth consisting of molten sulfur, stabilizing agents, film-forming materials, and blowing agents was prepared above the melting point of sulfur. This froth can be dispensed into molds or can be sprayed into cavities in which it solidifies on cooling. Since the specific gravity of sulfur is approximately twice that of other rigid plastic foam ingredients, the achievement of lower densities is more difficult. The density range of sulfur foams is 10 to 25 lb/ft^3.

Film-forming agents are tricresyl phosphate, dipentene dimercaptan, and silicones. Stabilizers like talc distribute themselves in the film cell walls and add strength and stability to the formulation. Their physical action is largely a function of the platelike particle shape. Aluminum flake is the second most promising stabilizing material. A combination of aluminum flake and talc results in smaller cell sizes than those obtained by using either one of the ingredients alone.

Many systems producing the blowing agents can be used; the combination of calcium carbonate and phosphoric acid has proved to be by far the best blowing agent. Two typical formulations for sulfur foams are listed in Table 6 [95].

TABLE 6

Typical Formulations for Sulfur Foams[a]

Component	Parts by weight
Sulfur	100
Phosphorus pentasulfide	3
Styrene[b]	3
Dipentene dimercaptan[c]	3
Tricresyl phosphate	0.25
Talc	7.5
Calcium carbonate	3
Orthophosphoric acid	2.55

[a]Data from Ref. [95].

[b]Used in styrene-based formulation only. Resultant foam has a density of 12 to 14 lb/ft^3.

[c]Used in dipentene dimercaptan-based formulation only. Resultant foam has a density of 10 lb/ft^3.

Unlike other plastic-based foam materials that can be foamed in place, sulfur foams are favored by low temperature since it speeds the solidification of the sulfur-foam froth, thereby ensuring a more uniform product. Compressive strength seems to be adequate for a variety of structural applications. A typical stress-strain curve for rigid sulfur foam is shown in Fig. 2. This type of stress-strain curve is desirable for materials used in shock-isolation applications [96]. The thermal conductivity of sulfur foam is between 0.3 and 0.4 Btu/(h)(ft^2)(^{o}F/in.) [97]. Low moisture absorption and low water-vapor permeation are observed in sulfur foams with intact skins (Table 7) [95]. The foam core, however, is much more absorbing and permeable.

For producing sulfur foams a machine consisting of two reaction tanks heated by gasoline burners and provided with agitation is used. An air-cooled gasoline engine that operates an air compressor and a hydraulic pump completes this portable unit, which has been used by the Southwest Research Institute to pour as much as 30 ft^3 of foam at a time.

In spite of the potentially low cost of sulfur foams, no commercial applications have yet been developed for them.

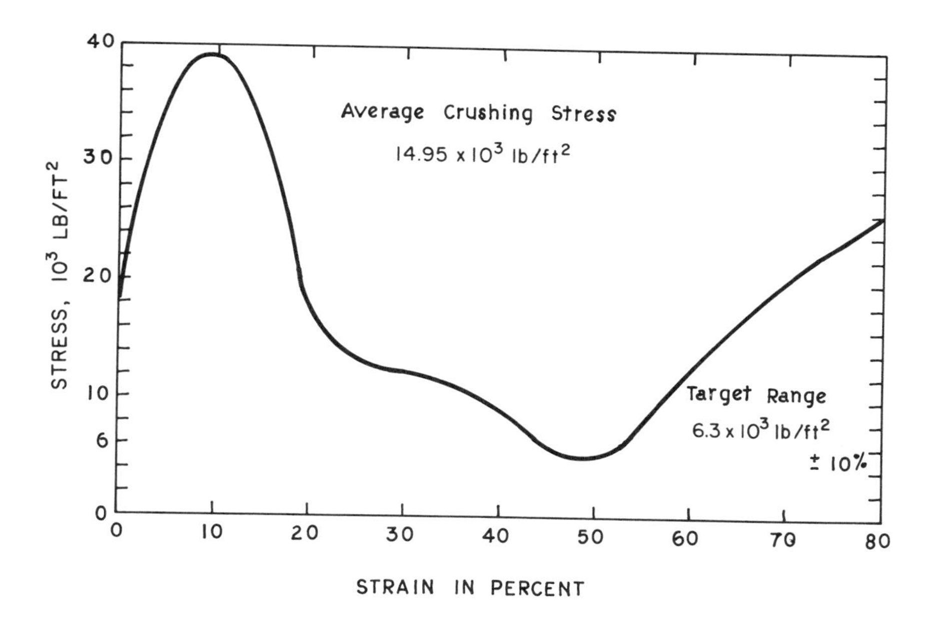

FIG. 2. Typical stress-strain curve for rigid sulfur foam. Reprinted from Ref. [96].

TABLE 7

Moisture Absorption and Water-Vapor Permeability of Sulfur Foams[a]

Property	Dipentene dimercaptan base		Styrene base	
	With skin	Without skin	With skin	Without skin
Water-vapor transmission, grains/ft^2-h	0	1.3	1.3	3.0
Permeance, perms	0	2.83	2.83	6.54
Permeability, perm.-in.	0	4.25	5.66	9.82

[a]Data from Ref. [95].

For producing sulfur foams a machine consisting of two reaction tanks heated by gasoline burners and provided with agitation is used. An air-cooled gasoline engine that operates an air compressor and a hydraulic pump completes this portable unit, which has been used by the Southwest Research Institute to pour as much as 30 ft^3 of foam.

In spite of the potentially low cost of sulfur foams, no commercial applications have yet been developed for them.

REFERENCES

1. J. H. Connelly and R. V. Harrington (to Corning Glass Works), U.S. Pat. 3,174,870 (1961).
2. W. O. Lytle (to Pittsburgh Plate Glass), U.S. Pat. 2,215,223 (1941).
3. R. A. Miller and W. O. Lytle (to Pittsburgh Plate Glass), U.S. Pat. 2,233,631 (1941).
4. B. Long, U.S. Pat. 2,123,536 (1938).
5. W. D. Ford (to Pittsburgh Corning), U.S. Pat. 2,544,954 (1951).
6. F. Schill, et al U.S. Pat. 3,163,512 (1964).
7. D. D'Eustachio (to Pittsburgh Corning), U.S. Pats. 2,775,524 (1956) and 2,860,997 (1958).
8. W. A. Weyl, Glass Ind., 24, 17 (1943).
9. J. P. Hyde and S. D. Stookey (to Corning Glass Works), U.S. Pat. 2,564,978 (1951).

10. M. J. Shoemaker (to Pittsburgh Corning), U.S. Pat. 2,582,852 (1952).
11. W. D. Ford (to Pittsburgh Corning), U.S. Pats. 2,596,669 (1952) and 2,600,525 (1952).
12. M. J. Udy (to Carborundum), U.S. Pat. 2,741,822 (1956).
13. J. A. Winterburn (to Thermal Syndicate), U.S. Pat. 3,396,043 (1968).
14. W. D. Ford (to Pittsburgh Corning), U.S. Pats. 2,377,076 (1945) and 2,620,597 (1952); J. T. Littleton (to Corning Glass Works), U.S. Pat. 2,536,192 (1951); A. H. Baker (to Pittsburgh Corning), U.S. Pats. 2,544,947 (1951) and 2,725,680 (1955); O. W. Wiley (to Pittsburgh Corning), U.S. Pat. 2,595,115 (1952).
15. E. O. Schylz (to VEB Schaumglaswerk Taubenbach), U.S. Pat. 3,403,990 (1968).
16. Anon., Chem. Eng. News, 24, 37 (1948).
17. W. D. Ford (to Pittsburgh Corning), U.S. Pat. 2,691,248 (1954).
18. G. Slayter and R. W. Soltis (to Owens-Corning Fiberglas), U.S. Pat. 3,207,588 (1962).
19. D. D'Eustachio et al. (to Pittsburgh Corning), U.S. Pat. 3,354,024 (1967).
20. R. F. Shannon (to Owens-Corning Fiberglas), U.S. Pat. 3,325,341 (1964).
21. A. L. Burwell (to University of Oklahoma Research Institute), U.S. Pat. 2,466,001 (1949), (to Soc. Anon. des Manufactures des Glaces and Produits Chimiques de Saint-Golbain Chauny and Cirey), U.S. Pat. 1,945,052, (1934).
22. Pittsburgh Corning Corp. Tech. Bull. FB-113, March 1966.
23. M. Hubscher, Silikatechnik, 5, 243 (1954).
24. Pittsburgh Corning Corp. Tech. Bull FC-100, March 1966.
25. D. D'Eustachio et al. (to Pittsburgh Corning), U.S. Pat. 2,596,659 (1952).
26. E. H. Haux (to Pittsburgh Plate Glass), U.S. Pat. 2,237,032 (1941).
27. B. Sosnick, U.S. Pat. 2,434,775 (1948).
28. B. Sosnick, U.S. Pat. 2,553,016 (1951).
29. J. Bjorksten and L. F. Yntema, Construction Specifier, 14, 50 (1961).
30. S. O. Fiedler et al. (to Lor Corp.), U.S. Pat. 2,937,938 (1960); U.S. Pat. 2,979,392 (1961).
31. W. S. Fiedler (to Lor Corp.) ,U.S. Pat. 3,214,265 (1965).
32. J. C. Elliott (to Lor Corp.), U.S. Pat. 3,005,700 (1960).
33. J. F. Pashak (to Dow Chemical), German Pat. 1,201,559 (1965).
34. J. A. Bjorksten (to Lor Corp.), U.S. Pat. 3,305,902 (1967).
35. H. Henkel (to Ringsdorff-Werke), German Pat. 1,015,228 (1957).
36. M. F. Grandey (to General Electric), U.S. Pat. 2,917,384 (1959).
37. J. F. Pashak (to Dow Chemical), U.S. Pat. 2,935,396 (1960).
38. F. R. Hensel (to P. R. Mallory & Co.), U.S. Pat. 2,450,339 (1948).
39. Mallory Battery Co. of Canada Ltd., Netherlands Pat. 6,604,091 (1966).

40. D. M. Albright et al. (to U.S. Steel), U.S. Pat. 3,360,361 (1967).
41. D. Yarnton, Eng. Mater. Des., 9, 83 (1966).
42. Ipsen Industries, Inc., Annual Report N-65-10839, G. C. Marshall Space Flight Center, Huntsville, Ala., Sept. 1964.
43. J. E. Hughes, New Scientist, 20, 268 (1963).
44. J. C. Elliott (to Lor Corp.), U.S. Pat. 2,3983,597 (1961).
45. Anon., Chem. Eng., 66, 120 (1959).
46. I. G. Farbenindustrie AG, British Pat. 497,844 (1938).
47. F. V. Lenel (to General Motors), U.S. Pat. 2,191,936 (1940).
48. W. N. Redstreake, Iron Age, Oct. 25, 1962, pp. 101-103.
49. J. M. Knote (to Quigley Co.), U.S. Pat. 2,122,288 (1938).
50. M. Wismer and J. F. Bosso Ind. Eng. Chem. Prod. Res. Develop., 5, 282 (1966).
51. M. Calis and J. Peyssou (to Commissariat a l'Energie Atomique), Belgian Pat. 615,236 (1962).
52. W. Hallett et al., Ipsen Industries, Inc., "Research on Method for Production of Thermantic and/or Heat Shielding Materials" (Phase 1), IR-8-117(1), contract No. AF33 (657)-11286, Rockford, Ill., July 1963, p. 19.
53. S. Kohn (to Office National d'Etudes et de Recherches Aerospatiales), U.S. Pat. 3,124,542 (1964).
54. E. I. duPont de Nemours & Co., British Pat. 1,012,197, (1965).
55. P. Leroy and R. Simon (to Compagnie des Ateliers et Foyes de la Loire), U.S. Pat. 3,322,867 (1967).
56. J. W. Newsome, H. W. Heiser, A. S. Russel, and H. C. Stumpf, Alcoa Tech. Paper No. 10, 1960, pp. 26, 27.
57. E. P. Flint Mater. Sci. Res., 2, 193 (1965).
58. J. L. Stein (to General Refractories Co.), U.S. Pat. 3,341,339 (1967).
59. J. Magder (to Horizons, Inc.), U.S. Pat. 3,330,675 (1967).
60. A. J. Montvala and H. H. Nakamura (ITT Research Institute), "Development of Lightweight Thermal Insulation Materials for Rigid Heat Shields," IITRI-G-600 2-27, contract No. NAS 8-11333, June 1964, p. 2.
61. E. J. Baker, Jr., in Proc. 12th Natl. Sample Symposium Soc. Aerospace Mater. and Process Eng., 12 (1967), Section I-4.
62. E. W. Blocker and R. D. Paul, Hamilton Standard Division of United Aircraft Corp., "Development of Alkali Metal Peroxide and Superoxide Blown Ceramic Foam," N66 31363, contract No. NAS-8-20089, April 1966, p.2.
63. Babcock & Wilcox Co., Refractories Division, PR-4010.
64. Babcock & Wilcox Co., Refractories Division, PR-3309.
65. Norton Co., Refractories Division, P-3, pp. 1-17.
66. Ipsen Industries, Inc., Specification Sheet.
67. Ipsen Industries, Inc., Specification Sheet.

68. Ipsen Industries, Inc., Specification Sheet.
69. L. K. Lindell, J. B. Cloud, et al., Ipsen Industries, Inc., "Research on Method for Production of Thermantic Structural and/or Heat Shielding Materials" (Phase 2), IR-8-117(2), contract No. AF33(657)-11286, Rockford, Ill., October 1963, p. 34.
70. Babcock & Wilcox Co., Refractories Division, R901-3001.
71. W. C. Allen, National Beryllia Corp., "Research on Thermal Transfer Phenomena," W. C. Allen, N66 23512, contract No. NASW-1197, Haskell, N. J., January 1969, abstr.
72. L. K. Lindell et al., Ref. 69, p. iii.
73. L. K. Lindell et al., Ref. 69, p. 16.
74. E. L. Strauss, Martin Co., Space Systems Division, in Summary of the Fifth Refractory Composition Working Group Meeting, III, RTD-TDR-63-4131, Project No. 7381, November 1968, p. 682.
75. W. C. Allen, Ref. 71, p. 2.
76. W. B. Urmston, U.S. Pat. 3,236,925 (1966).
77. P. E. Starnes (to John Laing & Son Ltd.), U.S. Pat. 2,635,052 (1953).
78. M. Dabour, French Pat. 1,135,021 (1957).
79. Anon., Engineer, 221, 812 (1966).
80. Anon., Engineering, 201, 1093 (1966).
81. G. Thiessen (to Koppers), U.S. Pat. 3,021,291 (1962).
82. R. C. Sefton (to Koppers), U.S. Pat. 3,257,388 (1966).
83. R. C. Sefton (to Koppers), U.S. Pat. 3,214,393 (1965).
84. L. E. Rivkind, J. Cell. Plastics, 3, 329 (1967).
85. J. L. Hanold, Concrete, 64, 32 (1956).
86. Elastizell Corp. of America, British Pat. 928,987 (1963).
87. A. Sommer, U.S. Pat. 3,141,857 (1964).
88. J. Pfanner, Silikattech, 6, 396 (1955).
89. Chemieprodukte G.m.b.H., German Pat. 1,124,415 (1962).
90. Elastizell Corp. of America, Alpena, Mich., A.I.A. File Numbers 4-E-6/4-E-13/4G, 1960.
91. R. C. N. Valore, Amer. Concrete Inst. J., 50, 775, 817 (1954).
92. F. C. McCormick, Amer. Concrete Inst. J., 64, 104 (1967).
93. R. W. Kluge, Amer. Concrete Inst. J., 28, 375 (1956).
94. J. M. Dale et al. (to Southwest Patents, Inc.), U.S. Pat. 3,337,355 (1967).
95. J. M. Dale and A. C. Ludwig, U.S. Army Materiel Command, Cold Regions Research and Engineering Laboratory, Preparation of Low Density Sulfur Foam, CRREL Tech. Rep. 206 (AD-661,315), Hanover, N.H., September 1967.
96. J. M. Dale and A. C. Ludwig, U.S. Army Natick Laboratories, Airdrop Engineering Laboratory, Foamed Sulfur for Airdrop Cushioning, Tech. Rep. 68-67 AD, Natick, Mass., May 1968.
97. J. M. Dale and A. C. Ludwig, Sulphur Inst. Journal, 2 (3), 6 (1966).

Chapter 17

EFFECTS OF CELL GEOMETRY ON FOAM PERFORMANCE

R. H. Harding

Union Carbide Corporation
South Charleston, West Virginia

I. INTRODUCTION

Any material that can be processed as a fluid and returned or converted to the solid state is potentially the base of a stable foam. Chapter 2 describes the complex colloidal variables that must be balanced to provide useful results, and succeeding chapters show how frequently and ingeniously this delicate balance has been achieved in practice.

Chapters 3 through 16 classify foams naturally on the basis of the constituent solids, whose chemical and physical properties set limits on propsective manufacturing processes and service conditions. But nearly every foam--flexible or rigid, unicellular or reticulated--is derived from a dispersion containing some expanding vapor as the (sometimes temporarily) discontinuous (internal) phase in a fluid matrix. The solidified continuous (external) phase then normally accounts for minor and major fractions of the final volume and weight, respectively, and so the product's qualities also depend on the characteristics of its gas phase.

The properties and relative amounts of its components further circumscribe without defining a foam's performance. The chemistry and technology employed to produce a specific foam jointly determine its internal physical structure, which varies significantly among systems and among processes, especially at the relatively low densities of greatest commercial interest. Cell geometry is therefore the bridge relating composition quantitatively to performance. Several physical parameters (i.e., density, relative size, and openness of cells) are regularly measured and controlled for their obvious economic significance. Others (e.g., cell shape, within-cell distribution of solid) are not. This is because their equally important effects are not only less generally recognized (and indeed are often attributed directly to the solid) but are also at present more difficult to measure [1] and control [2].

After a brief review of the physical structures predicted by surface chemistry and observed in practice, this chapter summarizes their known effects on selected properties of a variety of foams. Although significant progress has been made, the thorough understanding needed to optimize performance by controlling these variables invites further work. Many unanswered questions provide opportunities for noteworthy contributions to foam technology and the design of composite foam-containing assemblies.

II. FOAM STRUCTURES

A. Theory

Figures 1, 2, and 3 illustrate predicted unit-cell, internal, and cell-packing geometries, respectively.

Because it minimizes interfacial surface and capillary pressure, the spherical bubble is inherently the most stable configuration in Fig. 1. However, close-packed monodisperse spheres can occupy no more than 0.74% of the available space, and the volume fraction of solid in most commercial foams is far below the 0.26% remainder. Monodisperse spheres deform into regular polyhedra as they expand to occupy more than three-fourths of the system's total volume, favoring shapes that simultaneously minimize interfacial area and balance capillary pressures.

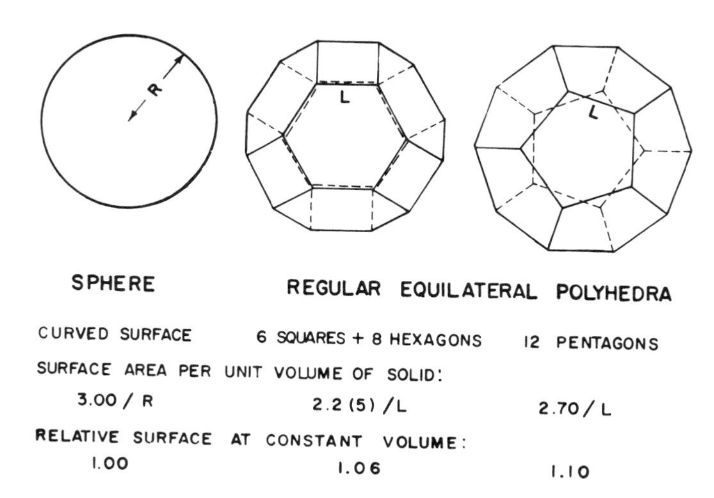

FIG. 1. Several ideal cell geometries. Reprinted from Ref. [1], p. 4, by courtesy of the American Society for Testing and Materials.

Kelvin's tetrakaidecahedron [3] in the center of Fig. 1 approximates the first, but not the second, criterion, so the dodecahedron on the right becomes a more probable structure because its equiangular geometry (faces intersect at 120-degree angles to form an edge, and edges intersect at 109.5-degree angles to form a corner) optimizes "wet-foam" stability.

These line drawings show thin-walled cells from essentially zero-density foams. Figure 2 illustrates some effects of bringing three identical bubbles together progressively to produce a three-dimensional cell edge. The top sketch represents the solid between close-packed spheres. The second merely adds weight to the dodecahedron of Fig. 1 and cannot exist because capillary pressure is infinite within the liquid phase. The third sketch portrays an expected closed-cell-foam geometry [4]: the relative amounts of solid and gas are functions of bulk density, and the actual distribution of solid between cell faces and edges depends on force balances imposed by viscosity, capillarity, and surface tension--force balances that change progressively with time during the foaming process.

Cell faces necessarily thin as expansion continues and as capillary pressures drive liquid toward their edges. This process can ultimately cause faces to begin pinholing, at which point the foam may collapse, adjacent bubbles may coalesce, or the faces might simply begin retracting or tearing, as indicated in the fourth sketch. The probability of obtaining each successive result increases with the fluid's viscoelasticity at that moment, the last alternative producing completely open-celled foams when one-sixth or more of the faces have ruptured. The bottom sketch illustrates the improbable limit where all faces have withdrawn completely into the strut.

Cells must occupy all space within a foam, and Fig. 3 projects several idealized possibilities in two dimensions. The top left sketch represents a close-packed array of uniform spheres whose minimum bulk density is one-fourth of the solid's density independently of cell size. Foam density can be reduced by displacing interstitial solid with smaller spheres, as shown at top right, but this structure is very unstable during foaming because smaller bubbles generate higher capillary pressures, which promote their assimilation into larger neighbors.

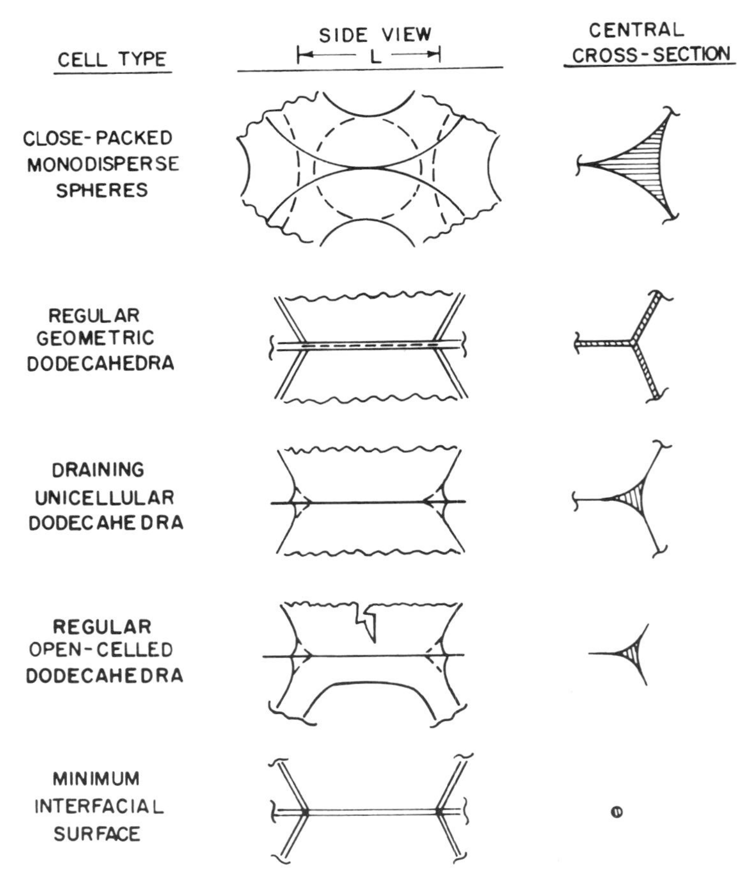

FIG. 2. Potential geometries of the solid rib (strut or Plateau border) between adjacent cells. Reprinted from Ref. [1], p. 6, by courtesy of the American Society for Testing and Materials.

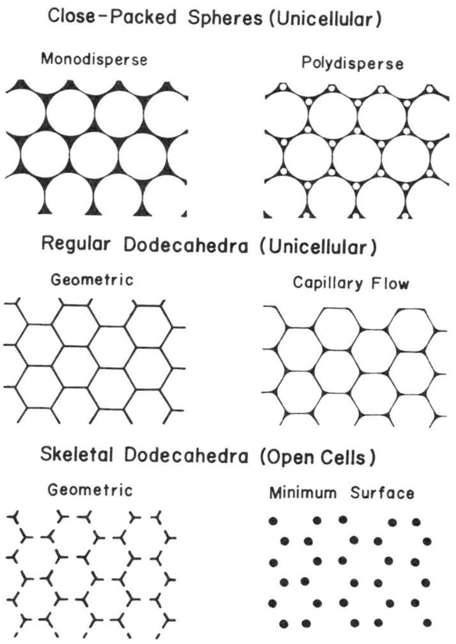

FIG. 3. Potential geometries of packed foam cells. Reprinted from Ref. [2], p. 385, by courtesy of Technomic Publishing.

The remaining sequence of sketches in Fig. 3 is directly analogous to the last four in Fig. 2. Since all are based on identical dodecahedra, it should be noted that this polyhedron must distort slightly to fill all space in these close-packed arrangements. Kelvin's tetrakaidecahedron packs to fill space without distorting but unbalances capillary pressures in wet foams. Williams' tetrakaidecahedron [5] has two quadrilateral, eight pentagonal, and four hexagonal faces; it fills space with equal angles, no distortion, and a dodecahedron's surface-to-volume ratio. Although it may therefore be the structure preferred by nature, the preceding observations generally apply to either of these regular polyhedral structures.

B. Practice

All large-scale commercial foaming processes generate a gas by chemical reaction, mechanical distribution, or vaporization; distribute it uniformly throughout a liquid; and finally stabilize the system by cooling or chemical reaction (e.g., polymerization). Operations are conducted to produce foams in a wide range of shapes and sizes, and specific techniques are normally required for each solid-liquid-foam combination. Although a corresponding broad range of physical structures results, empirical observations in a variety of systems [6-8] tend to confirm the expected geometries, introduce additional variables, and provide the basis for several interesting and useful generalizations [1, 2].

Useful densities range from 0.33 lb/ft^3 for some phenolics to 50 lb/ft^3 for some organosilicone and inorganic foams, with 1 to 4 lb/ft^3 including the dominant share of commercial production. Higher densities are generally associated with load-bearing service, whereas lower densities meet requirements for cushioning, insulating, structural composites, and many special applications. Constituent solids range from very soft through brittle mechanically. Although cell structure can theoretically be controlled independently of these parameters, thick walls naturally promote the formation of closed cells at high density, whereas dimensional stability may require open cells in extremely-low-density or resilient foams.

Ideally uniform and dependent only on the number of nucleating sites and the stability of interfacial surfaces, cell sizes are actually statistically distributed with standard deviations approximating ± 25% of the average unit's volume. Although coarse-celled foams contain as few as 1000 units per cubic inch, typical products contain 20,000 to 200,000, and microcellular foams contain 1 million cells per cubic inch.

The internal structures of unicellular and open-celled foams commonly approximate the third and fourth sketches in Fig. 2, respectively. A larger fraction of the solid phase is normally found in cell edges as foam density increases or the solid's hardness decreases. Thermoplastic materials tend to distribute themselves more uniformly between faces and edges than do thermosets, presumably because their effective in-process viscosities are higher and they are therefore more resistant to capillary drainage. By the same token, ruptured faces are usually cracked in rigid, but partially retracted in flexible, systems.

Although most cell faces are pentagonal, the smaller and roughly equal numbers of tetragons and hexagons normally observed tend to substantiate the presence of some tetrakaidecahedral structures. These faces are often approximately equilateral but seldom equiangular. Mechanical restraints imposed on expanding foams cause cells to elongate along lines of least resistance, normally the preferred local rise direction. Average height-to-width ratios can range from 1 in isotropic cells to 2 in fully elongated equilateral cells, and still higher near external foam surfaces, where flow patterns are irregular, stress gradients larger, and cell-edge lengths more variable.

The extent to which these anomalies occur depends on the complex effective force-energy balance associated with a specific manufacturing process. Techniques that simultaneously minimize frictional drag and gradual changes in the expanding liquid's physical properties, combined with maximum wet-foam stability, approximate the ideal structure most closely. We shall see, however, that in many applications intentionally distorted cell shapes can enhance foam performance and economy.

C. Measurement

Advances in our understanding of and ability to control cell structure depend on the availability of suitable analytical tools. Although careful microscopical inspection has contributed significantly to this objective, defining a foam's three-dimensional structure remains an infant art because few physical methods suitable for routine application have been developed.

The volume fraction of solid in a cellular material can be calculated from bulk densities measured by a number of standard tests (see Chapter 8). For example, ASTM Test D1940 for porosity of rigid cellular plastics employs the air-displacement principle as a relative measure of cell interconnection. Successive refinements in the original approach [9] define the volume fraction and average size of any foam's closed cells [10, 11] and promise to provide a direct measure of cell elongation [12]. A visually calibrated, dimensionless mechanical analogy [13] predicts that elongation will approximate

$$E = \frac{R + 2}{3},$$

where E is the average cell height-to-width ratio and R is the ratio of compressive yield strengths measured along the same mutually perpendicular axes in that foam specimen.

In relatively open-celled foams air-flow rates can be measured [14, 15] to compare permeability variations resulting from combined density, cell-size, and face-retraction effects. No other physical tests have been developed specifically to help define foam morphology.

III. CORRELATIONS BETWEEN PHYSICAL STRUCTURE AND PROPERTIES

Foam density or volume fraction of solid, cell size and shape, the solid's distribution between cell faces and edges, and the degree of cell interconnection can vary independently of each other and of raw materials within broad limits. In any specific case they are fixed by a combination of physical, chemical, mechanical, thermal, and time factors [16] whose exact effects and interactions are not yet well defined. The final result can, however, be noted and used to explain the relative contributions of solid, vapor, and their geometric arrangement in space to overall foam performance.

Resistance to chemicals, solvents, bacteria, and fungus; adhesive qualities; stability on exposure to various types of radiation; transition temperatures; and other properties are associated exclusively with the constituent solid. Ignition temperature and flammability depend primarily on the solid, although burning rates are generally much higher in foamed than in bulk form (more obviously so for open-celled foams, whose large internal surfaces contact air directly, and for unicellular foams containing combustible vapors, than for closed-cell products containing nonflammable gases). Odor, corrosiveness, and similar qualities might be due to the choice of solid and/or blowing agent. Such characteristics are peculiar to specific systems; they have been described in the appropriate chapters and will not be discussed here.

The literature also contains numerous references to the effects of density on properties, but usually with no assurance that other morphological parameters were held constant or that related changes were monitored. This approach is quite proper in the purely contemporary economic sense of describing what is available, but it sheds no light on useful improvements that might be realized by planned modifications of physical structure or on the sometimes large and unexpected effects of manufacturing the "same" foam by a different process. Accordingly subsequent comments attempt to distinguish the "direct" effects of solid and gas phases, density, and other physical variables within the limits of data currently available for adequately defined foams.

A. Mechanical

1. Static

Mathematical analyses [17-21] indicate that the conventional elastic properties of low-density foams containing regular uniform cells are direct functions of the corresponding mechanical properties of the constituent solids and complex functions effectively ranging between the first and third powers of solid content. Since all foams are heterogeneous, however, constituent solids are stressed unevenly and provide lower strengths than might be expected by linear extrapolation from the bulk. The more detailed analyses predict that large-celled foams will be stiffer than fine-celled materials and will yield or fail with less deformation at higher loads. Moduli are expected to decrease progressively as the solid distribution shifts from cell faces toward edges.

Data cited in other chapters certainly confirm the anticipated density effects. Empirical studies of cell-size effects on foam strengths have produced conflicting results, probably because solid distributions changed concurrently: the most recent work suggests that tear and tensile strengths increase with cell size but compressive strengths do not [8, 10], whereas industry results with flexible polyurethane foams generally indicate the opposite.

Removing its cell faces apparently increases an open-celled foam's tensile and tear strengths [8], probably due to reducing the flaws, while reducing its compressive strength [8, 14]. Opening the cells of predominantly unicellular materials can also depress compressive stress-strain curves appreciably [22] when deflections are large enough to compress entrapped gases.

The mathematical models omitted one morphological variable whose well-known effect is second only to that of the solid itself. Figure 4 illustrates the importance of cell elongation, and the relative direction of loading, to mechanical performance. Height-to-width ratios greater than 1 represent stress application parallel to the local foaming direction (the long-cell dimension), whereas ratios smaller than 1 describe tests on the same foams in a perpendicular orientation. Although the urethane's strengths increased in proportion to the 7/4 power of density, the effective strength of a 3-lb/ft^3 foam loaded perpendicularly can be matched at half the cost by redesigning to compress a highly anisotropic 1.5-lb/ft^3 foam in the parallel direction.

Similar cell-shape relationships apply to other mechanical properties [2]. A 3:1 parallel-perpendicular compressive-strength ratio in a given foam is usually accompanied by a similar tensile-strength ratio (lower if the solid is relatively brittle or cell size relatively heterogeneous), a 4:1 elastic modulus ratio, and 2:1 ratios in flexural and shear strengths and modulus of rigidity. Equal parallel and perpendicular strengths normally anticipate 1:1 ratios in other mechanical properties of the same foam.

Such essentially isotropic materials provide an ideal basis for separating true effects of density, solid composition, and other physical variables. They are, however, seldom made. For mechanical properties at least, one study [23] indicates that the geometric mean of strengths measured along the three mutually perpendicular major axes provides an equivalent result. The same work suggests that the relative importance of cell structure decreases, and that of solid-phase properties increases, as the service temperature rises.

Coefficients of thermal expansion are essentially independent of the structures of open-celled and medium-density (>3 lb/ft^3) unicellular foams [2]. However, since the pressure of gases trapped within closed cells varies with temperature, a pneumatic stress is superimposed on the solid's

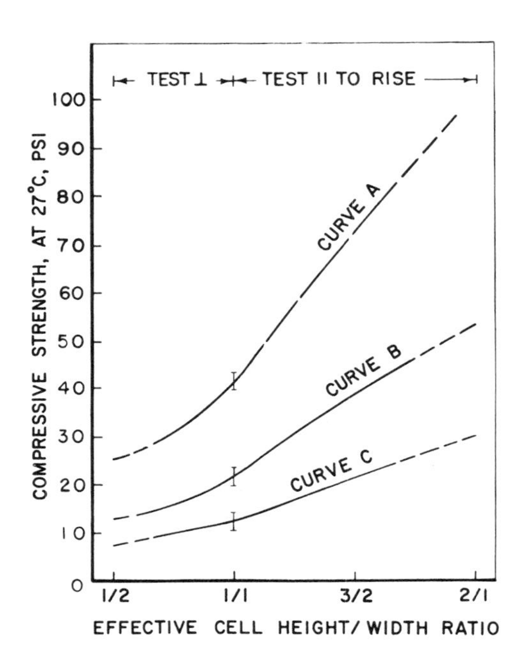

FIG. 4. Typical effect of cell shape on rigid-foam strength by ASTM test D1621. Curve A, 3-lb/ft^3 urethane, 2-lb/ft^3 polystyrene; curve B, 2-lb/ft^3 urethane; curve C, 1.5-lb/ft^3 urethane, 2-lb/ft^3 phenolic. Reprinted from Ref. [1], p. 10, by courtesy of the American Society for Testing and Materials.

inherent thermal sensitivity. Table 1 shows how this effect increases the apparent coefficients of low-density unicellular products, especially those containing elongated cells.

2. Dynamic

Foams bonded to sheet materials rapidly damp resonant vibrations by absorbing, distributing, and dissipating the energy of surface impacts and of cyclic stresses like those associated with transportation or nearby machinery. Large closed cells maximize damping efficiency by combined stiffness and pneumatic mechanisms [24]. Efficiency also increases with polymer stiffness and foam density. The results achieved in specific cases are functions of imposed frequencies.

Resilient foams are used primarily as cushioning materials. Flex fatigue and creep resistances are therefore as important as conventional mechanical properties. Comfort requires essentially open cells for air circulation, and resilience increases at least twofold [14] as the material in perforated faces retracts toward cell edges to minimize internal pneumatic effects. Fatigue resistance improves as cells become smaller [25] and as cell faces, whose progressive tearing contributes substantially to apparent changes with repeated flexing [14], retract more completely during manufacture. Since creep is inversely proportional to the foam modulus appropriate for anticipated use conditions, structural changes that increase static moduli decrease creep [2].

Shock-absorbent-packaging applications are best filled by firm unicellular foams, produced with minimum capillary drainage to enhance their stiffness and pneumatic cushioning capabilities. Relatively large cells, elongated in the design stress direction, optimize durability and performance [22].

TABLE 1

Thermal Expansion of Rigid Foams
(-62 to +60° C by ASTM D-696)

Cell structure		Density	Linear expansion coefficient
Type	Shape	(lb/ft^3)	((in./in.-° C) x 10^5)
Open	Any	1-7	6.3
Unicellular	Any	3-7	7.2
	Ideal	1	9
	Elongated	1	16.2

B. Mass Transfer

Assuming that hydrostatic pressures are not sufficient to rupture closed cells and that solids are not attacked chemically or physically, liquids will only penetrate an immersed foam's interconnected cells.

Gases and liquids move readily through open-celled foams under the influence of moderate pressure or concentration gradients. Resistance to transfer by either mechanism increases as density rises or cell size decreases, but the most significant factor is the fractional cell-face area occupied by foam solids. Reticulating a normal open-celled foam can reduce its air-flow resistance fivefold [14]. Restoring the integrity of half its cells (closing) can reduce moisture-vapor diffusion rates threefold [2].

Molecular diffusion is the only mechanism by which gases and vapors can move through unicellular foams (again assuming that pressure differentials will not rupture cell faces), which are therefore always much less permeable than open structures. At the same time their performance is far more sensitive to the variable composition and total pressure of contained fluids: atmospheric gases (air, moisture) attempt to enter these cells while blowing agents (carbon dioxide, fluorocarbons, hydrocarbons, steam, nitrogen) tend to diffuse out, each at its own characteristic rate.

The balance of this section will consider the effects of morphology on the diffusion-sensitive properties of essentially unicellular foams. But it is important to note here that, although transfer potentials exist in all cellular materials, overall gas composition and pressure cannot change when such products are isolated from their environment by efficient barriers.

1. Theory

Diffusion principles have been applied to develop mathematical models for foams containing dodecahedral [13], generalized volumetric [26], and cubical cells [27]. Constituent solids provide the major resistance to gas transfer, whereas thermal and/or concentration gradients provide the driving force since any specific vapor then exerts differential partial pressures on opposite sides of a cell face.

Fick's law defines the net rate at which a specific vapor diffuses across the membrane separating cell layers i and i + 1 below the foam surface i = 0 as

$$Q_m = \frac{K_p A_m (P_i - P_{i+1})}{X_m},$$

where K_p is the coefficient of permeability of cell-face material to that vapor at T° C, normally an exponential temperature function, A_m is the membrance area between adjacent cell layers, P_i is the partial pressure of vapor in the ith cell layer below the reference foam surface, and X_m is the cell-face thickness. An internal material balance [26] indicates that

$$P_{i,j+1} = \frac{P_{i-1,j} + P_{i+1,j}}{2}$$

and

$$\Delta = \frac{X_m V_x}{2RK_p A_m (T + 273)},$$

where P_{ij} is the partial pressure of specified vapor in the ith cell layer of a foam aged for j time increments, Δ is the duration of the standard time increment j for the specific vapor-solid combination at T° C, V_x is the gas-phase volume associated with one cell layer, and R is the universal gas constant (or appropriate modification for vapor).

The rate at which a gas penetrates a unicellular foam clearly depends as strongly on physical structure as on that gas-solid combination's inherent permeability. For example, the time required to achieve a given pressure change increases linearly with membrane thickness and thus with foam density when all other factors are constant.

Solving this numerical matrix for isothermal semiinfinite foam slabs cut to different thicknesses [26] indicated that the total amount of a vapor entering or leaving is constant only for a very short time: its net diffusion rate begins declining more rapidly as soon as partial pressure changes in the cell layer farthest removed from an external transfer surface. Increasing foam thickness or reducing cell size effectively retards overall composition changes; a vapor's mean partial pressure is directly proportional to thickness (expressed as a number of cell layers) at very short times, and to thickness squared at intermediate ages. Bonding a perfect barrier to one surface doubles the cut slab's effective thickness, and covering both faces stops diffusion completely. Bonding either or both surface(s) to permeable barriers retards composition and pressure changes significantly by increasing surface resistance.

Open cells naturally let gases bypass this "cascade" effect created by repeated diffusion through cell walls and across vapor-filled cavities. The solid's final distribution across each continuous cell face also influences rates because capillary flow reduces thickness much faster than transfer area. Cell elongation slowly increases the total area and reduces the thickness of the membranes within a unit volume of foam, thereby tending to accelerate diffusion, but its major effect results from larger changes in the number of layers per unit foam thickness. Twice as many cell walls oppose diffusion perpendicular to the direction of rise as resist parallel diffusion in the same thickness of one foam containing fully elongated equilateral cells.

The practical importance of diffusion phenomena is easy to illustrate in unicellular products.

2. Moisture-Vapor Permeability

The moisture vapor that permeates our atmosphere contributes to the premature failure of so many products (by promoting corrosion and mildew, water-logging low-temperature insulation, etc.) that designers routinely consider its effects. Steady-state water-vapor permeability, the yardstick commonly used to appraise the utility of a given material or composite, can be measured as a function of temperature and ambient humidity.

Table 2 shows that, like fibrous or granular materials, open-celled foams offer negligible resistance to the passage of moisture and that resistance generally increases with density in any product family.

TABLE 2

Vapor Diffusion through Various Materials

Material	Physical State	Density (lb/ft^3)	23° C WVP[a] (perm. -in.)
Air	Stagnant gas	0.08	120
Mineral wool	Loose fiber	6-18	116-100
Gypsum wall-board	Compact granular	47-49	13-19
Corkboard	Bonded cellular chips	7-11	10-2
Glass foam	Unicellular, friable	9-10	~0
Foamed polymers	Open cells, rigid-flexible	1-3	120-60
	Unicellular, tough	1-3	6-0.3

[a]Water-vapor permeability = grains of water per hour per square foot (vapor-pressure differential per inch of thickness).

Table 2 also illustrates the "hundredfold" water-vapor-permeability reduction associated with closing a foam's interconnected cells. Among essentially unicellular foams, low fractions of open cells increase water-vapor permeability less rapidly than predicted by linear interpolation [28]. Solid-phase chemistry, temperature, morphology, and transfer direction have relatively more influence on results. As might be expected, foams based on very hydrophobic molecules are least permeable when all other factors are constant. Water-vapor permeability rises with temperature at rates determined by the solid's properties.

Figure 5 isolates several morphological effects. It shows that doubling one unicellular foam's density reduced water-vapor permeability by about 1 perm. -in. It also indicates that a unicellular foam's permeability can vary significantly with cell size, cell elongation, and relative test direction.

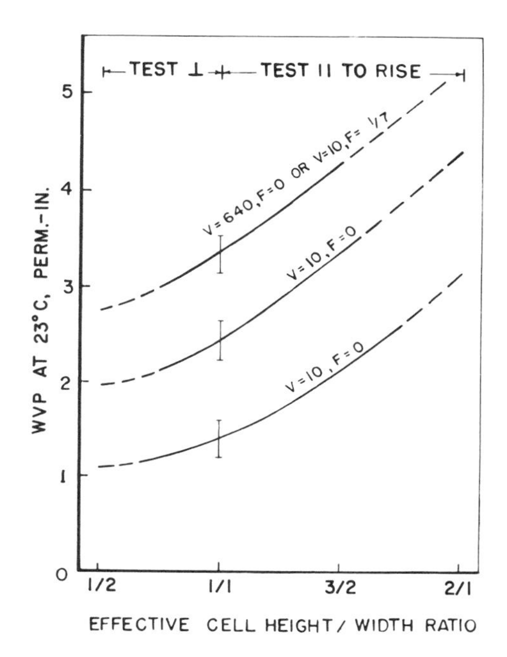

FIG. 5. Effects of morphology on the water-vapor permeabilities (WVP) of uniform specimens cut from foams based on one rigid urethane polymer. Top curve, 1.5-lb/ft^3 foam; middle curve, 1.5-lb/ft^3 foam; bottom curve, 3-lb/ft^3 foam. V is the average cell volume, 10^{-6} in.^{3}; F is the fraction of cells open. Reprinted from Ref. [1], p. 11, by courtest of the American Society for Testing and Materials.

Opening one-seventh of a foam's cells or increasing the mean diameter fourfold raised water-vapor permeability by about 1 perm.-in. Elongated cells transmitted vapor up to 2 perm.-in. more rapidly parallel than perpendicular to the rise direction.

Small cells impede moisture transfer more effectively than larger units. In open-celled products the larger number of finer edges functions mainly by blocking convection currents. Calculations show that the number, area, and thickness of regular geometric faces should vary to maintain water-vapor permeability independent of cell size in unicellular products. Since laboratory data showed that foam solids were actually less effective barriers in larger closed cells, they imply that solids drain more readily toward the edges of potentially thicker faces.

These observations were based on the diffusion of one common, important, easily measured vapor. Although other gases diffuse through foams more or less readily at rates that may differ by several orders of magnitude, the same principles and therefore the same design considerations apply to them as well.

3. Dimensional Stability

Total pressures within closed foam cells rise and fall with both gas diffusion and temperature. Differences between internal and atmospheric pressures establish pneumatic forces promoting dimensional change. Since these driving forces are contained by the solid matrix, the stability of unicellular foams generally improves with higher density and inherently stiffer constituent solids. Open-celled foams are not subject to pneumatic stresses, while partially interconnected structures relieve them more quickly than their unicellular counterparts.

Closed-cell foams are first exposed to pressure differentials when they cool and tend to shrink immediately after production. Commercial extruded sheet [29] and cut slabstock [30] both show spontaneous post-expansion as air subsequently diffuses inward. Foam volume may increase several percent during a period lasting minutes to months. This stress-relaxation process is most noticeable in foams containing elongated cells, which try to resume their preferrred isotropic configuration. Dimensional changes perpendicular to the foam-rise direction are then larger than those that are parallel.

We have noted that ideal blowing agents and atmospheric gases diffuse out and in, respectively, at rates determined by their respective permeability coeffficients and partial pressure differentials. Figure 6 demonstrates that this phenomenon can generate surprisingly large total pressure differentials, and especially gradients, as specimens age isothermally. A pressure maximum develops quickly just below exposed surfaces, generally growing larger and moving toward the center as time passes. Internal pressures ranged from 0.67 to nearly 2 atm. Differentials across individual cell faces, the forces most conducive to rupture and instability, tended to be highest near foam surfaces; they reversed direction in time but were quite persistent.

The blowing agent in these examples diffused more slowly than air, thereby generating internal total pressures higher than atmospheric. Blowing agents that diffuse more rapidly than air leave a temporary vacuum by the same mechanism. The pressures within air-blown foams are unbalanced by heating or cooling; differentials are therefore relatively small (less than 3 psi per 55° C change) and begin decaying the instant they are established. Similar considerations apply to any foam sealed within impermeable barriers, which prevent the diffusional decline of pressure gradients but normally stabilize the foam mechanically, unless larger differentials are produced by cooling below the point at which the blowing agent's partial and vapor pressures are equal, so the blowing agent begins condensing.

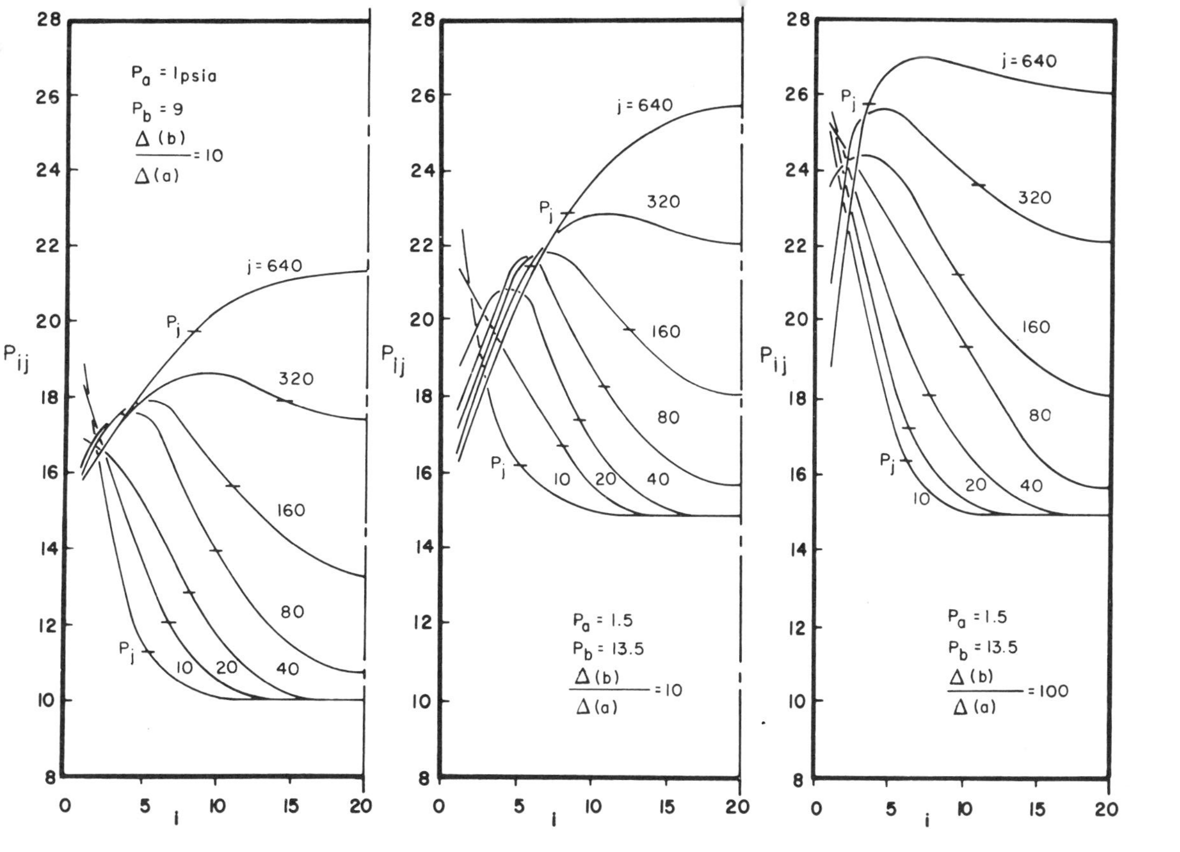

FIG. 6. Total pressure of gases in unicellular foam slabs cut 40 cell layers thick and aged in dry air (15 psia) at constant temperature. P_{ij} is the total absolute pressure in the ith cell layer below the cut surface of foam aged j standard time increments for air; P_a is the original pressure of air within the foam, psia; P_b is the original pressure of the blowing agent within the foam, psia; $\Delta(a)$ is the duration of the standard time increment for air--about 1 day for a typical rigid 2-lb/ft^3 urethane foam at room temperature; $\Delta(b)$ is the duration of the standard time increment for the blowing agent. Reprinted from Ref. [26], p. 228, by courtesy of Technomic Publishing.

On a different time scale, and with variable rather than steady-state gradients, these driving forces are opposed by the morphological factors that enhance mechanical performance and influence water-vapor permeability. Cell faces too weak (low density or highly drained) or brittle may relieve pressure by rupturing, whereas cells based on "tougher" solids may swell, shrink, or even collapse under sufficiently large imposed differentials.

Cut-foam specimens are commonly aged hot and cold in laboratory tests. Cold-aging can be a fairly realistic "go-no go" test in which unicellular foams are stable above some threshhold density-temperature combination but shrink abruptly or rupture cells at lower levels. Humid-aging at elevated temperatures induces progressive changes that can only be used to compare similar foams [30]. The results obtained from any such test clearly vary with manufacturing conditions, formulations, specimen morphology, and age, and do not constitute desirable accelerated-aging tests because short times under vigorous conditions are generally not equivalent to longer times under more realistic conditions. Great care must be exercised in interpreting results for any purpose other than quality control. Product durability is best evaluated by mathematical analysis or by testing prototypes of the assembled composite.

C. Heat Transfer

Since thermal insulation provides a major stimulus for many applications of low-density-foam products, factors affecting this variable quality have been studied extensively [2, 13, 26, 31-37]. Figure 7 shows that the 75° F thermal conductivity coefficient (k factor) can range from 0.09 to 0.28 Btu/(h) (ft^2) (° F/in.), respectively, for low-density unicellular fluorocarbon-blown systems and for air-filled materials (at 75° F, k = 0.06 and 0.18 Btu/(h) (ft^2)(° F/in.) for trichlorofluoromethane and for air, respectively). Calculations assumed that a 22° C/in. gradient was imposed on 1.5- and 3.0-lb/ft^3 specimens containing regular undrained geometric cells. The pairs of solid and broken lines respectively represent typical and large cells, the lower line in each pair corresponding to the lower density. Group A lines describe freshly fluorocarbon-blown foams, whose k factors drift through group B and slowly approach group C as the cut specimens age at room temperature. Group C represents equilibrium conditions (literally infinitely old, unicellular, vapor-blown foams) and is therefore more realistic for closed-cell air-blown products and for open-celled foams whose faces cracked or pinholed without retracting. Group D describes air-filled foams whose cell faces either have retracted or offer negligible resistance to radiant-heat transmission.

Figure 7 then indicates that open-cell content, cell size, and foam density are morphological parameters that dominate thermal conductivity. Figure 8 further illustrates the cascade, diffusion-controlled, k-factor

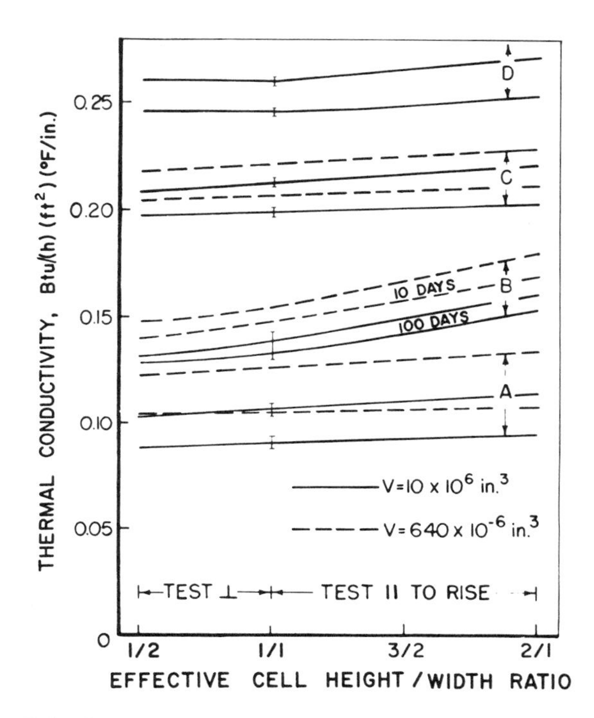

FIG. 7. Calculated effects of cell structure on the thermal conductivity (k factor) of cut 1-in.-thick foam specimens at a mean temperature of 75° F. Reprinted from Ref. [1], p. 12, by courtesy of the American Society for Testing and Materials.

drift phenomenon in slabs cut from one foam to different thicknesses. Similar results would have been obtained if the number of cell layers per specimen were changed by varying cell size or shape.

The same foam becomes the core of a lightweight composite insulating panel, potentially the wall of a refrigerated rail car or truck trailer, in Fig. 9. With diffusion through the cool face only, the thin skin that forms naturally at molded surfaces limits calculated 20-year k-factor drift to that obtained when a slightly thicker cut specimen aged 400 days in the laboratory. A more realistic panel, faced on the cool side with plastic sheeting to prevent mechanical damage, drifts negligibly during 20 years of service; its in-place k-factor at the end of this period approximates the initial conductivity of a cut 1-in.-thick laboratory test specimen.

The effective thermal conductivity of any unicellular foam can increase dramatically if the material is installed in such a way that moisture vapor enters its structure more rapidly than it diffuses out [35, 36]; water may then condense at low temperatures to "short-circuit" the foam's insulating

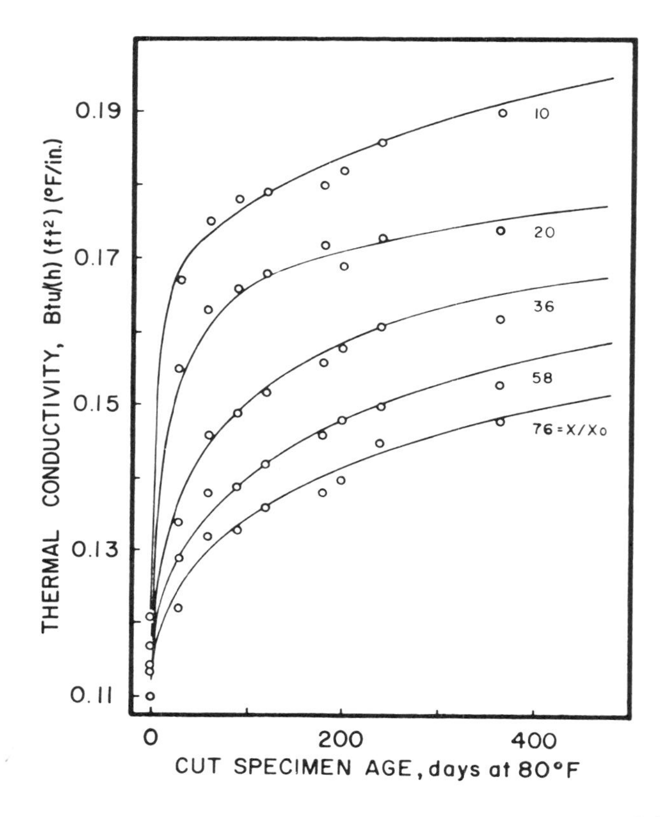

FIG. 8. Effective thermal conductivities at 75° F of a 2-lb/ft^3, fluorocarbon-blown, rigid urethane foam. X/X_0 = specimen thickness in cell layers (39 layers per inch); points are experimental data, and curves are calculated from the heat transfer-diffusion model. Reprinted from Ref. [37], p. 208, by courtesy of Technomic Publishing.

capacity (at 75° F, k = 0.13 and 4.2 Btu/(h) (ft^2) (° F/in.) for water vapor and for liquid, respectively). This problem can and must be avoided by properly designing the foam-containing composite. Conductivities generally rise with temperature as a function of foam composition. Since many blowing agents are low-boiling-point liquids rather than perfect gases, their partial and vapor pressures can balance at reasonable temperatures. Like water, they begin condensing within closed cells if cooled below this point, progressively increasing the relative amount of more conductive air in the gas phase and increasing the k-factor to approach the upper limit set by an air-filled foam at very low temperatures [2, 32]. Aside from the reduced

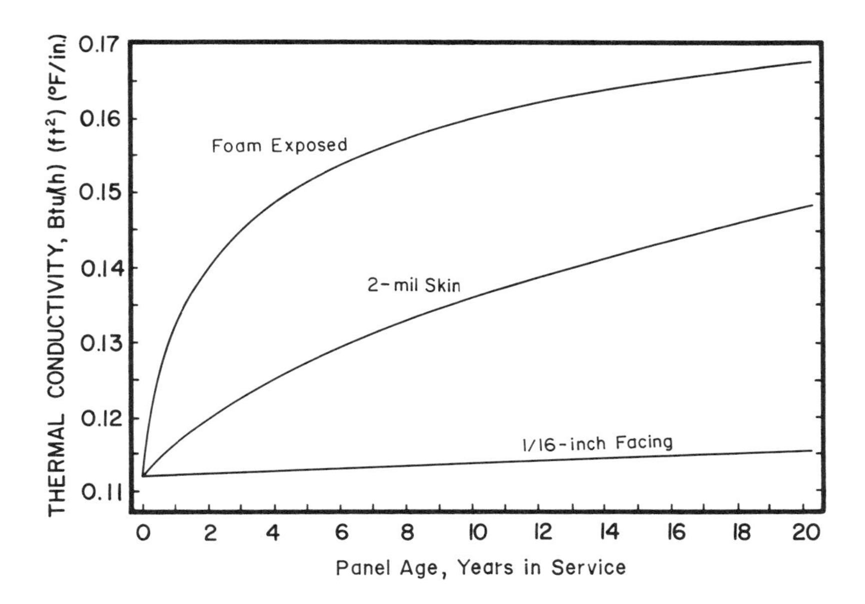

FIG. 9. Thermal conductivities at 80° F (mean) of 1.8-in.-thick, fluorocarbon-blown, rigid-urethane-foam cores in panels aged with an impermeable skin on the warm, 120° F side and diffusion through the cool, 40° F face. Reprinted from Ref. [37], p. 211, by courtesy of Technomic Publishing.

rate constants and partial pressures associated with the low temperatures at which they might occur, neither of these potentially important design considerations has any direct effect on diffusion or the morphological factors that dominate it in unicellular foams.

Thus, as was the case for dimensional stability, we see that the performance of low-density, unicellular, vapor-blown foams is significantly better in service than in most conventional laboratory tests. Changing internal gas composition is responsible for k-factor drift, and changing internal gas pressure promotes dimensional instability, but both diffusional mechanisms are controlled by structural design, service temperature, and foam morphology as well as the inherent properties and composition of the gas-solid combination foamed.

D. Miscellaneous

Relatively little is known about the direct effects of morphological variables on other useful properties of cellular materials. Studies of acoustical properties [38] did indicate that 1-in.-thick layers of foams provide noise-reduction coefficients in the range of commercial sound-absorbing products. Foam performance was reasonably insensitive to density, but small open cells were superior to larger or closed ones. A thin surface film further improved the efficiency of open-celled flexible foam.

IV. DISCUSSION

This review of factors affecting performance indicates that each morphological parameter affects every foam property. For example, mechanical strengths and moduli, vapor-diffusion rates, and thermal conductivities are all largest parallel to the rise of foams containing highly elongated cells. Transfer rates are much higher and dimensional stability much better in open-celled than in unicellular products. Dimensional stability is also enhanced by isotropic structures in closed foam cells.

Density is the only morphological variable that affects product cost directly, so its effects are relatively well defined. Although cell size, shape, and interconnection all influence foam properties, their relative contributions to specific properties vary widely. Within-cell distribution of solids is a parameter whose important effects are poorly understood, largely because it is difficult to measure and control.

The practical significance of standard-laboratory-test data is inherently limited by the fact that they are normally obtained from cut specimens foamed by different processes than those used to make specific products. Under such conditions they are useful for comparing the relative quality of similar foams and provide a basis for tentative product-design considerations. But cellular materials generally become internal elements of composite assemblies, where their morphology may be different and their properties complement those of the other components present. Proper interpretation and application of measured properties therefore offer significant challenges to both the foam technologist and the design engineer.

There is considerable incentive for continuing and progressively more sophisticated studies of foam morphology and its interactions with composition and manufacturing processes. The optimum balance can greatly improve the performance of a specific product at lower overall cost.

REFERENCES

1. R. H. Harding, in Resinography of Cellular Plastics, ASTM STP 414, American Society for Testing and Materials, Philadelphia, 1967, pp. 3–17.
2. R. H. Harding, J. Cell. Plastics, 1, 385 (1965).
3. W. T. Kelvin, Phil. Mag. J. Sci., 24, 503 (1887).

4. G. Gioumousis, J. Appl. Polymer Sci., 7, 947 (1963).
5. R. E. Williams, Science, 161, 276 (1968).
6. R.E. Wright, in Resinography of Cellular Plastics, ASTM STP 414, American Society for Testing and Materials, Philadelphia, 1967, pp. 19-27.
7. A. R. Ingram, R. R. Cobbs, and L. C. Couchot, in Resinography of Cellular Plastics, ASTM STP 414, American Society for Testing and Materials, Philadelphia, 1967, pp. 53-67.
8. E. A. Blair, in Resinography of Cellular Plastics, ASTM STP 414, American Society for Testing and Materials, Philadelphia, 1967, pp. 84-93.
9. W. J. Remington and R. Pariser, Rubber World, 138, 261 (1958).
10. R. H. Harding, Mod. Plastics, 37, 156 (1960).
11. D. M. Rice and L. J. Nunez, SPE Journal, 18, 321 (1962).
12. G. W. Schael, J. Appl. Polymer Sci., 11, 2131 (1967).
13. R. H. Harding, Ind. Eng. Chem. Proc. Des. Develop., 3, 117 (1964).
14. R. E. Jones and G. Fesman, J. Cell. Plastics, 1, 200 (1965).
15. D. Finlayson and A. H. Radcliffe, J. Cell. Plastics, 4, 474 (1968).
16. C. J. Benning, J. Cell. Plastics, 3, 174 (1967).
17. A. N. Gent and A. G. Thomas, J. Appl. Polymer Sci., 1, 107 (1959).
18. A. N. Gent and A. G. Thomas, J. Appl. Polymer Sci., 2, 354 (1959).
19. V. A. Matonis, SPE Journal, 20, 1024 (1964).
20. W. L. Ko, J. Cell. Plastics, 1, 45 (1965).
21. R. Chan and M. Nakamura, J. Cell. Plastics, 5, 112 (1969).
22. C. J. Benning, J. Cell. Plastics, 5, 40 (1969).
23. R. H. Harding and C. J. Hilado, J. Appl. Polymer Sci., 8, 2445 (1964).
24. A. N. Gent and A. G. Thomas, Rubber Chem. Technol., 36, 597 (1963).
25. B. Beals, F. J. Dwyer, and M. Kaplan, J. Cell. Plastics, 1, 32 (1965).
26. R. H. Harding, J. Cell. Plastics, 1, 224 (1965).
27. E. F. Cuddihy and J. Moacanin, J. Cell. Plastics, 3, 73 (1967).
28. C. J. Hilado and R. H. Harding, J. Appl. Polymer Sci., 7, 1775 (1963).
29. L. Leese and D. G. Gray, Brit. Plastics, 40, (3), 103 (1967).
30. R. H. Harding and C. J. Hilado, Proc. SPI Cellular Plast. Div. Tech. Conf., 7th, New York, April 1963, Section 3A, The Society of the Plastics Industry, New York.
31. R. E. Skochdopole, Chem. Eng. Progr., 57, (10), 55 (1961).
32. G. A. Patten and R. E. Skochdopole, Mod. Plastics, 39, (11), 149, 191 (1962).
33. D. J. Doherty, R. Hurd, and G. R. Lester, Chem, Ind., 1340 (1962).
34. F. O. Guenther, SPE Trans., 2, 243 (July 1962).

35. W. Schmidt, Kunststoffe, 53, 413 (1963).
36. M. M. Levy, J. Cell. Plastics, 2, 37 (1966).
37. R. H. Harding, J. Cell. Plastics, 2, 206 (1966).
38. G. L. Ball, M. Schwartz, and J. S. Long, Off. Digest, 425, 817 (1960).

Chapter 18

THERMAL DECOMPOSITION AND FLAMMABILITY OF FOAMS

Paul E. Burgess, Jr.
and
Carlos J. Hilado

Union Carbide Corporation
South Charleston, West Virginia

I. INTRODUCTION

In recent years cellular plastics have received considerable attention and widespread acceptance in many market areas because of the advantages they offer in thermal insulation, cushioning, packaging, flotation, and decorative molded pieces. Rapidly growing emphasis on safety requirements, motivated by a high degree of safety consciousness among researchers and regulatory bodies alike, has resulted in tremendously increased efforts to improve the flammability characteristics of plastic materials designed for use in some of these areas. These efforts have resulted in record sales of flame retardants to the plastics industry.

A cellular plastic is an organic material and exhibits some degree of fire hazard under certain conditions. Some plastics ignite and burn more readily than others, but all plastics can be modified to impart some degree of flame resistance. The performance of a cellular plastic in specific fire tests is often a major factor in its acceptance for a particular application, and the lack of adequate flame retardance is thought to be the largest single barrier to extensive new markets for cellular plastics. Consequently considerable effort has been expended in the study of the flammability characteristics of plastics.

The purpose of this chapter is to present briefly the various combustion- and flame-retardance concepts that have been advanced. These concepts evolved from the many excellent papers describing the flammability characteristics and flame-retardance requirements of various polymeric materials that have been presented at various institutional and society functions or published in the literature [1-106]. Individual papers will be referred to in the discussions that follow. Application of these basic principles to specific systems will be found in other chapters of this book and the references cited here.

II. THERMAL DECOMPOSITION AND FLAMMABILITY ASPECTS

A. Fire-Hazard Characteristics

The risk of fire is an ever-present threat to all who fabricate, handle, store, or otherwise employ plastic materials. The most important fire-hazard characteristics of cellular plastics [17, 21-23, 29, 32, 34, 38, 55, 70] are the following:

1. Thermal instability.
2. Ease of ignition.
3. Flame spread.
4. Fire endurance.
5. Fuel contribution.
6. Smoke density.
7. Products of pyrolysis and combustion.

The thermal instability of a cellular plastic may be defined as the ease by which heat produces changes in the chemical structure of the polymer network. These changes may involve simple bond-rupturing dissociations or reaction reversals and provide more volatile components, or they may result in extensive pyrolysis and fragmentation of the polymer. This characteristic provides a measure of fire hazard in that a more thermally stable polymer is less likely to ignite and contribute to a conflagration than a less stable one.

Ease of ignition may be defined as the facility with which a cellular plastic, its volatile components, or its pyrolysis products can be ignited under given conditions of temperature and oxygen concentration. This characteristic provides a measure of fire hazard in that a material whose ignition temperature is significantly higher than that of another would be less likely to contribute to a conflagration, all other factors being the same in both cases.

Flame spread may be defined as the rate of flame-front travel under given conditions of burning. This characteristic provides a measure of fire hazard in that surface flame spread can transmit fire to more flammable materials in the vicinity and thus enlarge a conflagration, even though the transmitting material itself contributes little fuel to the fire. Surface flame spread has in recent years been the flammability characteristic that has received the most attention.

Fire endurance may be defined as the resistance offered by a cellular plastic to the passage of fire, normal to the exposed surface over which flame spread is measured. This characteristic provides a measure of fire hazard in that a material that will contain a fire represents more protection than one that will give way before it, all other factors being equal. For example, a cellular plastic that develops an intumescent coating or strong char on exposure to fire would resist flame penetration much longer than one that melts away, both materials being equal in ease of ignition, surface flame spread, and fuel contribution.

Fuel contribution may be defined as the heat produced by the combustion of a given weight or volume of cellular plastic. This characteristic provides a measure of fire hazard in that a material that burns with the evolution of little heat per unit quantity burned will contribute appreciably less to a conflagration than a material that generates large amounts of heat per unit quantity burned. The actual quantity of generated heat is a function of the heat developed per unit quantity burned (fuel contribution) and the quantity of material burned; the latter is a function of the area of exposed surface burned (flame spread) and the extent to which fire has penetrated into the material (fire endurance).

Smoke density may be defined as the degree of light or sight obscuration produced by the smoke from the burning material under given conditions of combustion. This characteristic provides a measure of fire hazard in that an occupant has a better chance of escaping from a burning structure if he can see the exit, and a fireman has a better chance of putting out a fire if he can see where it is. The effect of smoke on visibility is the subject of increasing interest.

The products of the pyrolysis and combustion of cellular plastics are usually volatile gases, entrained solid particles (smoke), and solid carbonaceous char, or residue. A liquid phase can result if melting occurs to an appreciable extent before charring takes place, and ignition of a liquid increases the flame-spread hazard. Complete disintegration of the foam can be a significant hazard if it is relied on for structural strength. The toxicity of the combustion gases can be a serious hazard.

B. The Combustion Cycle

The combustion cycle of cellular plastics can be described as a five-step destructive process comprised of heatup, dissociation or degradation, volatilization, ignition, heat transfer, and propagation [21, 28, 32, 38, 89, 91] . The combustion cycle is shown diagrammatically in Fig. 1.

In the first step of the combustion cycle heat from an external source progressively increases the temperature of the cellular material. The rate of temperature increase varies with the density, heat capacity, and thermal conductivity of the cellular polymer. Cellular polymers, which are normally good thermal insulators, tend to promote a very rapid temperature increase at the affected area, thus producing a more rapid progression to the second step of the combustion cycle.

The second step of the combustion cycle involves dissociation or chemical reversals, decomposition or degradation, and volatilization. The products may be combustible or noncombustible, and they may be gases, liquids, entrained polymer fragments, and carbonaceous char. Volatiles and evolved gases expand into adjacent areas, where sufficient oxygen or oxidizing agents may be present to effect the third step of the combustion cycle.

The third step of the combustion cycle involves ignition of the combustible gases. If sufficient oxygen is present, combustion begins and the heat of combustion provides the source for the fourth step of the combustion cycle.

The fourth step of the combustion cycle involves heat transfer. After ignition, the heat of combustion increases the temperature of polymer in the decomposition or degradation area. Heat transfer to adjacent polymer may be effected by conduction, by convection created by the expanding combustion and evolved noncombustible gases, by diffusion, and in extreme cases by radiation from char residue heated to incandescence.

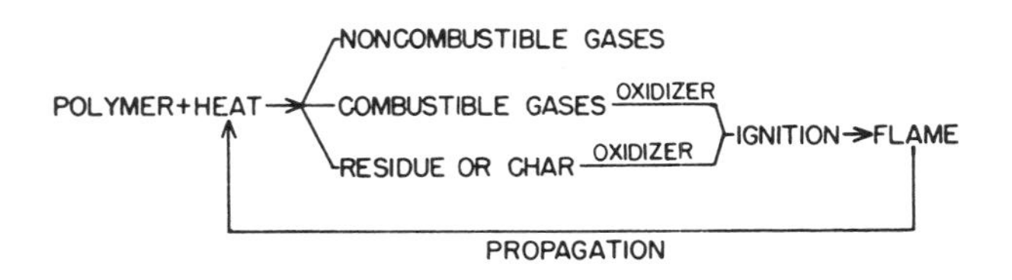

FIG. 1. The combustion cycle.

The fifth step of the combustion cycle is flame propagation. The heat transferred from the combustion area to adjacent areas produces a repetition of the first four steps of the combustion cycle, and combustion continues until a step of the cycle is interrupted or the cellular material is consumed.

C. Considerations of Polymer Composition

Foamed polymers are characterized by a relatively low mass per unit volume, highly developed surface area, and low thermal conductivity. The combustibility of foamed polymers is governed by the thermal stability of the polymer and by the flammability of thermal degradation products [16]. Therefore polymer combustibility is the product of all the energy factors of the molecule. The most important energy factors related to polymer combustibility are cohesive energy, bond-dissociation energy, and heat of combustion [17, 21, 23, 32, 34, 38].

1. Cohesive Energy

Cohesive energy is concerned with the secondary bond forces of molecular aggregates in liquid or solid phases. It is the total energy necessary to remove a molecule from a liquid or solid material. Low cohesive energies mean low-melting-point and easily volatilized molecules, and both factors contribute to combustibility. Polar groups and increases in molecular weight contribute to greater cohesive energy. Secondary bonding between hydrogen and strongly electronegative atoms like oxygen strengthens the intermolecular forces of polymers, thus increasing the cohesive energy. When the cohesive energy exceeds primary bond energies, molecules decompose before they volatilize. The cohesive-energy values for some typical structural units of foamed polymers are presented in Table 1.

TABLE 1

Molar Cohesive Energy of Polymer Structural Units[a]

Group	Cohesive energy (kcal/mole)
Hydrocarbon ($-CH_2-$)	0.68
Ether (-O-)	1.00
Ester (-COO-)	2.90
Aromatic ($-C_6H_4-$)	3.90
Amide (-CONH-)	8.50
Urethane (-OCONH-)	8.74

[a] Data from Ref. [17].

2. Bond-Dissociation Energy

Bond-dissociation energy is the energy required to rupture structural chemical bonds. A high bond-dissociation energy is desirable for the basic structural units of the polymer in order to prevent or delay the degradation or breakdown of the polymer molecule. The bond-dissociation energies of some typical structural linkages commonly encountered in foamed plastics are presented in Table 2.

Clearly, polymers with C=C or C=O structural units are decidedly more thermally stable than polymers containing C–C and C–O structural units. However, since all organic plastic polymers exhibit some degree of combustibility, it may be desirable to have such structural units as C–Br and C–Cl, which have relatively low bond-dissociation energies and can be made available readily to interrupt combustion mechanisms.

3. Heat of Combustion

The heat of combustion is the heat or energy that is liberated during combustion. A burning polymer liberating energy in excess of that required for ignition will propagate, but when less energy than required for ignition is liberated, the fire will extinguish itself. The heats of combustion of some typical structural units commonly encountered in foamed plastics are presented in Table 3.

Other structural characteristics that have been related to polymer combustibility are crosslink density [18, 34, 38, 87, 89] and aromaticity [32, 34, 38, 89].

III. CONCEPTS OF FLAME RETARDATION

Flame retardants are of two general types, additive and reactive. The additive type, particularly the filler variety, is used extensively in noncellular polymers, but the reactive type is rapidly becoming the workhorse of foam polymer systems [13, 18, 19, 30, 88]. In reviewing the history of flame retardants it becomes obvious that the bulk of flame retardants being used today are derived from the same elements as were the flame retardants employed at the turn of this century. Basically these are the elements of groups V and VII of the periodic table. The group V elements most frequently found in flame retardants are nitrogen, phosphorus, antimony, and to some degree the heavier elements. The group VII elements most frequently found in flame retardants are fluorine, chlorine, and bromine. From the group III elements, compounds of boron are frequently employed as flame retardants.

A. General Theories

Research and development on flame retardants has been and continues to be largely empirical. On this basis the most effective flame retardants

TABLE 2

Bond-Dissociation Energies of Typical Structural Units of Polymers[a]

Bond	Bond-dissociation energy (kcal/mole)
C–N	49–60
C–C	59–70
C–O	70–75
N–H	84–97
C–H	87–94
O–H	101–110
C=C	100–125
C=O	142–166
C–Br	54
C–C1	67

[a]Data from Ref. [17].

TABLE 3

Heats of Combustion of Chemical Bonds[a]

Type of bond	Heat evolved (kcal/mole)
C≡C	203.2
C=C	121.8
C–C	52.9
C–H	52.7
C–O	15.0
C=O	0

[a]Data from Ref. [17].

have been found to be compounds of phosphorus, halogen, and antimony or compounds containing two or more of these elements. Some general theories [1, 19, 21, 27, 88] for retarding the combustion of flammable materials have been advanced.

1. Coating

The coating theory suggests the formation of a layer or coating capable of either excluding or significantly limiting the access of oxygen or oxidizing atmosphere. A flame retardant may fuse and form an impervious coating, or it may itself intumesce or promote polymer intumescence to give a carbonaceous or foamed structure that excludes or limits access of oxygen and also effectively insulates the underlying polymer against further thermal degradation. Examples of coating materials are metallic oxides, inorganic salts, mineral fillers, and condensed phosphorus acids and anhydrides. In addition to coating, the flame retardants may contribute such gases as carbon dioxide, water vapor, and ammonia to the flame zone and thus aid in interrupting the combustion cycle.

2. Thermal

The thermal theory suggests agents capable of dissipating heat from the affected area by virtue of endothermic changes or by conduction. The flame retardant may undergo phase transitions or endothermic changes in fusion, sublimation, or decomposition which result in the dissipation of energy from the affected area. If used in sufficient quantity, the flame retardant may serve to conduct heat from the source and to distribute it over a large area, thus interrupting the heatup cycle of combustion.

3. Gas

The gas theory suggests agents that will decompose at elevated temperatures to produce non-combustible gases to dilute the flammable gases and thus limit access of oxygen, or to blanket the combustion region, thus reducing or limiting the access of oxidizing atmosphere. Quenching [18, 62] action through rapid gas evolution is an integral part of the gas theory. Examples are oxides of carbon and sulfur, nitrogen, ammonia, halogen compounds or halogen acids, and water vapor.

4. Chemical

The chemical theory suggests agents that are capable of producing species that can influence or inhibit the combustion cycle. Individual species and mechanisms are as uncertain as the complex reactions involved. One proposed mechanism involves changes that alter the course of oxidation to favor a thermochemical path less conductive to flame propagation by promoting the formation of carbon monoxide rather than carbon dioxide [1, 21]:

Reaction	Heat of reaction (kcal/mole)
$C + O_2 \rightarrow CO_2$	94.4
$C + 1/2O_2 \rightarrow CO$	26.4

Yet another mechanism involves the destruction of the very-high-energy ·OH radical species in the combustion region, replacing it with a radical of lower energy which is less effective in continuing the chain reactions of the combustion process [19]:

Propagation:

$$CO + \cdot OH \longrightarrow CO_2 + H\cdot$$

$$H\cdot + O_2 \longrightarrow \cdot OH + O\cdot$$

Inhibition:

$$\cdot OH + HX \longrightarrow H_2O + X\cdot$$

$$X\cdot + RH \longrightarrow HX + R\cdot$$

A third mechanism assumes catalysis of reactions leading preferentially to carbonaceous chars, as in the Lewis-acid-catalyzed dehydration of cellulose.

Although the chemical and gas theories are the most widely accepted, it is possible that all of these concepts can operate simultaneously.

B. Mechanisms

The mechanisms of flame retardance are not completely understood. However, on the basis of empirical results, various concepts have been advanced to describe the mechanisms involved in flame retardance or inhibition. These proposed mechanisms will be discussed individually as they have been applied to the elements most generally considered to provide the most effective flame retardance.

1. Phosphorus-Based Flame Retardants

Phosphorus is believed to perform most of its flame-retardant function in the condensed or solid phase. If a phosphorus compound is present in a cellular plastic and is capable of chemical transformation to phosphorus oxides and condensed phosphorus acids by combining with water vapor from decomposing hydrocarbon moieties, then it may influence the mechanisms of combustion as follows [1, 16, 21, 28, 38, 44, 88]:

1. By coating the surface of hot, decomposing materials, thus limiting access of oxygen.

2. By favoring the formation of carbonaceous char by redirection of the oxidation of carbon by an oxidation-reduction route whereby carbon is oxidized to carbon monoxide at the expense of an activated phosphorus acid, which is reduced. The reduced phosphorus species can recombine with available oxygen at the carbonaceous surface and provide a heat-transfer route whereby the energy from oxidation of the low-valency phosphorus species can be dissipated to the surrounding atmosphere. Perhaps the oxidized phosphorus species can then coat adjacent surfaces where repetition of the oxidation-reduction cycle can occur. The overall effect is that less energy is expended at the polymer surface or combustion region and the combustion cycle is interrupted.

3. By serving as a halogen-transport agent. If halogen is also present in the polymer, then phosphorus may influence the mechanisms of combustion by serving as a halogen-transport agent or catalyst for halogen-atom formation at the surface of decomposing polymer. Activated species containing phosphorus-halogen bonds or phosphorus oxyhalides are distinct possibilities.

2. Halogen-Based Flame Retardants

Halogens are believed to perform most of their flame-retardant function in the gaseous phase, either by quenching or by redirection or termination of the chemical reactions involved in combustion. The exact mechanisms are uncertain, but the following theories have been proposed:

1. The theory of flame quenching has been related to the complete halogen-containing molecule and the particular electron-capture efficiency of the molecule [4, 10, 26, 62] and also to local high-concentration regions of halogen products of pyrolysis which serve to snuff out portions of the local flame front, thus altering the total heat balance and favoring further quenching.

2. The theory of flame inhibition in vapor flames [2, 3, 19, 26, 33, 62] postulates that the active atoms and radicals responsible for flame propagation are removed by halogen species by producing less active species:

Free-radical combustion mechanism:

$$\text{fuel} + O_2 \xrightarrow{\text{fire}} CO + CO_2 + H_2O$$

Propagation:

$$CO + \cdot OH \rightarrow CO_2 + H\cdot$$

$$H\cdot + O_2 \rightarrow \cdot OH + O\cdot$$

Inhibition:

$$\cdot OH + HX \rightarrow H_2O + X\cdot$$

$$X\cdot + RH \rightarrow HX + R\cdot$$

3. Another theory proposes a mechanism characterized by catalyzed oxidation and halogenation followed by dehydrations and dehydrohalogenations to produce double bonds, water, and hydrohalogen acids until most of the hydrogen in the polymer is expended, thus avoiding the formation of flammable gases and leaving a residual structure of increasing aromaticity until a graphitelike state is achieved [44, 88, 93].

4. It is generally conceded that organic bromine compounds are more effective than the corresponding organic chlorine compounds in reducing polymer flammability. Likewise aliphatic and alicyclic halogen compounds are generally more effective than aromatic halogen compounds [16, 18, 30, 32, 38, 50, 88, 91]. These differences are governed by the energies of dissociation of bonds [26, 50, 70, 82] between the halogen atom and its parent compound atoms, usually carbon. The bond-dissociation energies of some halogen compounds are given in Table 4.

For greatest effectiveness the flame retardant should be available to exhibit its greatest activity at the precise condition of time and space at which the combustion process is most vulnerable to its particular mechanism of extinction. It could be expected from differing bond-dissociation energies that mixtures of flame retardants or compounds containing more than one halogen atom might promote controlled availability of the halogen, thus enhancing and extending combustion-retarding performance.

3. Synergism

Synergism is defined as the capacity of distinct species to react cooperatively to produce a total effect that is greater than the sum of two effects taken independently. Synergism has been employed frequently in the art of reducing the flammability of polymers. Flame-inhibiting synergists can generally be classified into two groups: (a) the group V elements of the periodic table, particularly the heavy-metal oxides and sulfides, and (b) free-radical inhibitors. The economic value of synergists is readily apparent.

TABLE 4

Bond-Dissociation Energies[a]

Organic radical	Bond-dissociation energy (kcal/mole)			
	Halogen atom			
	F	Cl	Br	I
H–		103	88	
CH_3–	108	81	67	54
C_2H_5–		81	67	
CH_2=CH–		104	87	
nC_3H_7–		82	68	
$(CH_3)_2CH$–		73	59	
$(CH_3)_2C$–		78	64	
CH_2=CH–CH_2–		58	48	
C_6H_5–	145	86	71	57
$C_6H_5CH_2$–		68	51	39
C_6H_5CO–		73	57	
CCl_3–	102	68	49	
CF_3–	121	83	65	

[a]Data from Refs. [26], [50], and [107].

The oxides and sulfides of antimony and bismuth are perhaps the oldest flame-inhibiting synergists in common use, and by far the most important of these are the oxides of antimony. However, antimony trioxide by itself is ineffective [18, 35, 80, 82, 88] in reducing flammability, perhaps because it has too high a melting point to effectively serve as a fluxing agent or coating medium and also does not have the capacity to dehydrate polymers as do the condensed acids and oxides of phosphorus. The mechanism by which antimony oxide and halogens perform their flame-retarding function has not been firmly established. The formation of antimony halide or oxyhalide intermediates [38, 44, 52] is generally accepted, but it is not known whether the flame-retarding function proceeds through inhibition of the free-radical mechanisms of flame propagation [52, 80] or by blanketing or coating [19, 52] the affected polymer area, or whether these possible routes are simultaneously operative.

In addition to the heavy-metal oxides and sulfides of group V, synergism has been attributed to phosphorus and to a lesser degree to nitrogen [18, 75]. Combinations of phosphorus and halogen are considered by many authors to exert a synergistic effect [16, 26, 31, 38, 53]. The number of commercially available organophosphorus-halogen flame retardants lend credence to the idea of enhancement by combining halogen and phosphorus in the same molecule [50].

Synergists classified as free-radical initiators are generally thermally unstable organic compounds, such as peroxides and nitroso and azo compounds [15, 18, 19, 26, 30, 31, 33, 38, 50, 62, 80]. It has been suggested that free-radical initiators inhibit flame propagation by (a) accelerating the decomposition of halogen compounds to produce flame-retardant species under favorable conditions within the combustion region; (b) causing a lowering of the active oxidation temperature of the polymer, favoring simultaneous decomposition of the polymer and halogen flame retardant; (c) accelerating the decomposition of polymer and subsequent reactions between polymer fragments and halogen flame retardant, thus lengthening the residence time of active flame-retardant species in the combustion region.

IV. FLAMMABILITY TESTS AND GEOMETRIC CLASSIFICATION

The flammability characteristics of cellular materials are more difficult to define than those of solid materials, because the physical structure of the solid-gas combination is extremely important. Pyrolysis and combustion in the direction perpendicular to the exposed surface become major factors because of the greater thickness of material involved. The appreciable gas content of cellular materials gives them a low specific heat per unit volume and a low thermal conductivity, resulting in a tendency to concentrate heat at the exposed surface instead of dissipating it to underlying material or to a substrate. Because the heat applied may destroy the cellular material before it can be dissipated, cellular materials in thicknesses sufficient to provide thermal insulation can exhibit distinctive flammability behavior. This behavior determines the flame-extinction mechanisms that must be employed.

Because no two accidental fires are identical in all respects, predictions of actual fire performance of these materials must of necessity be based on their behavior in tests that will, it is hoped, provide some correlation with actual fire exposure. Most small-scale tests have an inherent deficiency in that they fail to reproduce the massive effect of heat present in a large-scale fire, yet small scale is a necessity if a test is to be applicable to the small quantities in which experimental polymers must be made.

Surface flame spread is an important characteristic and is often the major factor determining actual fire hazard or fire-protection requirements.

TABLE 5

Flammability Tests for Cellular Materials

Test[a]	Specimen dimensions	Applied heat source	Angle θ (degrees)	Refs.
ASTM D1692, UL 94	6 x 2 x 0.5	Burner, 60-sec ignition	0	[108-110]
Monsanto Ridel 117/2	6 x 2 x 0.5	Burner	0	[111]
ASTM D635	5 x 0.5	Burner	0	[112, 113]
Federal Standard 406, Method 2021	5 x 0.5	Burner	0	[113]
ASTM D757	5 x 0.5 x 0.125	Globar element, 950°C, 3-min ignition	0	[114]
British Standard 476	36 x 9	Radiant panel, 500°C	0	[115]
ASTM E108, UL 790, NFPA 256	40 x 52 or 136	Burners, wood brands	0-45	[116-119]
National Aniline CF-TM-11	8 x 2.5 x 0.25	Globar element	90	[120]
ASTM D568	18 x 1	Burner	90	[121]
Federal Standard 406, Method 2022	18 x 1	Fusee or benzene drop	90	[122]
Hooker HLT-15	8 x 0.5 x 0.125	Burner	90	[123]

Vertical bar	18 x 2 x 2	Burner, 60-sec ignition	90	[124]
Butler chimney	10 x 0.75 x 0.75	Burner, 960°C, 10-sec ignition	90	[125]
UL Bull. 55	6 x 0.5	Burner, two 10-sec ignition	90	[110]
Schlyter	12 x 31 (two)	Burner, 37 or 291 Btu/min, 3-min ignition	90	[126]
ASTM C209	12 x 12	Alcohol, 1 ml	135	[127]
Monsanto Rideal 118/2	6 x 6 x 2	Alcohol, 0.3 ml	135	[111]
U.S. Navy (Vallejo)	6 x 1 x 0.5	Candle	135	[128]
Pittsburgh-Corning tunnel	30 x 3.875	Burner, 90 Btu/min	150	[129, 130]
Upjohn tunnel	30 x 3.875	Burner, 940°C	150	[131]
Monsanto tunnel (Vandersall)	24 x 4	Burner	152	[132]
ASTM E286	96 x 14	Burner	174	[133-135]
Federal Spec. SS-A-118b	36 x 36	Burner, 28,000 or 60,000 Btu	180	[136]
UL 181	18 x 18	Burner	180	[137]
ASTM E84, UL 723, NFPA 255	300 x 20	Burner	180	[117, 138-143]
FM calorimeter	54 x 60	Burners	180	114, 145

TABLE 5 (cont'd.)

Test[a]	Specimen dimensions	Applied heat source	Angle θ (degrees)	Refs.
ASTM E162	18 x 6	Radiant panel, 670°C	240	[146-150]
Federal Standard 406, Method 2023	5 x 0.5 x 0.5	Electric heating coils, 55 A	270	[151]
ASTM D1929	3 g	Heated air		[152,153]
ASTM E136	1.5 x 1.5 x 2	Heated air, 860°C		[154]
Rohm and Haas XP2 smoke	1 x 1 x 0.25	Burner		[155,156]
Monsanto smoke (Cass)	0.2-0.4 g	Globar element, 950°C		[157]
U.S. Bureau of Mines 6366	6 x 6 x 1	Burner, 1050°C		[158]
U.S. Bureau of Mines 6837	6 x 6 x 1	Burner, 1180°C		[159]
FM heat damage	16 x 16	Burner		[160]
ASTM E119, UL 263, NFPA 251		Burners		[117,161-166]
ASTM E152, UL 10(b), NFPA 252		Burners		[167-169]
ASTM E163, UL 9		Burners		[170,171]
FM "White House"	1200 x 240	Wood cribs		[172]

[a]Abbreviations: ASTM, American Society for Testing and Materials; UL, Underwriters' Laboratories; NFPA, National Fire Protection Association; FM, Factory Mutual Engineering Division.

Consequently many tests provide a measure of this characteristic. The position of the surface exposed to fire testing can be expressed as the angle formed by the exposed surface with the horizontal, which will be referred to as surface angle θ. If θ is 0 for a flame front moving on the upper side of a horizontal surface, as in ASTM D1692, rotating this surface about the origin until θ becomes 90 degrees gives a burning vertical surface in which the flame front travels upward. Further rotation until θ becomes 180 degrees gives a burning horizontal lower surface, as in ASTM E84.

The value of θ is irrelevant for small-scale tests in which specimen dimensions are too small to provide a measure of surface flame spread, and large-scale tests in which combustion travels in more than one direction.

Some of the flammability tests that have been employed for evaluating cellular materials are presented in Table 5. In addition, thermogravimetric and differential thermal analyses have been very useful tools for preliminary evaluations of polymers and components 24, 28, 32, 72, 77, 78, 89, 90, 94 .

REFERENCES

1. R. W. Little, ed., Flameproofing Textile Fabrics, Reinhold, New York, 1947.
2. W. A. Rosser, J. H. Miller, S. H. Inami, and H. Wise, Mechanisms of Flame Inhibition, final report, Phase II, Stanford Research Institute, June 1958.
3. W. A. Rosser, H. Wise, and J. Miller, Seventh Symposium (International) on Combustion, Butterworths, London, 1959.
4. E. C. Creitz, J. Res. Natl. Bur. Stand., 65A, 389 (1961).
5. K. W. Rockey, Plastics, 26, 103 (1961).
6. N. E. Boyer, Plastics Technol., 8 (11), 33 (1962).
7. R. H. Dahms, Hydrocarbon Proc. Petrol. Ref., 41 (3), 132 (1962).
8. S. S. Feuer and A. F. Torres, Chem. Eng., 69, 138 (1962).
9. H. N. MacFarland and K. J. Leong, Arch. Environ. Health, 4, 39 (1962).
10. T. G. Lee, J. Phys. Chem., 67, 360 (1963).
11. J. J. Anderson, Ind. Eng. Chem. Prod. Res. Develop., 2, 260 (1963).
12. S. S. Penner, Chem. Eng. News, 41, 74 (1963).
13. J. H. Saunders and K. C. Frisch, Polyurethanes, Chemistry and Technology, Interscience, New York, 1964.
14. N. E. Boyer and A. E. Vajda, Trans. SPE, 4, 45 (1964).
15. J. Eichhorn, J. Appl. Polymer Sci., 8, 2497 (1964).
16. H. Piechota, J. Cell. Plastics, 1, 186 (1965).
17. E. A. Dickert and G. C. Toone, Mod. Plastics, 42, 197 (1965).

18. P. Robitschek, J. Cell. Plastics, 1, 395 (1965).
19. W.G. Schmidt, Trans. J. Plastics Inst., 33, 247 (1965).
20. M. Kaplan et al., J. Cell. Plastics, 1, 355 (1965).
21. J.H. Saunders, "Thermal Degradation and Flammability Characteristics of Urethane Polymers," Wayne State University Polymer Conference Series, May 1965.
22. A.A. Briber, "Flammability Requirements for Cellular Plastics as Materials of Construction," Wayne State University Polymer Conference Series, May 1965.
23. E.A. Dickert, M. Kaplan, and G.C. Toone, "Chemical Factors Affecting Flame Retardance of Rigid Urethane Foam," Wayne State University Polymer Conference Series, May 1965.
24. J.K. Backus, W.C. Darr, P.G. Gemeinhardt, and J.H. Saunders, J. Cell. Plastics, 1, 178 (1965).
25. L. Nicholas and G.T. Gmitter, J. Cell. Plastics, 1, 85 (1965).
26. C.T. Pumpelly, in Bromine and Its Compounds (Z.E. Jolles, ed.), Academic Press, New York, 1966, Chapter 6.
27. R.H. Essenhigh and J.B. Howard, Ind, Eng. Chem., 58, 14 (1966).
28. J.H. Saunders and J.K. Backus, Rubber Chem. Tech., 39, 461 (1966).
29. A.A. Briber, J. Cell. Plastics, 2(2), 112 (1966).
30. P. Robitschek, "Flammability Characteristics of Polymeric Materials," Wayne State University Polymer Conference Series, June 1966.
31. R.R. Hindersinn, "The Effect of Phosphorus and Chlorine on the Flammability of Polymeric Materials," Wayne State University Polymer Conference Series, June 1966.
32. I.N. Einhorn, "Flammability Characteristics of Cellular Plastics," Wayne State University Polymer Conference Series, June 1966.
33. J. Eichhorn, "Self-Extinguishing Systems for Vinyl Aromatic Polymers," Wayne State University Polymer Conference Series, June 1966.
34. J.J. Anderson, "The Relatiosnhip of Crosslinked Density to the Flammability Characteristics of Polyurethane Foams," Wayne State University Polymer Conference Series, June 1966.
35. R.C. Nametz, "Self-Extinguishing Polyester Resins," Wayne State University Polymer Conference Series, June 1966.
36. B.J. Bremmer, "Synthesis of New Halogen Containing Epoxy Resins," Wayne State University Polymer Conference Series, June 1966.
37. W.G. Woods, "Boron Compounds as Flame Retardants in Polymers," Wayne State University of Polymer Conference Series, June 1966.
38. J. Tilley, "Fire Resistance of Plastics Equated to Formulations with Respect to Chemical and Physical-Chemical Composition," Wayne State University Polymer Conference Series, June 1966.

39. I. N. Einhorn, "Small Scale Test Methods for Evaluation of the Flammability Characteristics of Plastics," Wayne State University Polymer Conference Series, June 1966.
40. A. Bortosic, "Smoke Density," Wayne State University Polymer Conference Series, June 1966.
41. E. A. Boettner, "The Identification of the Combustion Products of Plastics," Wayne State University Polymer Conference Series, June 1966.
42. A. F. Robertson, "Recent Developments in Flammability and Smoke Measurements at the National Bureau of Standards," Wayne State University Polymer Conference Series, June 1966.
43. D. Mitchell, "Correlations Between Small Scale and Full Scale Fire Tests," Wayne State University Polymer Conference Series, June 1966.
44. P. M. Hay, "Flameproofing of Polymeric Materials for the Textile Industry," Wayne State University Polymer Conference Series, June 1966.
45. H. C. Vandersall, "Fire Retardancy through Phosphorus-Catalyzed Intumescence," Wayne State University Polymer Conference Series, June 1966.
46. D. Mitchell, "Dust Hazards in the Manufacture of Plastics," Wayne State University Polymer Conference Series, June 1966.
47. C. P. Fenimore, "Flammability of Polymers," Wayne State University Polymer Conference Series, June 1966.
48. A. Singh, L. Weissbein, and J. C. Mollica, Rubber Age, 98, 77 (1966).
49. R. C. Anderson, J. Chem. Educ., 44, 248 (1967).
50. Z. E. Jolles, Trans. J. Plastics Inst., 35, 3 (1967).
51. C. T. Pumpelly, Trans. J. Plastics Inst., 35, 11 (1967).
52. K. M. Bell, 35, 27 (1967).
53. J. K. Jacques, Trans. J. Plastics Inst., 35, 33 (1967).
54. J. I. Jones, Trans. J. Plastics Inst., 35, 17 (1967).
55. C. J. Hilado, Ind. Eng. Chem. Res. Develop., 6, 154 (1967).
56. M. M. Levy, J. Cell. Plastics, 3, 168 (1967).
57. J. D. Downing and R. M. Anderson, J. Cell. Plastics, 3, 236 (1967).
58. M. A. Conway, R. J. Gabler, and D. E. Jackson, J. Cell. Plastics, 3, 547 (1967).
59. C. E. Miles and J. W. Lyons, J. Cell. Plastics, 3, 539 (1967).
60. O. A. Krueger, D. E. Jackson, and K. C. Lyle, J. Cell. Plastics, 3, 497 (1967).
61. R. C. Nametz, Ind. Eng. Chem., 59 (5), 99 (1967).
62. P. Volans, Trans. J. Plastics Inst., 35, 47 (1967).
63. C. F. Culles and D. J. Smith, Trans. J. Plastics Inst., 35, 39 (1967).
64. D. J. Rashbash, Trans. J. Plastics Inst., 35, 55 (1967).

65. C. D. Sneddon, Trans. J. Plastics Inst., 35, 63 (1967).
66. M. A. Denney, Trans. J. Plastics Inst., 35, 67 (1967).
67. B. J. Howthorne, Trans. J. Plastics Inst., 35, 73 (1967).
68. R. A. Blease, Trans. J. Plastics Inst., 35, 79 (1967).
69. K. Fischer, Trans. J. Plastics Inst., 35, 85 (1967).
70. C. J. Hilado, P. E. Burgess, and W. R. Proops, J. Cell. Plastics, 4, 67 (1968).
71. I. N. Einhorn, J. Cellular Plastics, 4, 188 (1968).
72. J. K. Backus, D. L. Bernard, W. C. Darr, and J. H. Saunders, J. Appl. Polymer Sci., 12, 1053 (1968).
73. F. J. Martin and K. R. Price, J. Appl. Polymer Sci., 12, 143 (1968).
74. P. G. Pape, J. E. Sanger, and R. C. Nametz, J. Cell. Plastics, 4, 438 (1968).
75. G. C. Tesoro, S. D. Sello, and J. J. Willard, ACS Polymer Chemistry Division – Symposium, April 1968.
76. J. A. Cengel, J. Cell. Plastics, 4, 309 (1968).
77. J. N. Tilley, H. G. Nadeau, H. E. Reymore, P. H. Waszeciak, and A. A. R. Sayigh, J. Cell. Plastics, 4, 22 (1968).
78. N. N. Tilley, H. G. Nadeau, H. E. Reymore, P. H. Waszeciak, and A. A. R. Sayigh, J. Cell. Plastics, 4, 56 (1968).
79. Y. Iwakura, K. Uno, and N. Kobayashi, J. Polymer Sci., 6, 2611 (1968).
80. R. F. Lindemann, Ind. Eng. Chem., 61 (5), 70 (1969).
81. C. K. Lyon and G. Fuller, Ind. Eng. Chem. Prod. Res. Develop., 8, 63 (1969).
82. C. J. Hilado, Flammability Handbook for Plastics, Technomic, 1969.
83. J. W. Crook and G. A. Haggis, J. Cell. Plastics, 5, 119 (1969).
84. A. G. Walker, British Plastics, 42, 131 (1969).
85. C. J. Hilado, Fire Technol., 5 (2), 130 (1969).
86. M. Narkis, M. Grill, and G. Lesser, J. Appl. Polymer Sci., 13, 535 (1969).
87. J. DiPietro, "An Overview of Flammability of Polymeric Materials, Yesterday, Today and Tomorrow," University of Detroit Polymer Conference Series, June 1969.
88. A. D. Delman, "Recent Advances in the Development of Flame-Retardant Polymers," University of Detroit Polymer Conference Series, June 1969.
89. I. N. Einhorn, "Thermal Degradation and Flammability Characteristics of Urethane Polymers," University of Detroit Polymer Conference Series, June 1969.
90. R. W. Mickelson, "Kinetics Pertaining to the Thermal Degradation and Combustion of Polymers," University of Detroit Polymer Conference Series, June 1969.

91. R. C. Nametz, "Flame-Retarding Synthetic Textile Fibers," University of Detroit Polymer Conference Series, June 1969.
92. S. S. Hirsch, "Factors Affecting Thermal Oxidation and Fire Resistance of Polymers," University of Detroit Polymer Conference Series, June 1969.
93. C. E. Hathaway, J. M. Butler, J. B. Cross, and R. P. Quill, "Highly Chlorinated Polyols for Flame-Retardant Polyurethane Foams," University of Detroit Polymer Conference Series, June 1969.
94. J. D. Seader, "Use of Thermogravimetric Analysis in Polymer Degradation Studies," University of Detroit Polymer Conference Series, June 1969.
95. W. S. Perkowski and R. G. Cheatham, "Flammability and Design Considerations for Commercial Airplane Interior Materials," University of Detroit Polymer Conference Series, June 1969.
96. J. R. Gaskill, "Analysis of Smoke Development in Polymers During Pyrolysis or Combustion," University of Detroit Polymer Conference Series, June 1969.
97. E. Boettner, "Analysis of Volatile Degradation Products of Polymer Combustion," University of Detroit Polymer Conference Series, June 1969.
98. H. H. Cornish, "Toxicological Aspects of Flammability and Combustion," University of Detroit Polymer Conference Series, June 1969.
99. D. E. Cagliostro, "Polymeric Composites and Flame Suppression," University of Detroit Polymer Conference Series, June 1969.
100. A. Briber, "Testing and Evaluation of the Flammability Characteristics of Polymers and Plastics," University of Detroit Polymer Conference Series, June 1969.
101. C. J. Hilado, in Polyurethane Technology (P. F. Bruins, ed.), Interscience, New York, 1969, Chapter 6.
102. G. W. Miller and J. H. Saunders, J. Appl. Polymer Sci., 13, 1277 (1969).
103. J. K. Fincke and G. R. Wilson, Mod. Plastics, 47, 108 (1969).
104. A. E. Boettner, G. Ball, and B. Weiss, J. Appl. Polymer Sci., 13, 377 (1960).
105. C. P. Fenimore and G. W. Jones, J. Appl. Polymer Sci., 13, 285 (1969).
106. S. S. Hirsch, J. Polymer Sci., 7, 15 (1969).
107. T. L. Cottrell, Strength of Chemical Bonds, Butterworths, London, 1954.
108. American Society for Testing and Materials, ASTM Stand., 14, 556 (November 1967).
109. C. J. Hilado, J. Cell. Plastics, 3, 280 (1967).

110. L. M. Kline, Underwriters' Lab. Res. Bull., No. 55, February 1964.
111. D. A. Carpenter, Brit. Plastics, 38, 284 (1965).
112. American Society for Testing and Materials, ASTM Stand., 14, 183 (November 1967).
113. U.S. General Services Administration, Federal Test Method Standard 406, Method 2021, October 1961.
114. American Society for Testing and Materials, ASTM Stand., 14, 285 (November 1967).
115. British Standards Institution, British Standard 476, Part 1, July 1953.
116. American Society for Testing and Materials, ASTM Stand., 14, 388 (November 1967).
117. Building Officials Conference of America, BOCA Basic Building Code, Appendix G, 406, 1965.
118. National Fire Protection Association, Natl. Fire Codes, 4, 251-1 (1965).
119. Underwriters' Laboratories Standard UL 790, September 1958.
120. National Aniline Division, Allied Chemical Corporation, National Aniline Test Method CF-TM-11, September 1959.
121. American Society for Testing and Materials, ASTM Stand., 27, 160 (June 1967).
122. U.S. General Services Administration, Federal Test Method Standard 406, Method 2022, October 1961.
123. Hooker Chemical Corp., Intermittent Flame Test, September 1955.
124. C. J. Hilado, Ind. Eng. Chem. Prod. Res. Develop., 6 (3), 154 (1967).
125. O. A. Krueger, K. C. Lyle, and D. E. Jackson, J. Cell. Plastics, 3, 497 (1967).
126. J. A. Wilson, ASTM Special Tech. Publ. No. 301, February 1961.
127. American Society for Testing and Materials, ASTM Stand., 14, 50 (November 1967).
128. F. W. Breuer, paper presented at the 12th Annual Conference SPI Cellular Plastics Division, October 1967.
129. M. M. Levy, Fire Technol., 3 (1), 38 (1967).
130. M. M. Levy, J. Cell. Plastics, 3, 168 (1967).
131. G. W. Schael, Upjohn Company, North Haven, Conn., private communication, November 1966.
132. H. L. Vandersall, Monsanto Special Report No. 6717, May 1966.
133. American Society for Testing and Materials, ASTM Stand., 14, 475 (November 1967).
134. H. D. Bruce and V. P. Minuitti, Forest Products Lab. Publ. No. 2097, November 1957.
135. C. C. Peters and H. W. Eickner, ASTM Special Tech. Publ. No. 344, October 1962.

136. U.S. General Services Administration, Federal Specification SS-A-118b, April 1959.
137. Underwriters' Laboratories Standard UL 181, November 1961.
138. American Society for Testing and Materials, ASTM Stand., 14, 358 (November 1967).
139. International Conference of Building Officials, Uniform Building Code, 3, Section 4202, 1964.
140. National Fire Protection Association, Natl. Building Codes, 4, 255-1 (1965).
141. A. J. Steiner, Underwriters' Lab. Res. Bull., No. 32, September 1944.
142. Underwriters' Laboratories Standard UL 726, March 1965.
143. C. H. Yuill, ASTM Special Tech. Publ. No. 344, October 1962.
144. Factory Mutual Engineering Division, Laboratory Standards for Classification of Insulated Metal Roof Deck Constructions, (January 1961).
145. N. J. Thompson and E. W. Cousins, Natl. Fire Prot. Assoc. Quart., 52, 186 (1959).
146. American Society for Testing and Materials, ASTM Stand., 14, 444 (November 1967).
147. D. Gross and J. J. Loftus, ASTM Bull., No. 230, 56 (May 1958).
148. D. Gross and J. J. Loftus, National Bureau of Standards Report No. 7325, August 1961.
149. A. F. Robertson, D. Gross, and J. Loftus, ASTM Proc., 56, 1437 (1956).
150. A. F. Robertson, ASTM Special Tech. Publ. No. 344, October 1962.
151. U.S. General Services Administration, Federal Test Method Standard 406, Method 2023, October 1961.
152. American Society for Testing and Materials, ASTM Stand., 14, 661 (November 1967).
153. N. P. Setchkin, J. Res. Natl. Bur. Stand., 43, 591 (1949).
154. American Society for Testing and Materials, ASTM Stand., 14, 414 (November 1967).
155. F. J. Rarig and A. J. Bartosic, ASTM Special Tech. Publ. No. 422, 1967, p. 106.
156. Rohm and Haas Co., Natl. Fire Prot. Assoc. Quart., 57, 276 (1964).
157. R. A. Cass, J. Cell. Plastics, 3, 41 (1967).
158. D. W. Mitchell, J. Nagy, and E. M. Murphy, U.S. Bureau of Mines Report No. 6366, 1964.
159. D. W. Mitchell, E. M. Murphy, and J. Nagy, U.S. Bureau of Mines Report No. 6837, 1966.
160. Factory Mutual Engineering Division, Susceptibility to Heat Damage Test, November 1966.
161. American Society for Testing and Materials, ASTM Stand., 14, 397 (November 1967).

162. J. A. Bono, ASTM Special Tech. Publ. No. 344, October 1962.
163. International Conference of Building Officials, Uniform Building Code, 3, Section 4302, 1964.
164. National Fire Protection Association, Natl. Fire Codes, 4, 251-1 (1965).
165. Southern Building Code Congress, Southern Standard Building Code, Section 1001, 1965.
166. Underwriters' Laboratories Standard UL 263, June 1964.
167. American Society for Testing and Materials, ASTM Stand., 14, 427 (November 1967).
168. National Fire Protection Association, Natl. Fire Codes, 4, 252-1 (1965).
169. Underwriters' Laboratories Standard UL 10(b), February 1962.
170. American Society for Testing and Materials, ASTM Stand., 14, 456 (November 1967).
171. Underwriters' Laboratories Standard UL 9, June 1962.
172. Factory Mutual Engineering Division, Insulated Metal Roof Deck Fire Tests, May 1955.

Chapter 19

FOAMS IN TRANSPORTATION

M. Kaplan and L. M. Zwolinski

Allied Chemical Corporation
Buffalo, New York

I. INTRODUCTION

Many different types of foams based on a wide variety of organic polymer backbones have gained considerable growth in the transportation industry during the past decade. This increased usage of plastic foams in transportation is attributed to a combination of factors, which include economics, processability, ease of application and physical properties.

The outstanding physical properties of (a) cushioning (for seating and safety), (b) thermal insulation, and (c) high strength-to-weight ratio will continue to extend the use of foam in automobiles, aircraft, trucks, trailers, railroads, and public transportation applications.

II. MARKETS

Considerable marketing data on the use of foam plastics in transportation applications have been published [1-12].

In general, urethane foams dominate the market for flexible, semirigid, and rigid foams used in transportation. Flexible latex foams are used to a significant extent, and vinyl and olefin foams are finding use in specialty semirigid-foam applications. Rigid polystyrene foam, in expanded or bead form, is used extensively for board-stock insulation.

A detailed survey on the use of rigid cellular plastics by transportation-equipment manufacturers (e.g., insulated truck vans, trailer vans, truck tanks, trailer tanks, tank cars, and freight containers) indicates that materials and techniques used to a large extent include polystyrene-foam board stock, urethane-foam insulation [13].

III. FLEXIBLE AND SEMIRIGID FOAMS

The use of flexible cellular materials in transportation ranges from such applications as seating cushions to oil-resistant gaskets to exterior automobile bumpers. Designers have used the remarkable engineering properties of expanded flexible polymers to fill the needs of modern vehicle comfort and performance requirements. A variety of materials are now used in high volume and in specialized critical applications in all modes of transportation.

Certain use areas are common to several forms of conveyance. Hence, these areas, such as interior trim, seating, exterior bumpers, and filters, are presented as separate topics and related to specific automobile, air, rail, and marine utilization.

Historically the anticipation of using expanded flexible polymers for comfort cushioning can be attributed to Charles Goodyear, who, in discussing the use of "Gum-Elastic" sponge rubber in 1885, wrote, "the seats of all vehicles without springs may be rendered easy and comfortable at a very trifling cost." From this beginning, uses of flexible cellular materials had progressed to the point where the automobile industry in 1947 used 7 lb of sponge rubber per car. By 1953 the automobile industry consumed 43 million lb of chemically blown elastomer [14]. Introduction of new polymers accelerated flexible-foam uses in transportation, and of urethane foam alone a total of 35 million lb was used in 1962. This increased to 110 million lb in 1966 [10], and by 1970 this market reached 260 million lb. Increased emphasis on vehicle safety padding and cushioning comfort ensures an excellent growth rate for flexible foams in all transportation.

Several foam-polymer types have been adopted widely because of special property or processing characteristics. Molded latex foam rubber and urethane foam are used in large quantities for seat cushioning because of their resiliency and load-bearing characteristics. Extruded and molded foam rubber is used for weatherstripping and for seals. Semirigid urethanes are used for interior trim parts, such as safety crash pads, arm rests, and sun visors. Fabric-backed expanded vinyls find use as upholstery materials, and injection-molded vinyl steering-wheel horn hubs provide decorative safety padding. Other flexible cellular materials based on polyethylene, polypropylene, polychloroprene, and ethylenepropylene rubbers have filled specialized needs.

An insight into the utility of foams in transportation may be gained by examining the amounts of urethane foam used in a 1968 Chrysler Plymouth Fury (Table 1).

TABLE 1

Flexible-Urethane-Foam Usage in a 1968 Chrysler Plymouth Fury[a]

Application	Quantity (lb)
Fillers and inserts	3.0
Front-seat cushions and backs	3.0
Rear-seat cushions and backs	3.5
Arm rests	0.75
Crash pad	1.0
Head restraints	1.0
Total	12.25

[a]Reprinted from Ref. [15] by courtesy of Cahners Publishing Co., Inc.

It has been reported that the amount of urethane foam used in Chrysler products varies from about 8 lb in Plymouth Valiant and Dodge Dart to 18 lb in Chrysler Imperial and New Yorker models [15]. The larger models have some latex foam substituted in seat cushions, backs, and trim pads, and these figures include the total of both foams. Other sources report that a middle-line car uses an average of 12 to 14 lb of flexible urethane foam [9, 16].

Although automotive uses account for the major share of flexible foams used, other forms of transportation use foam products extensively. Cushioned seating is perhaps the most common outlet for foam in automobiles, trucks, buses, airplanes, passenger railroad cars, and pleasure craft. However, polyethylene foams as truck mat underlays, fabric-backed expanded vinyl upholstery for aircraft and boats, polychloroprene-foam journal-box wheel-bearing lubricators for railroad cars, and a host of other uses add up to a substantial consumption of flexible foams. Innovations are still being made at a rate that will further stimulate the use of flexible foam in the transportation industry in the years ahead.

A. Seating

The first cellular material to win acceptance for seat cushioning was foam rubber. The appeal of this material was that seat cushions could be produced to a shape corresponding specifically to the designers' requirements, with trimming considerably simplified. Subsequently the development of urethane foams resulted in a displacement of foam rubber in some seating applications due to improved economics [17]. By 1965 urethane foam supplied 42% of the automotive padding and seating market, with consumption at 60 million lb. Latex foam rubber and various fibers supplied the remaining 58% [18]. At present, urethane foam dominates the automobile-seat market, with latex foam rubber being used primarily in the premium models. Similarly flexible urethane foams are now used in about 95% of all passenger aircraft seating [19].

Automobile-seat construction has evolved through the years to a now widely used low-profile design. This type of seat has a stamped metal frame strung with square-sine-shaped wire springs. Over the springs a burlap pad is placed and covered by molded flexible foam or cut-to-shape slab-stock urethane foam. The fabric or fabric-backed vinyl upholstery is draped over this assembly and "hog" ringed in place to the metal frame [20]. In ground vehicles the seat springs take up the main deflection under load while the flexible foam provides localized support for passenger-weight distribution and adjusts to individual body differences [21].

Service requirements and economics determine the spring-support design and choice of soft cellular padding. In automobiles the bench-type seats are usually padded with a molded-urethane-foam topper pad 1 to 2 in. thick. Where the shape is simple, this topper pad is more economically cut from slab-stock urethane foam. Bucket-seat pads are generally molded urethane foam in thicker cross sections, and in some models molded latex foam rubber. The foam used must be sufficiently firm to prevent a passenger from "bottoming out" (striking through to the base) when going over rough roads.

The urethane padding used for automobile seats usually has a density of 1.6 to 2.0 lb/ft^3 and is of medium firmness (30-40 $lb/50in.^2$ at 25% indentation load). With molded foam-rubber pads load-bearing characteristics are controlled by formulation (density, fillers) and by core hole size and distribution. The back-support pad in seats needs only to provide comfort cushioning rather than heavy load bearing and hence is made of softer, lighter foam.

City-bus seats are much firmer, and springs of heavier duty construction are required to endure greater wear. These firmer, shallower seats allow easier passenger interchange. Long-wearing, tough bus seats are made by slush-molding a cover from vinyl plastisol and then filling with mechanically foamed vinyl [22]. In long-distance buses the coaches are upholstered more luxuriously, with head rests added for head and neck support. In aircraft and railroad seating foam cushioning is used principally for static seating comfort because there is little need for vibration-shock absorption [21].

Boat seats with a vacuum-formed vinyl skin, embossed in a dielectric unit, have been made by molding a prepolymer foam into the vinyl cover [23]. In another operation upholstered boat seats are made with a vinyl-foam-vinyl sandwich, dielectrically sealed in a variety of patterns. This construction uses vinyl-impregnated, open-pore urethane foam to impart good loft to the quilting. Finished seats have an attractive appearance and withstand weather, saltwater, and marine abuse [24].

A greater emphasis on vehicle safety, primarily automotive, has had an effect on seat design and materials of construction. National Highway Safety Bureau and General Service Adminstration regulations made the addition of head restraints mandatory in all front seats of automobiles produced after January 1969 [25, 26]. Head rests are made in most cases of semiflexible urethane foam molded in a vinyl cover or with a cut-and-sew fabric-backed vinyl cover added after foam molding (Fig. 1).

For further passenger protection some automobiles have front seats with an impact-absorbing seat top. These seats have a corrugated metal top structure covered by semirigid-foam padding and regular upholstery. Under rear-passenger impact, in the so-called second collision, the structure collapses and absorbs impact energy [16, 25].

Seat height is critical in passenger safety. Experimental rigidly mounted seats with high integral seat backs were shown to protect passengers against injury in rear-end collisions at speeds of over 30 mph [27] (see Fig. 2).

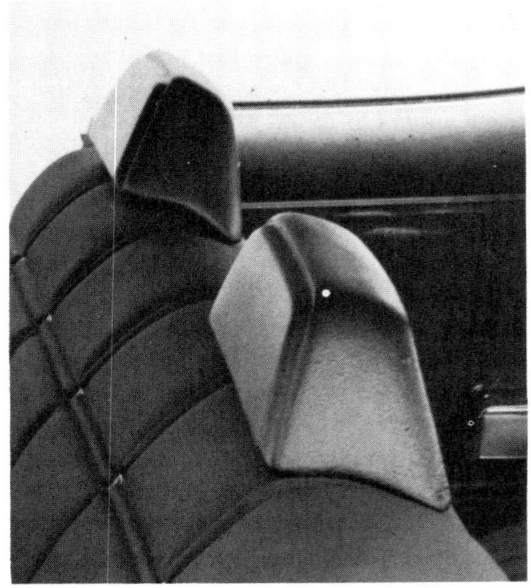

FIG. 1. Polyurethane-foam-filled head restraints, now a standard item in all American automobiles, provide protection against neck-injury for front-seat passengers. The part shown at the left was made from a molded, firm, flexible-urethane-foam core and a cut-and-sew fabric-backed vinyl cover. The head rest on the right features a rotationally cast vinyl plastisol cover into which the urethane foam was directly molded. Courtesy of Ford Motor Co.

As another safety consideration, the flammability of automobile-interior-trim material has received more attention. The Society of Automotive Engineers has contributed by establishing a recommended practice for determining the burning rate of components used in such items as seats, arm rests, crash pads, etc. [28] . Fire resistant polychloroprene foams with densities of 3.7 and 5 lb/ft^3 have been commercially offered for seating cushioning. The properties of this material compare favorably with those of natural latex foam rubber, and, in addition, the polychloroprene foam has superior oxidation and oil resistance [29] . A modified polychloroprene foam is made by adding a blend of undistilled isocyanate to the frothed rubber latex. The foam rapidly develops high wet-gel strength, and low densities are obtained. This material can be used for transportation seating and truck-cab mattresses [30] .

Although the bulk of foam in seating is used as support padding, considerable quantities of expanded vinyl, backed with fabric, are used for upholstery. The soft hand and luxurious appearance of leather that this material possesses offer good styling possibilities at a relatively low

FIG. 2. Padded high-back automobile bucket seats eliminate the need for a separate head restraint. Flexible urethane foam is used to provide occupant comfort and to achieve a finely styled custom appearance. The foam used along the top of the seat back is a firm material to give added protection to rear-seat passengers. Courtesy of Chrysler Corp.

cost [8, 31]. The composite is tough and can be dielectrically embossed with decorative patterns.

For greater styling effects thin sheets of urethane foam have been used with vinyl-coated fabric to achieve high-loft patterns in automobile upholstery. Pleated cover pads for sports-car bucket seats have been made using 0.25-in.-thick slab urethane foam needled with a combination of polyester and acetate fibers. After being slit into 1.25-in.-wide strips the foam is cemented to cotton sheeting. Surface nylon bodycloth or knit-backed vinyl is sewn to the cotton sheeting between the foam strips to form deep pleats [20].

Dielectric bonding has been used to manufacture pleated seat inserts by sealing 0.25-in.-thick vinyl foam or vinyl-impregnated urethane foam between knitted cotton backing and a leather-grained vinyl cover (Fig. 3). This process has the advantage of producing unique designs not possible by sewing [20]. Another process makes use of open-celled reticulated foam coated with vinyl. The use of the open-pore foam allows heat sealing with no dead spots when used between fabric backing and vinyl covering [32]. A new method that claims improvements on both the sewing and dielectric heat-sealing systems uses a special press and an adhesive. Vinyl cover, foam, and backing are coated with the adhesive prior to being put through the press. Press pressures of 2 to 4 psi and a maximum heating temperature of 95 C are used to obtain a material loft of up to 0.5 in. [32].

The use of all foams in seating will continue to grow as new seating concepts are introduced. It has been suggested that the present method of seat building with foam pads over springs may be supplanted by a design featuring a rigid-foam core, a flexible-foam top pad, and a covering [19].

B. Interior Trim and Accessories

The trend toward more luxurious, highly styled vehicle interiors and greater emphasis on safety has helped extend the use of foam-padded trim parts. Automotive uses consume most of the semiflexible and semirigid foams in such elements as crash pads, arm rests, and sun visors, but airplanes with padded overhead racks and arm rests, trucks, and buses contribute to the total amount of material used [34].

As already mentioned, government action dictated the inclusion of a number of safety features in automobiles made after January 1969. But even before then the value of padded instrument panels, sun visors, and arm rests for passenger safety was recognized, and these items were offered as added purchase options. It has been shown that such interior trim parts as dashpads must absorb smoothly the large forces acting on the passenger's body during a collision. Semirigid foam, in combination with a deformable metal backing, accomplishes this particularly well. At collision speeds of under about 20 mph a well-designed dashpad with semirigid urethane foam can provide excellent protection because the foam shows a behavior under stress that approaches that of an ideal deceleration material. On collision, the force acting on the body increases rapidly with a small deformation of the pad. There follows a pronounced plateau where an increase in deformation with a resultant absorption of shock energy does not cause a significant increase in the force acting on the body. This plateau is then followed by an increase in force after about 60% deformation. For a dashpad to function effectively the impact energy should therefore be dissipated by the time the padding material reaches about 50% deformation to avoid high stresses [35].

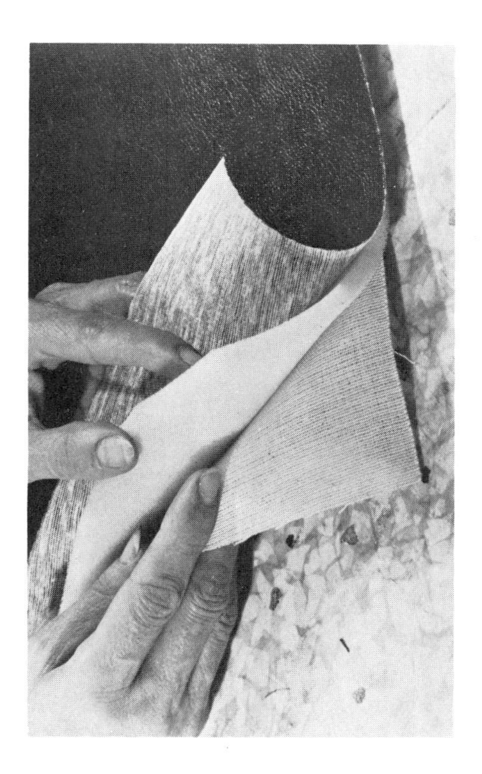

FIG. 3. Flexible urethane foam is used in fabric-backed vinyl-foam-fabric laminates to achieve high loft patterns in automobile upholstery. The composites can be sewn or dielectrically heat-sealed in desired styles. Illustrated is the use of resin-impregnated reticulated foam, which uniformly dielectrically heat-seals. Courtesy of Foam Division, Scott Paper Co.

A Federal Aviation Agency study has shown, however, that at collision speeds of 45 to 50 mph resilient materials must be used in combination with a backing, such as thin-gauge metal or crushable plastic, to provide progressive deceleration on impact [36] (see Fig. 4).

To determine the performance of dashpads and other trim components, a Society of Automotive Engineers impact test has been developed [37, 38].

Foam-padded parts are made by several methods. A construction using a preformed cover with a semirigid-urethane-foam core in which a metal or plastic insert is encapsulated is most common. The cover for such items as arm rests, horn hubs, and sun visors is made by slush or rotational fusion of polyvinyl chloride plastisol in electroformed, patterned

FIG. 4. Automobile instrument-panel pads are now a standard item in all cars sold in the United States. These pads feature a vacuum-formed, injection-molded or slush-molded plastic cover with a semirigid-urethane-foam core. A crushable metal or plastic insert is included to absorb collision energy. Courtesy of Ford Motor Co.

metal molds [39, 40]. The vinyl skin is then transferred to a second foaming mold, where it is filled with semirigid urethane foam. The metal backup plate serves to give the part rigidity and provides for easy screw mounting of the part to the car body. For dashpads the cover is made in most cases by vacuum-forming leather-grained vinyl-ABS (acrylonitrile-butadiene-styrene) stock to the desired shape prior to filling with foam. Both one-shot and prepolymer urethane systems are used for part molding. For some items it is more economical to drape a cut-and-sew cover, usually made of fabric-backed vinyl, over a premolded semirigid-foam core.

Urethane foam has been the favored material for padded-interior-part production because its good foam-flow properties enable it to fill complex shapes during part molding. Furthermore, the properties of the form can be tailored by formulation changes to meet a variety of performance requirements. Other polymers, however, have made inroads because of special properties. A recent Ford model featured a collapsible steering-wheel horn hub constructed with a rigid-urethane-foam core plus an outer layer of soft vinyl foam, both encased in a vinyl cover [36]. Other horn-hub pads are made from injection-molded foamed vinyl plastisol [41]. Polyethylene foam as a floor-mat underlay for trucks has proved to be a tough, resilient material [42]. In this application the foams excellent chemical stability, low water absorption, and flexibility over a wide temperature range have been used to good advantage.

In some Ford automobiles closed-cell crosslinked polypropylene foam has been used since 1966 to make sun visors. The high energy-absorbing properties of this material have allowed a sewn construction of 1/8-in.-thick polypropylene-foam sheets and chipboard to pass an impact test of 600 g using a 6.0-lb aluminum head form traveling at 15 mph [43, 44]. Its excellent flex properties, heat-sealable character, and resistance to degradation make this foam an attractive material for future use in door panels, seat backs, and arm rests.

Expanded vinyl covering in combination with urethane foam used as an American Motors headliner in the Ambassador model eliminated a cut-and-sew operation and provided increased passenger protection [31, 36]. Foamed latex rubbers and expanded ethylene-propylene rubber molded pads covered with fabric-backed vinyl have been used as arm rests [45].

A recent innovation in molding interior trim parts has been the introduction of integral-skin urethane foam. In this fabrication technique a single foam shot charged into a mold rapidly expands to fill the mold and at the same time forms a densified skin in contact with the mold surface. These "self-skinned" foam articles take the shape of the mold, with the durable skin exactly duplicating the mold surface. This process eliminates the need for a preformed plastic cover, and parts can be used directly as molded or painted to match automotive interiors [46-48]. Foams with densities of 7 to 15 lb/ft^3 have been used to make arm rests, instrument bezels, crash pads, and padded horn hubs for automobiles, and an integral-skin-foam arm rest is used in the new jumbo jet airplanes. A European modification of this process uses vacuum-formed 0.5-mm-thick polystyrene single-use mold liners to obtain a textured integral-skin-foam arm rest. After demolding, the thin liner is stripped from the part and discarded. Semigloss or high-gloss finishes are possible, depending on the type of film used [49].

C. Gaskets and Weatherstripping

Flexible cellular polymers are used extensively in vehicles as deck lid and door seals, heater gaskets, roof rail strips, and in a wide variety of other applications [45]. For best performance the materials should have the following properties:

1. Weather and ozone resistance.
2. Low compression set.
3. Low water absorption.
4. High-temperature and aging-deterioration resistance.
5. Low-temperature flexibility.
6. Good abrasion resistance.
7. Good adhesion to metal, where required.

Acceptable service is often obtained by using polymer blends and in some cases by coating the blown elastomer with a tougher, more durable polymer film.

The early use of chemically blown sponge rubber and closed-cell expanded rubber for automotive needs relied on natural, reclaim, and synthetic rubber polymers [14]. More recently newly developed polymers with improved properties have extended vehicle life and have reduced unit maintenance requirements. For example, open-cell sponge polyisoprene rubber has been molded into door weatherstripping, floor-board seals, truck weatherstripping, antisqueak strips, trim fillers, and vibration dampers [50]. The physical properties of this material are comparable to those of natural rubber, with the added advantages of excellent building tack during processing and good low-temperature flexibility in finished goods. In another application closed-cell polychloroprene foam is used as a dam with SBR-molded wedges for positioning backup lights and windshields [26]. Isocyanate-modified polychloroprene foam with outstanding resistance to oil and aging deterioration has been introduced for automobile- and truck-body seals [30]. The foam is made by adding crude tolylene diisocyanate to frothed polychloroprene latex foam. Parts are demolded after 15 to 30 min at room temperature, with the usual steam vulcanization eliminated. Finished parts are firmer than those made from regular polychloroprene foams and have improved flame resistance.

A newer type of weatherstrip made from a blend of SBR and polychloroprene is extruded as closed-cell sponge and provides improved appearance, low door-closing effort, and low water absorption [38]. Plastic clips embedded in the weatherstripping are inserted into sheet metal holes of the car body to fasten the weatherstripping. Above the belt line of the car the strips are cemented into a channel to prevent sagging.

High-performance sports cars with forced-air intakes require large circular seals between the hood air scoop and air cleaner. Satisfactory performance has been obtained with seals constructed from cellular polychloroprene extrusions of triangular cross section. Total seal weight was about 12 oz. Depending on the car model, the cellular rubber seal was attached to either the underside of the hood scoop or to the air-cleaner mounting [51].

Ethylenepropylene foamed elastomers with resistance to ozone attack, good low-temperature properties, and good general weatherability are well suited for sealing applications. Extrusion and molding techniques have been used to prepare such articles as heater gaskets, door seals, air-conditioner gaskets, and glass window channels. Low-cost mineral fillers and oil extenders have been successfully compounded into the rubber stock for preparing both open- and closed-cell sponge. Closed-cell sponge molded into sheets has good recovery and snap, a smooth natural skin, and low water absorption. Blends of ethylenepropylene and butyl rubbers have shown unique damping properties in sponge compounds and may offer superior performance in door-seal uses [45] (see Fig. 5).

Expanded thermoplastics, such as polyvinyl chloride, polyethylene, and polypropylene, are making a penetration into vehicle gasketing because of lower manufacturing costs and desirable properties. Lightweight polyethylene foams have excellent chemical resistance, low water absorption, and low moisture-vapor transmission. This material is well suited for low-pressure gasketing where good solvent resistance is required [42]. Polypropylene foams, die cut from expanded sheets, have also been used for gasketing. The crosslinked material is unaffected by acids, bases, alcohols, and ketones and has good oil resistance at temperatures of up to

FIG 5. Expanded-rubber weatherstripping is used to seal vehicle doors, trunk lids, and windows. Extruded in a variety of cross sections, the rubber is cured in heated ovens where chemical blowing agents effect expansion. Where improved oxidation resistance is required, the weatherstripping is often covered with a polyurethane coating. Courtesy of Inland Manufacturing Division, General Motors Corp.

160°C. In one instance a foamed-polypropylene oil gasket used in a power-brake system replaced a cork-rubber composition because of equivalent performance at a reduced cost [44, 52].

D. Automobile Bumpers

A revolutionary energy-absorbing, high-density urethane-foam bumper was introduced by the General Motors Corp. in the 1968 GTO-model Pontiac. The Endura bumper, as Pontiac called it, featured a color-stable urethane lacquer coating applied over a tough, resilient-molded urethane foam, which was backed by a 0.15-in.-thick steel frame [53] . The 44-lb/ft^3 foam, with a deflection of 0.5 in. under a 1000-psi load, was designed to show complete recovery in 24 h after depression by a 4000-lb load for 8 h. Bumpers mounted on automobiles were able to withstand impact tests with no damage at speeds of 4 mph--more than twice the speed conventional bumpers can withstand.

This approach to the manufacture of exterior automobile components has allowed a new freedom in styling. Stamped metal bumpers, which are subsequently chrome plated, have design limitations tied to metal-shaping technology. With foamed elastomer bumpers, complex shapes and deep contours are possible. Moreover aesthetic effects characterized by an elegance brought about by the absence of ornamentation can be achieved. Added benefits with urethane bumpers are reduced maintenance and elimination of corrosion.

An important step in the introduction of the dentproof bumper was the development of a suitable color-stable, fully reacted urethane laquer coating [54, 55] . The coating was based on an aliphatic diisocyanate and could be color matched to metallic and nonmetallic body paints. The applied coating featured good flexibility, excellent abrasion resistance, high gloss, and good weatherability. Drying proceeds by solvent evaporation, and finished films are polishable with cleaners and waxes.

In practice molds with highly polished plated-nickel surfaces are sprayed with a lacquer-based prime coating. The high-density foam, based on a mixture of polyether polyols and an aromatic diamine, reacted with tolylene diisocyanate (TDI) in the presence of a small and carefully controlled amount of blowing agent, is cast into the mold and allowed to expand and fill the mold. The molded bumper, color coated to match the rest of the car body, is mounted to the frame by means of extensions from the metal insert encapsulated in the foam [16, 53].

The popularity of the energy-absorbing bumper has led to greater use of high-density urethane elastomer foam in exterior automobile parts. The Pontiac GTO bumper has been increased in size and carried through the 1970 model year (Fig. 6). Other Pontiac models have featured urethane-foam center-nose sections in split front-grille designs and rear-bumper

FIG. 6. The 1970 Pontiac GTO model features a distinctive front end highlighted by an energy-absorbing "Endura" bumper. This bumper is molded in flexible microcellular urethane foam with a density of 40 to 45 lb/ft^3. Painted with an elastomeric urethane coating, the unit is able to withstand low-speed impact without damage and is free of corrosion, and road-stone pitting. Courtesy of Pontiac Motor Division, General Motors Corp.

foam appliques. It has been reported that other car manufacturers may also adopt the energy-absorbing bumpers in at least some models [21, 51, 56].

E. Thermal and Sound Insulation

Because of an upsurge in the number of vehicles equipped with air conditioning, the use of foam for passenger-compartment and air-conditioner insulation has increased sharply. Likewise manufacturer emphasis on quiet-riding automobiles has provided added stimulus to the adoption of trim structures that allow noise isolation as well as thermal insulation.

It has been estimated that 30% of the automobiles produced in 1968 were factory furnished with air conditioning. In models priced above $4000 the share was greater, with about 75% produced with air conditioning [19]. For a full benefit from such equipment improved insulation has been required. Polypropylene foam, for example, was utilized in making under-the-hood air-conditioner covers for the 1969 Ford Thunderbird. Here the foam's vacuum formability, heat resistance, and insulation properties made it the material of choice. In another application polypropylene foam was chosen to insulate a 3.5-ft-long ventilating-heating duct for buses. The 3/8-in. foam sheeting, adhered to the ductwork, provided both desired insulation and sound deadening [43, 44].

Urethane-foam-insulated door panels and headliners have been designed by automobile engineers to function both as decorative trim and as insulation. A molded headliner was installed by one company in hardtop and two-door sedan models. In this construction a composite of vinyl skin, resin-modified semirigid foam, and kraft paper was molded to shape in matched metal dies and installed, with a fiber pad as an antirattle strip, on the car roof. This padding provided thermal and acoustical insulation as well as passenger protection [20, 36]. In a similar manner automobile-door panels with vacuum-formed ABS skins, deep-loft urethane padding, and a fiber-glass-reinforced-polyester base have been produced. These units (Fig. 7) have built-in arm rests and give a luxurious appearance to the interior in addition to insulating the passenger compartment [16].

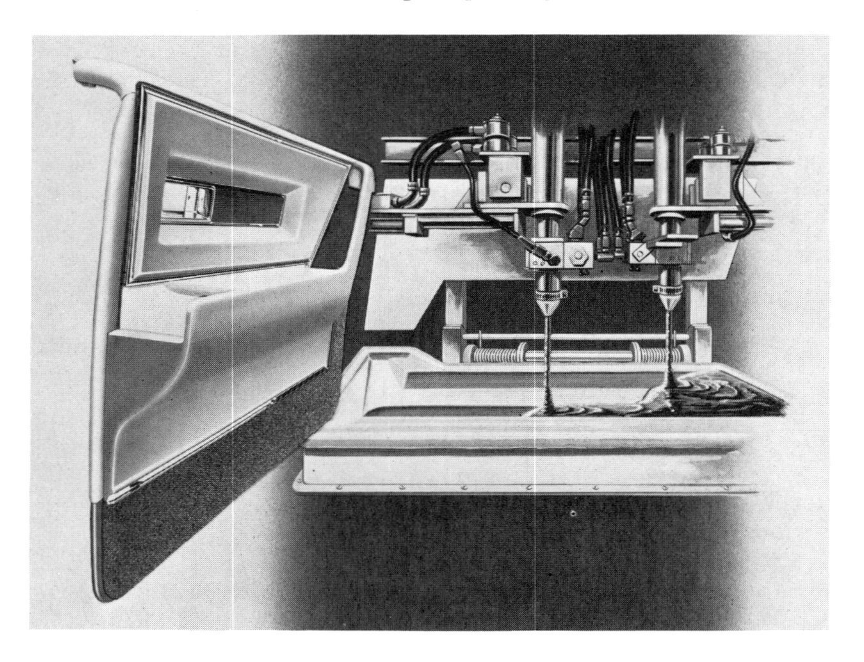

FIG. 7. Polyurethane door-trim panels for automobiles are manufactured by using a pour-in-place technique. The liquid-foam mix is poured into a vacuum-formed ABS plastic cover that is positioned in a mold. Courtesy of Fisher Body Division, General Motors Corp.

For protection in cold climates experimental truck cabs have been insulated with sprayed-in-place flexible urethane foam. These cabs allowed relief drivers to sleep in comfort on the bed shelf behind the driver in subzero weather.

F. Other Applications

Although cushioning and interior-trim applications account for most of the flexible foam used in transportation, several interesting smaller volume needs are met by expanded elastomers.

Experimental military vehicles and off-the-road construction equipment have been fitted with tires filled with molded-in-place flexible urethane foam. It was found that this design prevented major blowouts and extended the life of tires used in rough terrain. Operation of vehicles was possible even when sections of the filled tires had been gouged out. A recent experimental design for safety automobile tires featured an inner member of flexible foam. In normal use the vehicle load would be carried by the air-inflated portion of the tire. When a puncture deflated the outer portion, the vehicle load was transferred to the foam safety core, allowing the automobile to be brought to a smooth, safe stop [51].

Open-celled, or reticulated, foam produced by the controlled alkaline hydrolysis of flexible polyester-based urethane foam has gained acceptance for automobile-carburetor air filters (Fig. 8). The caustic-leached, membranefree foam has a density of about 1.5 lb/ft^3 and a porosity of about 45 pores per inch. The filters are made by hot-wire-cutting strips with the desired cross section from treated slabs of open-celled foam. After the ends of the strips are glued together to form a cylindrical element, the filter is impregnated with light oil, which aids in trapping and retaining dust particles. The filter is particularly effective because of the great dust-holding capacity, low pressure drop across the element, and high filtering efficiency. Filters can be easily cleaned periodically in gasoline to remove trapped dirt and then reused [57].

Military research with open-celled urethane foam led to the development of explosion-proof aircraft fuel tanks. It was found that metal tanks could be filled with cut-to-shape flexible reticulated polyester foam, with only a 3% reduction in tank volume. The foam acted like a fire screen, and when the tanks were punctured by bullets or shrapnel, the fine-mesh foam prevented fire and explosion propagation. Civilian use of this safety concept has been adopted for some racing-car and aircraft fuel tanks. The foam sharply reduces the hazards of crashes since gasoline will not readily spill from a ruptured tank. In addition, sloshing of fuel within the tank is eliminated, cutting "fuel ram" in the engine and the generation of explosive fumes [19, 58].

A further extension of the use of open-celled foam in fuel tanks has been introduced in Germany [59]. A technique was developed for forming

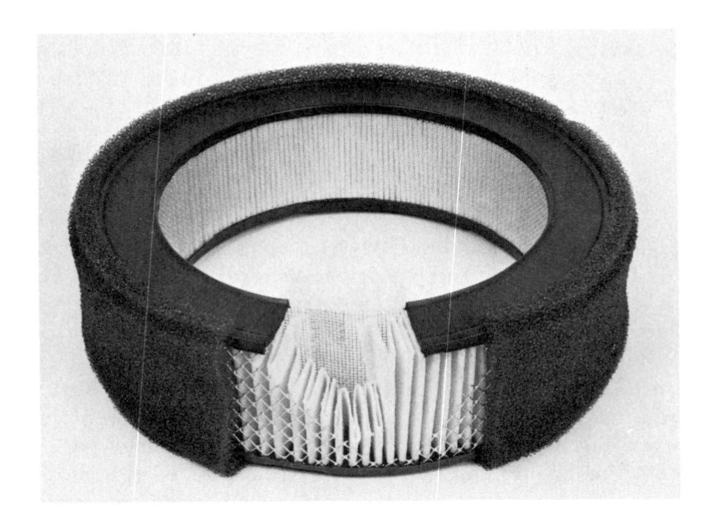

FIG. 8. Open-cell (reticulated) flexible polyester-based urethane foam is shown here as an element of a dual-stage air filter. Used in conjunction with the oiled-paper medium for heavy-duty vehicles, the outer polyurethane wrap may be removed for cleaning. Courtesy of AC Spark Plug Division, General Motors Corp.

reticulated polyether-based-urethane foam in situ in rotationally cast polyamide tanks. The advantages of this method are that seamless tanks, free of weak spots, are realized, and the open-celled foam can easily be formed in irregularly shaped tanks. The finished tanks have excellent heat and impact resistance, low fuel permeability, and outstanding explosion- and fire-resistance characteristics.

Expanded rubber has been used for a number of years as automobile shock mounts and heavy-duty vibration dampeners [14]. Now high-performance microcellular foams have been accepted as vibration isolators for suspension systems of light and heavy trucks and overload springs for trucks [54, 55] (Fig. 9). These microcellular foams can be made in densities of 20 to 50 lb/ft^3 and with load-bearing capabilities in excess of 1500 psi at 50% compression. In addition to good shock-absorbing properties, the microcellular foams, in contrast to solid rubber, exhibit little sideway bulge when compressed, and in installations in confined areas this could prove to be an important advantage (Fig. 10).

IV. RIGID FOAMS

Rigid cellular materials have found considerable use in the transportation industry. Applications for the foams include thermal insulation for

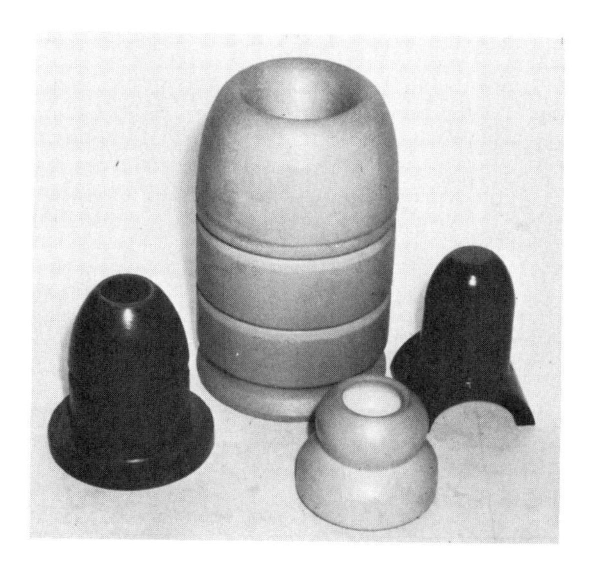

FIG. 9. Its controlled cushioning and high load-bearing properties make flexible microcellular urethane foam useful for shock absorbers, jounce pads, and auxiliary springs. The large element shown here is an auxiliary spring for a Mercedes-Benz bus. Smaller parts are automobile jounce cushions. Courtesy of Cellasto, Inc.

FIG. 10. Vehicle seals of flexible microcellular urethane foam. These elements seal running parts against road dust, and because of excellent oil resistance, good recovery from compression, and little sideways bulge, the parts have a long service life. Courtesy of Cellasto, Inc.

trucks and railroad cars, structural reinforcement for aircraft, and void fillers for ships. A wide variety of foamed polymers, including urethane, polystyrene, vinyl, ABS, and polyolefins, have been used.

Thermal insulation represents the largest single application for rigid plastic foams [2]. The high insulation efficiency and low weight-to-volume ratio are characteristic properties of the rigid cellular plastics. Table 2 shows the comparative efficiency of commercial foam insulants [1].

TABLE 2

Comparative Efficiency of Commercial Insulants[a]

Material	k factor (Btu/(h)(ft^2)(°F/in.))
Polyurethane:	
Foamed in place	0.11
Slab	0.15
Polystyrene:	
Expandable (beads)	0.24
Expanded	0.25
Fiber glass	0.24
Rigid polyvinyl chloride foam	0.17

[a]Adapted from Ref. [1].

The decision to use one foam instead of another is generally based on the economics and the k factor, as well as the requirements of specific end use. Such factors as maximum service temperature, dimensions, structural load bearing, production volume, and fabrication and installation techniques are important considerations.

Strength properties will vary depending on the polymer backbone of the cellular material as well as the density. Table 3 shows the comparative physical properties of rigid cellular plastics [1].

The overall structural strength of the finished vehicle will also depend to a large extent on the materials included in the composite structure; use of pour-in-place, spray, or froth technique with polyurethanes; and surface preparation and adhesives used (with board-stock foam).

Flame-retardance properties can be formulated into the rigid cellular plastic as required for the application.

TABLE 3

Comparative Physical Properties of Rigid Cellular Plastics[a]

Material	Density (lb/ft^3)	Compressive strength (psi)	Shear strength (psi)	Shear modulus (psi)
Polyurethane	1.5-2.0	20-60	20-50	250-550
	2.1-3.0	35-95	30-70	350-800
	3.1-4.5	50-185	45-125	500-1300
Expanded polystyrene	1.7-2.3	16-38	30-40	1000-1300
Expanded polystyrene (beads)	1.2-1.5	10-13	20-30	500-600
Rigid polyvinyl chloride foam	1.6	30	30	--

[a]Reprinted from Ref. [1] by courtesy of Wayne State University.

The rigid urethane foams are the most widely used of all the rigid cellular materials in the transportation field. Their characteristic combination of properties includes versatility in application (pour-in-place, slab stock, sprayed); highest thermal efficiency; excellent water, chemical, and solvent resistance; wide service range (below -175 °C to above +120 °C); low water permeability; and high structural strength. In 1965 the Cellular Plastics Division of the Society of the Plastics Industry, Inc., published a Guide for the Application of Rigid Foam in the Transportation Industry [60].

A. Trucks and Trailers

The use of rigid cellular materials as insulation in trucks and trailers for transporting a variety of cargoes has appeared in the literature [7, 61-63]. The characteristic low density of the foam results in a reduced truck weight, thus permitting increased cargo weight.

Truck and trailer bodies insulated with foamed-in-place urethane are considerably stronger than other types of construction since the composite structure distributes vibration and strains throughout the truck body. Previous methods of fastening with rivets and bolts produced localized stress at the points of attachment. The rigidity of the foamed-in-place core reduces the need for ribs and offers the possibility of reduced skin thickness [64]. Heffner [65] has reported detailed data on the performance and economics of foamed-in-place insulated trailers.

Rigid urethane foams, with their unique combination of engineering properties, offer many designs and applications in trucks and trailers. Insulated trailer tanks, which haul orange juice from Florida to New York without requiring a refrigeration unit, have been developed. The temperature of orange juice loaded at near freezing temperatures had increased only 1°C at time of unloading [61].

The Association of Food and Drug Officials of the United States has adopted a code requiring that a maximum of -18°C must be maintained in the shipment of frozen foods [66]. If the code is widely adopted, rigid-urethane-foam insulation may be a necessity. For example, high-performance urethane insulation is critical when cooling is supplied by evaporation of liquid nitrogen. Linde's Polar Stream Equipment [67] utilizes liquid nitrogen (bp -200° C) bled from a cylinder into the load space periodically. The Pure Carbonic Division of Air Reduction Corp. claims to have a similar system utilizing carbon dioxide [61]. Perkins [68] has developed detailed refrigeration requirements for protective vehicles for perishable goods.

Mechanical compressor systems have also been extensively used in conjunction with urethane-insulated trucks and trailers.

A new concept in the control of temperature and environmental atmosphere in the transportation of perishables is a development of the Tectrol

Division of Whirlpool Corp. Airtight leakproof containers have been molded from reinforced plastic with urethane-foam insulation (Fig. 11). Careful control of the temperature and gases inside the container can extend the storage life of fresh produce from three to five times their normal life [61].

Delivery trucks insulated with foamed-in-place urethane have been used to maintain the product quality of frozen and cream pies during delivery. The system eliminates the use of mechanical truck refrigeration by use of chilled plates and compartments that are connected to the plant refrigeration system prior to delivery [62].

Unrefrigerated trucks insulated with rigid urethane foam can also be used for a wide variety of ladings requiring constant temperature within the service limits of the foam. A truck can haul frozen-food containers one day, candy the next, perishable produce the next. When required, small heaters or Dry Ice can be used [7].

Sprayed rigid urethane foam has been suggested for use in protecting general freight commodities, such as paint, pharmaceuticals, cosmetics, and chemicals, from freezing during the winter months [63]. A foam thickness of 1.5 in. applied to a local-delivery van equipped with a propane-fired heater was reported sufficient to maintain above-freezing temperatures in below-freezing weather.

Hudgens reported [69] that reefer vans insulated with foamed-in-place urethane required considerably less maintenance than a fiber-insulated van after 4 years of operation. The vans, used for transporting fresh meat

FIG. 11. Airtight containers molded from reinforced plastic with rigid-urethane-foam insulation keep perishables fresh for long periods through controlled load-space environment. Shown here carried on a space-frame trailer, these containers can also be fork-lifted onto flat cars, lifted by overhead gantry and carried aboard ship, or transported by air. Courtesy of Litewate Transport Equipment Corp., and Tectrol Division of Whirlpool Corp.

and packinghouse products, were foamed with urethane, which formed a vapor seal and prevented the infiltration of fatty acids and condensates, which tend to deteriorate the nonfoam trailers very rapidly.

In addition to the excellent insulating characteristics of rigid urethane foam, the closed-cell structure prevents water absorption and buildup of deadweight load. With conventional fibrous insulation a reefer may absorb as much as 1000 lb of water. Truckers estimate that this deadweight costs $1 a year per pound in reduced earning capacity. Additional revenue loss is encountered when the trailer with fibrous insulation must be thawed and dried out, with the unit standing idle for 2 or 3 weeks. The closed-cell structure of the urethane foam keeps out meat juice and brine solution, which can cause residual odors and sanitation problems [7, 70-72].

In many instances refrigerated trucks are virtually custon made. They are produced to the specifications of the end user, which may differ radically. Therefore the type of construction will differ widely. In some cases foam-in-place processing can be highly efficient; for others construction with board-type insulation can be more efficient [66]. A sandwich construction is shown in Fig. 12.

The advantages of board-type (urethane or polystyrene) insulation include the following [73]:

1. Ease of handling and fastening because of rigidity.
2. Fixing in place with various adhesives or banding.
3. Fitting with hand tools.
4. Low labor cost due to ease of application.
5. Ease of repair.

In the case of urethane slabs a wide variety of solvent-based adhesives can be used. Contact types, which develop high initial tack, do not require shoring or jigs. Direct layup with polyester resins and fiber glass is useful in insulated transportation designs [74, 75]. Considerable data have been developed on the use of cellular plastic cores in sandwich panels. Structural design principles, including deflection data for thickness versus loading of polystyrene [76] and urethane [77-80] cores, have been reported.

An application utilizing the structural properties of rigid cellular materials is truck-radiator fan shrouds. Union Carbide's "Structural Foam" process produces a molded, foamed polypropylene part having a cellular core and an integral solid skin [81, 82]. Low-cost aluminum molds make this application practical for small-run truck productions. The polypropylene resin (specific gravity 0.905) is foamed to a specific gravity of 0.60. Fan-shroud units weigh from 2 to 5 lb and have an average wall thickness of 0.25 in.

This application has expanded to the point where a structural foam shroud is used in every GMC and Chevrolet truck over 3/4 ton. General Motors used 15 sizes and types, some in very low volume. Because of the low mold costs, the structural foam shrouds are claimed to be significantly

FIG. 12. Polystyrene foam is used as an insulating material in commercial trucks. The milk-delivery truck shown here has a sandwich construction with the rigid foam laminated between sheets of metal. In addition to its insulating value, the foam contributes to structural rigidity. Courtesy of Sinclair-Koppers Co.

cheaper than metal stampings, injection molding, or polyester compression molding in low volume [83-85] .

Rigid foam board based on polyvinyl chloride (produced by B. F. Goodrich and Johns-Mansville under license from Kléber-Colombes, France) has reportedly been used as a structural foam-core material in campers and truck trailers [86] .

B. Automobiles

The design capabilities of rigid plastic foams, primarily for structural strength properties, are of increasing interest to automobile engineers. In many cases it is too early to determine which uses will be commercialized to full scale and which will remain only speculative.

The structural design features and the weight reduction possible with rigid cellular materials offer considerable engineering latitude and economic benefits to automated mass-production techniques. In one unique application automobile manufacturers are evaluating a thin-gauge steel top that has a composite rigid-urethane and flexible-foam inner liner. The assembled unitized structure offers rigidity, insulation, vibration, dampening, soundproofing, and cushioning [87].

Further uses of rigid urethane as a structural reinforcement were demonstrated in a prototype station wagon. Poured-in-place rigid foam in steel sandwich panels used as decking resulted in an overall saving of $5 per vehicle, based on the use of lighter gauge steel, reduced manpower requirements, lower costs of raw materials, and elimination of fabrication steps [88].

Volkswagen is using rigid urethane foam as a poured-in core for hollow aluminum tubular framing. The foam reduces dust traveling through the structure, cuts vibration, deadens sound, and adds strength [88].

An experimental plastic car displayed by Bayer in Düsseldorf (October 1967) demonstrated new design concepts for use of plastic materials and associated production techniques in automobiles [89]. A principal novel design feature is a load-bearing base unit that completely replaces the conventional chassis or X frame. The base unit and front fenders are a sandwich construction, with outer shells of epoxy-fiber glass and the cavity filled with rigid urethane foam (6 lb/ft^3). The rear fenders are rigid-urethane-foam core-sandwich construction with thermoplastic facing. Urethane Duromer, a structural foam with solid walls and microporous interior (overall density 30-80 lb/ft^3) was used for the roof, hood, and trunk lid [90, 91].

Structural thermoplastic foams [85, 86] produced by injection molding of ABS have been suggested or are in production in automobile footrests, gear-shift knobs, and decorative scrolls and moldings. Marbon Chemical reports that the density can range from 30-60 lb/ft^3, with wall thickness of >200 mils. Finished parts can be glossy or dull, smooth or textured. By incorporating graining in the mold, a wood-grain appearance can be achieved. The closed-cell foamed ABS parts have excellent dimensional stability (even after immersion in water) and less tendency to warp and swell than wood, but they have the mass, sound, and feel of wood.

Structural polypropylene foam is also finding application in children's car seats. The structural foam seat and back support have replaced the metal pans. Hinges and bars are molded into the foam to reduce assembly costs, and a soft vinyl foam is used over the rigid foam for enhanced comfort and appearance [83-85].

Polypropylene foam has been used for the vacuum-formed air-conditioning insulator in the 1969 Thunderbird and a 3.5-ft-long ventilating heating duct for buses [85, 90]. Closed-cell polypropylene foam can be vacuum-formed or compression-molded, as well as die-cut, skived, and heat-sealed. Vacuum forming can be performed over a wide temperature range (e.g., 1 min at 170°C for 0.25-in. thickness). Lap and butt heat seals are made by heating the material at 205°C and pressing the surface together. Foam density is 4 to 6 lb/ft^3 (Haveg Industries, Inc., Minicell PPF type 5UM).

C. Railroad Cars

Rigid urethane foam, particularly the poured-in-place type, is the most widely used foam plastic for insulating railroad cars. In some cases styrene foam has been used in applications where insulation requirements are more moderate.

Railroad tank cars insulated with poured-in-place rigid urethane foam showed improved insulating efficiency, permitting 50% reduction in wall thickness, compared with conventional batt or block insulation. Tank-car capacity was increased, and the insulating job was completed in half the time [92]. Spray-on rigid urethane foam can be applied to a railroad tank car. Advantages claimed include ease of application, reduced painting maintenance, and ease of repair in case of damage. The chemical industry has been shipping temperature-sensitive chemicals in tank cars insulated with urethane foam. Similarly the food industry has been shipping hot cocoa butter and hot tomato paste in tank cars insulated with urethane foam without in-transit heating [93].

The increasing demand for rail transfer of food products has resulted in low-temperature insulated refrigerated box cars, as well as box cars that are insulated but not refrigerated. The objective of the latter type of design is to maintain the cargo at the same temperature at which it is loaded for 2 to 5 days. Construction of the insulated box cars varies widely. In some cases factory-produced, sized, and shaped sandwich panels with poured-in-place urethane-foam cores are assembled by gluing and conventional methods of construction [94].

Detailed techniques are reported [95] for reinforced polyester panels insulated with poured-in-place urethane foam. Froth foaming yielded foam densities of 2.5 lb/ft^3 and mold pressures of 7 psi. Comparison of the polyester sandwich construction with existing steel refrigeration cars showed the bending rigidity of the polyester sidewall to be approximately three times that of the steel car. The roof can resist 15% more compressive force, whereas the sidewalls can take equal forces of compression.

Another method is the use of poured-in-place urethane foam in the walls of a completely constructed box car [69]. For this type of application Fruit Growers Express [96] utilizes a heated, insulated building with sufficient space for foaming six 60-ft cars. Proper application of foamed-in-place insulation is reported to be dependent on correct temperature, bracing of car structure, accurate predetermined timed shots, and proper isolation of foam cavities. Airtightness tests on new mechanical refrigerator cars insulated with urethane foam showed a loss of only 45 ft^3/h in contrast to the 260-ft^3/h air loss generally considered acceptable. This airtight quality is particularly desirable for the controlled-atmosphere cars used to ship fresh produce. The insulation has also eliminated many structural members formerly required in cars using flexible insulation. Other benefits have been the elimination of lock nuts where the material surrounds bolt-and-nut connections and the waterproofing and sealing that are imparted. Connor and Sonderlind [96] report that a continuous-testing program is essential to maintain quality control. Foam insulation is checked periodically for density, stability, and moisture absorption. At specified intervals 4 x 4 x 1-in. samples are evaluated for volume change under high humidity (100% at 60 °C). Samples are tested for differences in volume change every 7, 14, and 28 days. Deep-freeze and k-factor tests are also made periodically.

A passenger train utilizing panels based on reinforced polyester and poured-in-place urethane foam was constructed in Germany [97]. The result was a 30% reduction in overall weight.

Preformed polystyrene-foam panels with mounting strips and vacuum-formed ABS plastic-sheet facings have been designed for refrigerated railroad cars [73, 98]. The plastic facings were designed with S-curved corrugations to provide air circulation within the car. The resulting interior surface is easily cleanable and resistant to abrasion, water, and impact. Conventional techniques can be used for mounting the polystyrene-foam (1.5 lb/ft^3) blocks and facings with nonconductive fasteners or adhesives. Steam expansion of large molded boxcar-door sections of preexpanded polystyrene beads has been demonstrated [66].

In 1963 the U.S. Steel Corp. developed the application of urethane-foam insulation for open gondola or hopper cars used to haul coal and iron ores. During the winter months the ore freezes solid in the uninsulated ore cars and must be thawed for unloading. The foam (500 lb per unit) is sprayed onto the outer surface of the car. It has been found that with the foam-insulated car the heat inherent in loading is sufficient to keep the load from freezing if the road haul and unloading are accomplished within 2 to 3 days [99-102]. Detailed temperature profiles of insulated railroad cars after exposure to low ambient temperatures for specific lengths of time reflected frost-line penetration. A test was made to compare an insulated hopper car (approx. 2 in. of 2lb/ft^3 sprayed urethane foam) with an uninsulated hopper car loaded with coal at an initial temperature of about 15°C. After 95 h at an ambient temperature of -15°C the insulated car emptied cleanly on unloading, but the uninsulated car could not be emptied [103].

D. Aircraft

The aviation industry represents a growing segment of the transportation industry for the application of rigid cellular materials. The important areas in air transportation are (a) in the structure of the aircraft and (b) in the containers or packaging for movement by air. Foamed plastics are finding application in the construction of airplanes as lightweight space fillers, moisture barriers, and to add rigidity and structure strength to the units in which they are applied. With respect to air cargo for shipment of perishable foods and flowers, foamed plastics offer low relative weight and excellent thermal insulation [104].

One of the first applications for rigid urethane foam was in aircraft wing tips, flaps, and ailerons for added structural strength without the addition of excessive weight. This application employs the urethane as a stiffening agent to bond in place the skins, thereby providing a continuous skin support and eliminating the need for ribs and internal strengthening members [73, 105, 106].

Sprayed urethane foam has been used in the pressurized portion of the interior cabin area of the Lear Jet as thermal insulation, panel dampener, and soundproofer [104].

Slab-stock urethane foam combined with a skin of fiber-glass impregnated with epoxy resin is used for the wing construction in all-plastic airplane by Windecker Research, Inc. (Midland, Texas). Except for the 290-hp Lycoming engine, controls, associated hardware, and the landing gear, the plane is 99.5% reinforced plastic and foam plastic [107].

Aircraft radomes utilize the transparency of urethane foams to waves of radar frequency. Radomes in Lockheed's F-94C Starfire planes are made of reinforced plastic-urethane foam sandwich construction [108]. Radomes are also used in commercial aircraft to protect the antennas from airstream, dust, and moisture [109].

Slab-stock rigid polyvinyl chloride foams laminated into sandwich panels are being considered as lightweight interior structural decks of aircraft [86]. Rigid polyvinyl chloride foams are said to retain 80% of their mechanical properties up to 95°C and can withstand unlimited exposure at temperatures of 150°C without permanent loss of properties [7].

Sandwich panels of ABS sheet-foam-sheet lamination are claimed to be potentially suitable for aircraft wing ribs [85] and side and divider panels [104]. Ease of molding to specific shapes as well as high resistance to heat distortion and impact are attractive to aircraft manufacturers considering this technique.

Rigid urethane foam (4 lb/ft^3) has been used as the core for helicopter blades, where the leading edge spar and the trailing edge skin are bonded. The blades have been flight tested for several thousand hours without cracking or deformation [73, 110]. Similarly the L-1049 Super Constellations utilize rigid urethane foam to modify propeller blades [111]. The increased surface area on the blade increases aircraft speed and improves engine cooling. The Avion 2-180 Gyroplane utilizes rigid urethane foam to fill the void in the rear annular duct to reduce resonance and metal fatigue [112].

E. Cargo Containers

The concept of containerized freight-transport units has become increasingly popular. The shipping cost advantage of not loading and unloading a cargo until it reaches its destination is attractive to the shipper, receiver, and carrier. Lower labor costs, more efficient packing, and reduced pilferage are among the advantages of this concept. The design of such shipping containers involves primarily the structural and insulating qualities of the rigid cellular materials.

Unrefrigerated containers insulated with 5 in. of rigid urethane foam bonded to reinforced plastic skins were used to ship cargoes frozen in liquid nitrogen. In the foam containers the cargo stayed frozen for up to 4 weeks at temperatures of 32°C without refrigeration. The containers were strong enough to permit seven-high stacking in ships' holds [113] .

Rapid growth in containerized shipments has occurred since 1958 [114] . As a means of facilitating efficient interchange between rail, marine, and highway carriers, the American Standards Association (MH-5 committee) has developed a modular series of container sizes. The nominal sizes established as American Standards for the van type containers are 10-, 20-, 30-, and 40-ft lengths and an 8 x 8-ft end configuration. In addition, modules that are 5 ft and 6 ft 8 in. long, 8 x 8 ft in. cross section, and with an optional height of 6 ft 10 in. have been included in the standard. The actual sizes are somewhat smaller in order to provide clearance between units when used in combination with each other to fit within the maximum 40-ft length [115] .

REFERENCES

1. R. L. Siren, "Markets and Economics of Cellular Plastics," Polymer Conference Series, Cellular Plastics Technology, Wayne State University, Detroit, Mich., 1967.
2. R. E. Skochdopole, in Encyclopedia of Polymer Science and Technology, Vol. 3, Wiley-Interscience, New York, 1965, pp. 80-130.
3. Anon., Plastics World, 25 (12), 38 (1967).
4. R. E. Tenhoor, Chem. Eng. News, 41, 242 (1965).
5. H. H. Noren, J. Cell. Plastics, 1, 242 (1965).
6. D. P. Shedd, paper presented at the SPI 7th Annual Technical Conference, 1963.
7. U.S. 1967-1968 Foamed Plastics Market and Directory, Technomic Publishing, Stamford, Conn., 1967, pp. 22-40.
8. Anon., Chem. Week, 93 (17), 171 (1963).
9. Anon., Urethane Industry Digest, 6, 17 (1968).
10. Anon., Mod. Plastics, 45 (5), 120 (1968).

11. K. Kawai, J. Cell. Plastics, 1, 234 (1965).
12. B. Simond, J. Cell. Plastics, 1, 229 (1965).
13. Anon., J. Cell. Plastics, 1, 371 (1965).
14. T. H. Rogers, Elastomeric Foam in the Automobile Industry, Chemical Research Association, Detroit, Mich., 1954.
15. P. D. Alexanders, Plastics World, 26 (7), 38 (1968).
16. Anon., Mod. Plastics, 45 (2), 89 (1967).
17. R. D. Byal, Chem. Eng. Progr. 63 (3), 94 (1967).
18. Anon., Plastics World, 25 (6), 7 (1967).
19. Anon., Mod. Plastics, 44 (5), 107 (1967).
20. G. M. Wolf and M. A. Marinetti, SPE Journal, 24 (7), 42 (1968).
21. E. W. Madge, Latex Foam Rubber, Interscience, New York, 1962, pp. 138-142.
22. Cellular Vinyl, Service Bull. G-26, B. F. Goodrich Chemical Co.
23. R. R. Scheinert and G. Mirr, J. Cell. Plastics, 2, 322 (1966).
24. Anon., SPE Journal, 24 (11), 72 (1968).
25. J. M. Callahan, Automotive News, Sept. 24, 1968.
26. R. Eshelman, Rubber Age, 100 (9), 68 (1968).
27. D. M. Severy, H. M. Brink, and J. D. Baird, SAE Journal, 77 (7), 20 (1969).
28. Flammability of Automotive Interior Trim Materials-Horizontal Test Method, Society of Automotive Engineers Technical Report J369, 1969.
29. E. L. Beidler and D. G. Walters, Rubber World, 157 (5), 51 (1968).
30. Anon., Chem. Eng. News, 46 (20), 46 (1968).
31. Anon., Mod. Plastics, 44 (7), 110 (1967).
32. Anon., Mod. Plastics, 45 (8), 106 (1968).
33. Anon., Mod. Plastics, 44 (3), 104 (1967).
34. J. H. Saunders and K. C. Frisch, Polyurethanes: Chemistry and Technology, Part II, Interscience, New York, 1964, pp. 164-168.
35. H. Wirtz, J. Cell. Plastics, 2, 324 (1968).
36. Anon., Mod. Plastics, 44 (2), 90 (1966).
37. Instrument Panel Laboratory Impact Test Procedure-Head Area, SAE J921a, Society of Automotive Engineers, November, 1967.
38. M. F. Garwood, Rubber World, 156 (3), 45 (1967).
39. Anon., Plastics World, 27 (2), 53 (1969).
40. A. B. Zimmerman, SAE Journal, 77 (4), 26 (1969).
41. Anon., Chem. Eng. News, 45 (4), 41 (1967).
42. Anon., Mater. Des. Eng. 63 (5), 87 (1966).
43. R. J. Fabian. Mater. Eng. 70 (5), 42 (1969).
44. H. H. Lubitz, J. Cell. Plastics, 5, 221 (1969).
45. L. Spenadel, Rubber World, 150 (5), 69 (1969).
46. C. R. Michaels and R. P. Kane, paper presented at the Automotive Engineering Congress, Society of Automotive Engineers, Detroit, Mich., 1968.

47. L. M. Zwolinski, SPE Journal, 25 (9), 24 (1969).
48. H. Wirtz, J. Cell. Plastics, 2, 324 (1966).
49. H. Wirtz, J. Cell. Plastics, 5, 305 (1969).
50. Compounding NS-11, Natsyn 200 Open Cell Sponge, Goodyear Tech. Book Facts, Goodyear Tire and Rubber Co.
51. R. Eshelman, Rubber Age, 101 (9), 66 (1969).
52. Anon., Mater. Des. Eng., 62 (4), (1965).
53. Anon., Motor Trend, February 1968.
54. M. E. Bailey, M. Kaplan, B. Taub, and G. Wooster, J. Cell. Plastics, 5, 237 (1969).
55. Anon., Prod. Eng., 40 (4), 61 (1969).
56. Anon., Chem. Week, 103 (13), 22 (1968).
57. Anon., Chem. Eng. News, 43 (23), 42 (1965).
58. Anon., Plastics World, 25 (3), 43 (1967).
59. G. Mueller and A. F. Bockman, J. Cell. Plastics, 5, 151 (1969).
60. A Guide for the Application of Rigid Urethane Foam in the Transportation Industry, Cellular Plastics Division, Society of Plastics Industry, Inc., 1965.
61. J. H. Reeves, paper presented at the SPI 7th Annual Technical Conference, Cellular Plastics Division, 1963.
62. Anon., Food Proc., 26 (3), 138 (1965).
63. H. J. Bruce, J. Cell. Plastics, 2, 309 (1966).
64. L. A. Nethmar, paper presented at the SPE 21st Annual Technical Conference, 1965.
65. G. Heffner, J. Cell. Plastics, 2, 310 (1966).
66. M. W. Riley, Mater. Des. Eng., 53 (3), 127 (1961).
67. Union Carbide Corp., Linde Polar Stream Bulletin, pp. 86-105.
68. W. E. Perkins, J. Cell. Plastics, 2, 318 (1966).
69. H. R. Hudgens, J. Cell. Plastics, 2, 317 (1966).
70. Anon., DuPont Magazine, Nov.-Dec. 1960, pp. 6-8.
71. Union Carbide Corp., Rigid Urethane Foam, 1964, p. 19.
72. Anon., Plastics World, 21 (9), (1963).
73. T. H. Ferrigno. Rigid Plastic Foams, 2nd ed., Reinhold, New York, 1967, p. 176.
74. Thurane Brand Insulation Board (Rigid Urethane Foam), Dow Chemical Co. Bulletin No. 171-167, p. 17, (1961).
75. J. F. Reeves, "Cellular Plastics in Transportation Applications: Cargo Containers Molded with a Reinforced Plastic with a Urethane Foam Core," Polymer Conferences, Wayne State University, 1965.
76. Dow Chemical Co. Bull. 171-144, 1960.
77. J. A. Hartsock, J. Cell. Plastics, 2, 332 (1966).
78. J. A. Hartsock, J. Cell. Plastics, 3, 81 (1967).
79. J. A. Hartsock, paper presented at the 2nd SPI International Cellular Plastics Conference, New York, 1968.

80. J. A. Hartsock, Design of Foam Filled Structures, Technomic Publishing, Stamford, Conn., 1969.
81. Anon., Mod. Plastics, 44 (2), 96 (1966).
82. Plastic Data Guide, Union Carbide Corp. Bull. J-2855, 1967.
83. J. H. Gehl, paper presented at the SAE Conference, Detroit, Mich., 1968.
84. R. G. Angell, J. Cell. Plastics, 3, 490 (1967).
85. R. J. Fabian, Mater. Eng. 70 (5), 43 (1969).
86. L. L. Scheiner, Plastics Technol., 14 (13), 39 (1968).
87. Anon., Mod. Plastics, 39 (2), 82 (1961).
88. Anon., Mod. Plastics, 44 (5), 107 (1967).
89. Anon., Brit. Plastics, 41 (7), 66 (1968).
90. H. H. Lubitz, J. Cell. Plastics, 5, 211 (1969).
91. Mobay Chemical Co., The Rolling Showcase for Engineering Plastics, 1968.
92. Mobay Chemical Co., Mobay Polygram, 2 (2).
93. U.S. Foamed Plastic Markets and Directory, Technomic Publishing, Stamford, Conn., 1965.
94. Anon., Railway Age, 157 (15) (1964).
95. A. Wethmar, SPE Journal, 22 (2), 101 (1966).
96. J. J. Connor and C. Soderlind, Plastics Design Proc., 9 (12), 23 (1966).
97. R. Vieweg and A. Höchtlen, Kunststoff-Handbuch, Vol. VII, Polyurethane, Hansen, Nunich, 1966, pp. 647-648.
98. Anon., Plastics Technol., 7 (9), 41 (1961).
99. Anon., Railway Age, 156 (7) (1964).
100. Anon., Chem. Week, 95 (26), 48 (1964).
101. Anon., Railway Locomotives and Cars, Oct. 1964.
102. Anon., Railway Age, 157 (18) (1964).
103. H. J. Bruce, J. Cell. Plastics, 2, 306 (1966).
104. A. P. Vance and R. M. Parks, J. Cell. Plastics, 2, 345 (1966).
105. J. P. Billington, Mod. Plastics, 45 (6), 76 (1968).
106. H. R. Moore, paper presented at the SPI 12th Annual Meeting, Reinforced Plastics Division, Chicago, 1957.
107. Anon., Prod. Eng., 39 (11), 157 (1968).
108. T. P. Dougan, Mater. Des. Eng., 49 (1), 86 (1959).
109. J. H. Saunders and K. C. Frisch, Polyurethanes: Chemistry and Technology, Part II, Interscience, New York, 1964, p. 277.
110. Mobay Chemical Co., Mobay Polygram, 2 (1).
111. Anon., Brit. Plastics, 31, 421 (1958).
112. C. H. Wheeler, in Foamed Plastics, Proceedings of a Conference, PB 181576, U.S. Department of Commerce, Office of Technical Services, Washington, D.C., 1963.
113. Anon., Plastics World, 19 (4) (1961).
114. Anon., Plastics World, 27 (11), 61 (1969).

115. F. Mueller, J. Cell. Plastics, 2, 340 (1966).
116. L. A. Harlander, J. Cell. Plastics, 2, 342 (1966).

Chapter 20

ARCHITECTURAL USES OF FOAM PLASTICS

Stephen C.A. Paraskevopoulos

University of Michigan
Ann Arbor, Michigan

I. PROPERTIES AND APPLICATIONS OF MATERIALS

The physical properties of foam plastics reflect those of the parent plastics, but modifications do occur as a result of the conversion process.

Closed-cell foam plastics offer excellent thermal insulation properties (with k-factors usually varying from 0.12 to 0.30 Btu/(h) (ft^2) (°F/in.)), low water absorption, and low moisture permeability. The best insulation value is achieved for each material at a specific density (usually between 2 and 3 lb/ft^3); beyond this point it decreases in each direction. The insulation properties are also closely related to the percentage of closed cells as well as to their geometry, size, and content.

As is the case with their parent materials, foam plastics are either thermosetting or thermoplastic. The former are cured through heat or chemical means into permanent form; because of the chemical crosslinking of molecules, they do not lend themselves to heat-forming techniques. The latter, in contrast, soften when heated and harden when cooled; by alternating the heating and cooling process they can be reshaped many times, since their change is substantially physical rather than chemical.

Virtually all thermosetting or thermoplastic resins can be foamed. The most common rigid foams are polystyrene, which is thermoplastic, and polyurethane, which is thermosetting. Because of their properties and their more advanced state of development, these are the only foam plastics that have found applications, other than experimental, in building. The discussion of foam plastics in this chapter is therefore concentrated on these two materials. However, the potential of other foam plastics, such as epoxy, polyvinyl chloride (PVC), and acrylonitrile-butadiene-styrene (ABS), for use in building merits further investigation.

From the standpoint of construction, plastics offer such advantages as light weight, ease of fabrication, relatively low cost, and a multiplicity of shapes and colors. On the other hand, they exhibit some definite limitations that relate to their nature as organic materials and their inherent mechanical properties. Compared with conventional building materials, plastics as a group have relatively low strength, low moduli of elasticity, high coefficients of thermal expansion, and high creep characteristics. Some are also susceptible to heat and to ultraviolet or hydrolytic deterioration.

The rather poor mechanical properties of plastics, which become even poorer in the case of foam plastics, should not inhibit their structural use. To be sure, poor mechanical properties impose limitations, but nevertheless much can be accomplished through proper design and application, as shown in the examples presented in this chapter.

So far, the main effort of the plastics industry has been to incorporate foam plastics into conventional building systems, and consequently, research regarding structural potential has been rather limited. Besides serving as roof and wall insulation, foam plastics have been used extensively as the core material in sandwich-panel construction. Many types of panels have been produced, but in most instances the skin (plywood, asbestos-cement, or metal) is the structural element, and the foam provides only insulation. One recent large-scale application has been at the CBS building in New York, where polyurethane foam was produced at the site by the frothing technique and deposited behind the granite slabs of the elevations.

Considerable work has been done by Dow Chemical Company to find construction applications for polystyrene foam. The foam has been successfully applied to the forming of concrete shell structures in which the form-giving material provides the roof insulation. One of the most significant developments introduced by this company has been its "spiral generation" process (this technique for the erection of spherical domes and other curvilinear surfaces is described later in this chapter).

Most pioneering work that sought a greater utilization of the structural properties of polyurethane foam has been instigated by the military. Good examples are the singly curved shell structures developed by the Army Corps of Engineers at Fort Belvoir, Virginia, for use in Greenland. The original structures were produced by spraying large molds with chopped glass fibers and polyester resin, closing the molds, and then filling them with polyurethane foam. After curing, the molded poured-in-place structural panels were assembled to form a complete structure. Subsequent structures were produced in a similar fashion except that the reinforced plastic skins were replaced by thin-gauge aluminum sheets.

On the civilian side, experimental structures made of molded or sprayed polyurethane components have been erected by various private concerns in this country and abroad. A number of conceptual ideas for the realization of structures through the use of polyurethane foam have also appeared in newspaper and magazine articles; more often than not they constitute fiction rather than fact.

Between 1962 and 1966 a research project was conducted at the University of Michigan [1] to investigate the structural potential of foam plastics. The activities and findings of this project are summarized in this chapter.

II. STRUCTURAL INVESTIGATION

One cannot design intelligently with any material unless facts are available as to its physical properties, the shapes and sizes in which it can be obtained, and the possibilities of assembling component parts into a final structure. Creative design imagination is not stimulated by lack of familiarity with a material, and the foam plastics are not familiar in the same sense as concrete, steel, and wood, the materials that architects and engineers have traditionally used.

The basic pattern of structural investigation followed by the University of Michigan project staff is shown in Fig. 1.

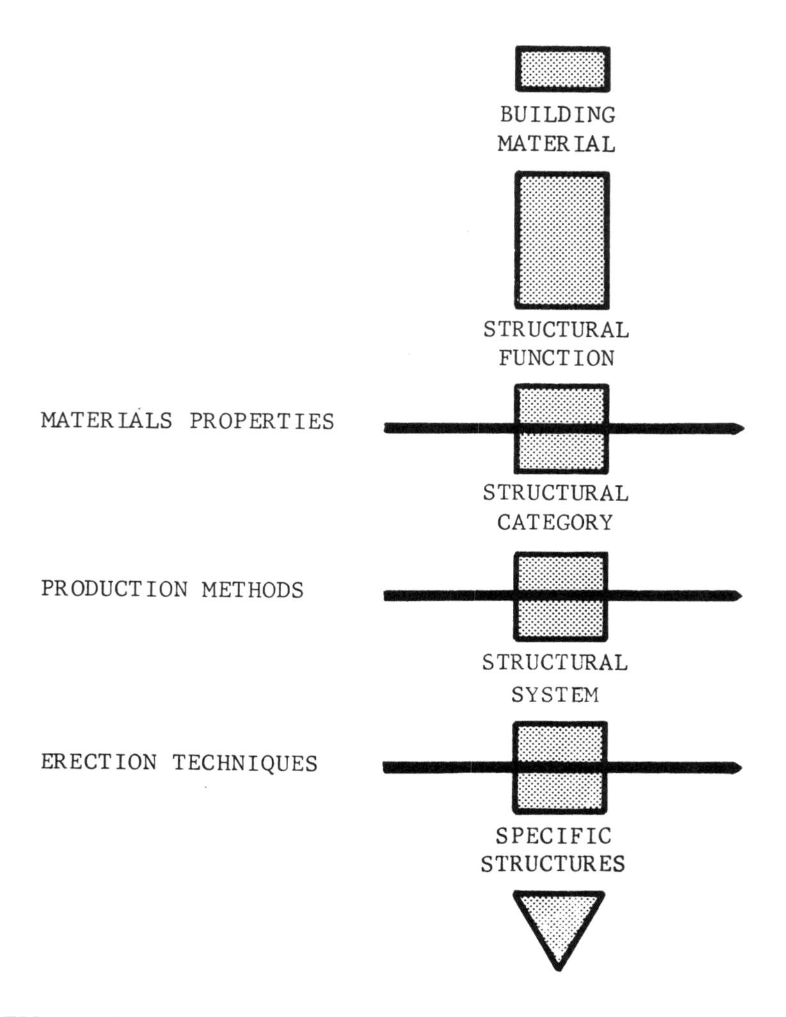

FIG. 1. Pattern of structural investigation.

Foam plastics were investigated in terms of their structural function as follows:

1. As a primary structural material that carries the principal loads and stresses of the structure.
2. As a secondary structural material that takes secondary stresses while allowing another material to carry the principal stresses.
3. As a contributing structural material, either as form-giving device that allows a production technique to be utilized or as a tertiary structural material that braces another material acting as a secondary element.

Structural development was approached from three directions:

1. Through investigation of the mechanical and other physical properties of foam plastics assumptions were drawn as to the kinds of structure that would be most logical from the standpoint of structural engineering.
2. Similar investigation of production methods revealed shapes and sizes of structures that may most readily be obtained.
3. Finally, various erection techniques that became possible through the use of these new materials also suggested ways in which structures could be developed.

The fundamental limitation in structural design with foam plastics lies in the low moduli of elasticity and the high creep characteristics exhibited by these materials. Because of their low E values, in order to eliminate excessive deformations, structural stiffness must be increased through geometry. This also leads to a reduction of stresses within the structure. It is essential to maintain low stress levels not only because of low material strength but also in order to control the effect of creep.

Structural solutions should be sought within the family of "surface structures," especially shells and folded plates, in which stress levels are kept low by the distribution of loads throughout the structure. Through further analysis, solutions that will be ideal from a purely structural standpoint can be obtained. However, to determine optimum solutions in terms of performance and cost, factors other than structural must also be considered.

Possibilities that are presented by various methods of producing foam plastics materials and the limitations imposed by such methods in terms of size and shape are major factors in determining optimum structural solutions. Finally, erection techniques should also be given due consideration. Foam plastics offer some unique possibilities for the erection of structures, and this can contribute further in reducing their cost in place.

The range of architectural applications of foam plastic structures is limited by structural shape and other limitations imposed by the nature of these materials. However, new architectural solutions are being made possible through the use of foam plastics. Their proper use can greatly contribute not only to the realization of new shell, folded-plate, and monocoque surface structures but also to a greater industrialization of the building process.

III. PRIMARY STRUCTURAL APPLICATIONS

A. General

In primary structural applications the foam plastics are viewed as constituting the building element or elements required for the overall stability of a structure. In addition to possessing sufficient strength, primary structures or structural components should be able to demonstrate limited time-related deflections and to maintain their properties under adverse conditions of exposure to use and the elements.

A major controlling factor in primary structural applications of the foam plastics is unit deformation, both elastic and plastic (creep). Two approaches are possible: (a) structural applications involving minimum uniform stress; (b) the use of structural shapes in which relatively moderate deflections can be accepted. The solution lies in the use of doubly curved shells, in which minimum stresses will occur as a result of sustained load, and where larger movements from those normally associated with building structures can be safely accepted.

It can be theorized that in optimum primary structural applications of foam plastics to shells the resulting shells must meet the following conditions:

1. The shell curvature and form must be such that it will result in acceptable stresses and corresponding deformations without distortion.
2. The shell boundary should be supported so as to maintain membrane stresses in the shell.
3. The resulting shell structures should be of constant strength.

To meet these conditions, the concept of minimum structure or structural membrane (where form is determined by forces rather than geometry) becomes a fundamental part of the general approach to the use of doubly curved foam plastic shells.

The use of foam plastics for principal structural components presupposes an economic use of such materials. Doubly curved shells for roof structures offer the possibility of capitalizing on the positive features of cellular plastics: their low weight, flexibility of application, and high insulating capabilities. Furthermore the negative features, such as creep, low absolute strength, and low service temperatures, can be accommodated without increasing the costs.

In general, the cost position of foam plastics versus concrete is quite favorable when these materials are used for the construction of doubly curved shells. The cost of such shells compares favorably even with standard wooden flat-roof construction on a square-foot basis.

The method of production and erection of shell structures becomes particularly important in obtaining overall economy. Two rigid foam plastics, namely, polystyrene and polurethane, were selected by the University of Michigan project staff for detailed exploration. Selection of materials was determined not only on the basis of physical properties and availability but also on a consideration of the systems available for the production and erection of shells. These systems were (a) the "spiral generation" system for polysterene and (b) the spray system for polyurethane.

B. Polystyrene "Spirally Generated" Dome

The "spiral generation" process [2] involves the use of a specially designed machine that places plastic foam in a predetermined structural shape. The equipment consists of a machine head mounted on a boom that revolves around a pivot mechanism. To erect a polystyrene foam dome the position of the pivot mechanism is fixed, and as the rotation process begins, foam board material is placed in the machine head, which then bends and heat-seals it, layer on layer, into a rising structural spiral.

The need for a golf-club structure in the Ann Arbor area offered the vehicle for a first architectural application of the "spiral generation" process. The structure, encompassing a dome 45 ft in diameter, was designed by the University of Michigan project staff and erected in collaboration with the Dow Chemical Co. in the summer of 1963. The erection process is shown in Fig. 2.

In erecting this structure a 1-ft-wide by 3-ft-deep circular trench, corresponding to the diameter of the dome, was first excavated. The trench was bridged at intervals by 2 x 4 wood boards, which supported a base ring made from angle iron. A starter strip of foam was attached to the base ring, and the generation process began. The dome, consisting of polystyrene foam boards 4 in. thick with an approximate density of 2 lb/ft^3, was constructed in less than 12 h.

After its completion the dome was lowered into the trench by 30 students using ropes connected to the base ring. After the dome had been lowered and the trench backfilled with earth (fully on the outside and partially on the inside), the openings were marked, cut out, and reinforced around the edges with glass-fiber tape and epoxy resin. The concrete floor slab was then poured, and its edge was anchored to the dome.

The dome exterior was painted with a mixture of latex paint and vermiculite. The interior was treated with a similar, but somewhat heavier, coating. Window glass for the exterior walls was cut and placed in wood mullions. The joint between the dome and the glass was made flexible by using a polyethylene-foam gasket, which was bonded with contact cement to the dome's interior surface.

The dome structure has performed extremely well; only minor repair and maintenance have been required over the years. A number of applications of the "spiral generation" process have since been made in domes with smaller as well as larger diameters. In most cases, however, their surfaces have been covered with concrete to meet fire-code requirements.

The "spiral generation" process is a unique method for constructing polystyrene-foam shells. It derives its principal advantage from a well-integrated use of relatively simple equipment and an economic and rational use of materials.

Fig. 2. Erection of a spirally generated polystyrene dome.

FIG. 2. (continued).

FIG. 2. (continued).

C. Sprayed Polyurethane Shells

Spray application of polyurethane foam offers another promising possibility for constructing shell structures. Transportability of equipment and ease of creating structural shells of varying shape and thickness are the basic advantages. The spray operation can either be conducted freehand or it can be mechanized. An unlimited number of shapes can be created through the hand-spray operation.

Nevertheless this operation is not satisfactory from the standpoint of quality control. Too much depends on the skill of the man who operates the spray gun. Since the material leaving the spray gun will expand usually 20 to 30 times, any imperfection will be magnified accordingly. Thus the results may be quite unpredictable.

An essential element of the spray application is the form-giving surface against which the foam components are to be sprayed. Air-inflated skins have been used for this purpose. Another possibility is offered through a lightweight wood lattice that was devised by the University of Michigan project staff as an armature to obtain doubly curved shell forms.

The armature is composed of wood slats bolted together as a uniform grid that can be folded. As such it can be preassembled and transported to the site. There it can be bent into a doubly curved form and manipulated to meet a range of shapes and dimensions.

A hand-sprayed shell was constructed by this method in the spring of 1964 in collaboration with Wyandotte Chemicals Corp., as shown in Fig. 3. A 27 x 27 ft armature was bent into a doubly curved shape anchored at four points describing a square, 21 ft on the side. The armature was then covered with a stapled-on nylon-reinforced paper skin and sprayed from the inside.

Since the armature is not capable of supporting the total weight of the structure, the sequence of spraying becomes an important consideration. First the buttresses and then the peripheral section were sprayed. Thus the foam began to work structurally as the armature was being sprayed.

The general form of this shell was based on construction rather than on geometry or forces. The edge support provided for the shell proved to be inadequate; this created high stress concentrations in the four buttress areas, and considerable deflection was experienced at the midpoint of the edge members. To correct this situation the shell was pulled back to its original shape and stiffening arches were installed at the edge line.

A folding-armature shell of this kind offers a reasonable basis for a primary architectural application of polyurethane foam, provided the armature is adequately stiffened at the edges when erected and tied effectively in a circumferential pattern. This is especially important since the consistency of the foam as a strcutural material when produced by freehand spraying is questionable.

To improve structural performance, the possibility of using the concept of minimum structure to determine the form of the structure was also subsequently explored, as shown in Fig. 4.

FIG. 3. Freehand-sprayed polyurethane shell.

FIG. 3. (continued).

A large-scale structural membrane of 1/4-in. flexible polyurethane foam supported at the corners and with simulated uniform surface loading was used to define the profile of the armature. The form obtained from the loaded membrane is in pure tension in the flexible stage. If the form is rigidized and reversed, it should result in a state of pure compression for a surface load. To do so a folding armature was formed to duplicate the profile of the membrane.

The armature was covered with the same nylon-reinforced paper. A sheet of 1/4-in. flexible polyurethane foam was then attached over this surface and secured at the edges. The armature was thus used to produce two shells: one sprayed from the top and incorporating the flexible membrane, and one sprayed from the bottom and incorporating the armature.

Testing of the two shells produced in this manner indicates that the wood armature contributes to the load-carrying capacity of the foam. It was further indicated that stress levels can be maintained within acceptable limits through the use of a proper form.

However, in spite of the careful spraying operation, inconsistencies in the material still occurred. Since it is questionable whether exacting design specifications can be met through the hand-spraying application, further effort in optimizing form does not appear to be warranted. If the rather crude appearance of a hand-sprayed surface is acceptable, the approach recommended in connection with the first sprayed armature appears to be the most promising.

The unpredictability and crudeness of texture associated with freehand spraying can be used to create a number of fun structures for playgrounds, etc., where structural integrity and economic use of materials may not be essential. If these qualities are essential, however, the mechanization of the spray operation appears to be a necessary prerequisite.

FIG. 4. Freehand-sprayed polyurethane shell.

Further experiments with the concept of minimum structure suggested the use of flexible polyurethane membranes as the form-giving device for the mechanized production of anticlastic umbrella shells (Fig. 5). A hexagonal type of anticlastic umbrella that is reasonably close to a surface of revolution indicated a method of mechanization comprising a rotating platform and spraying from a fixed position.

An operation for the fabrication of hexagonal shells was organized in the covered courtyard of the Architectural Research Laboratory in the summer of 1965. After initial experimentation seven shells, 12 ft in diameter, were produced and assembled to form a roof structure as shown in Fig. 6.

The hexagonal diaphragms consisted of 1/2-in. flexible polyurethane membranes that were stretched and attached to a wood frame constituting the final edge. Circular plywood disks were attached to the center of the diaphragms.

The rotating mechanism consisted of a hexagonal platform with a center element raised to the desired height of the umbrella shell. This carrousel was rotated around a central shaft at controlled speeds by a motor.

The hexagonal diaphragms were placed on the carrousel and stretched to the desired form. While the carrousel was rotating, polyurethane foam was sprayed from a suspended platform by an operator moving a gun along a metal guide. An initial layer was sprayed to rigidize the flexible foam, and then the thickness was gradually built up to the predetermined amount.

The completed shells were removed with a three-rope cradle and stacked one on top of the other. They were then coated with a liquid elastomer and individually assembled with pipe columns and bracing sprats. The umbrellas were anchored to the ground and attached to each other to form a pavilion.

The overall density of the sprayed foam was 6.6 lb/ft^3, and the thickness of the shells was designed to range from 2.5 in. at the edge to 4.5 in. at the crown. The weight of an assembled umbrella, including column and braces, was 480 lb, and its cost in place was computed at $2.50 per square foot of floor area.

The operation used to produce the hexagonal shells can be fully automated to produce structures meeting exacting specifications. Foam densities can be determined through speed of rotation and the distance at which the gun is located. In contrast to the freehand spray operation the surface texture is quite smooth and uniform. Also, considering the fact that a total roof system is provided, which includes structure, waterproofing, insulation, and acoustical treatment, the system is quite attractive from the standpoint of cost. More research is needed to establish the effective parameters of foam plastics when used as a primary structural material. However, research to date fully supports the thesis that certain foam plastics show great promise for the economic realization of doubly curved structures.

FIG. 5. Anticlastic umbrella.

IV. SECONDARY STRUCTURAL APPLICATIONS

A. General

In this case the foam plastics provide bracing or stiffening for other materials (metal, wood, or synthetic) that act in a primary structural capacity but are not proportioned to resist buckling on their own. The contribution of the strain in the foam plastic to the overall performance of the structure is assumed to be small, if not negligible, provided that

1. The primary structural material resists most or all of the direct stresses from the loads of the structure.
2. The stress in the foam resulting from the tendency of the primary material to buckle is neither large nor of long duration.

In contrast to primary structural applications, where forces in the foam plastics must be restricted to low levels of direct stress, the presence of a stronger material in secondary applications permits the consideration of higher stress concentrations and bending in the plane of the structure. Thus, with certain restrictions, the range of possible uses of foam plastics is extended to include singly curved shells, folded-plate structures, slab and panel structures, and monocoque tube structures of sandwich construction.

FIG. 6. Mechanized production of polyurethane shells.

FIG. 6. (continued).

FIG. 6. (continued).

The use of foam plastics as a core material in slab and panel structures is limited by the presence of permanent high shear stresses in the foam. In general an additional material must be introduced between the skins to resist this shear; otherwise, shear creep of the core material will produce intolerable deflections. The foam plastics serve then only to restrain buckling of the skins and to provide additional capacity for transient loads.

The application of foam plastics as the core element in monocoque tube structures of sandwich construction (as illustrated in the proposed filament wound structure) depends entirely on economic considerations. Where thermal efficiency is a major consideration, the use of foam plastics is indicated; otherwise the use of some other core material may be preferable.

The elements in singly curved shells and folded-plate structures are usually designed to resist stresses in the plane of the structure. Thus the restrictions for secondary structural applications of foam plastics are automatically satisfied. The use of foam plastics to stiffen skin membrane is appropriate for such structures, because the dead load is reduced while the primary structural material can still be maintained close to a theoretical minimum.

To prevent buckling the bond between the foam core material and the skins must be strong and continuous. Furthermore it must resist delamination under design stresses and changes in temperature and humidity. The cellular structure of the foam plastics provides uniform support and bracing to the primary structural elements. Foam plastics can be bonded with adhesives to a variety of skin materials, or as in the case of polyurethane and epoxy, which can be foamed in place, the bond can be achieved as the core material is produced between the skins.

By pouring polyurethane-foam components between glass-fiber and aluminum skins set into molds, a series of singly curved shell modules was produced by the U.S. Army Engineers Research and Development Laboratories at Fort Belvoir. By joining these modules a number of barrel-vaulted buildings were erected for military purposes.

B. Folded-Plate Structures

A variety of sandwich panels with polystyrene or polyurethane foam as the core material are commercially available and can be used for the erection of folded-plate structures. Normally, however, the contribution of the foam to the overall structural performance of the panel is very limited. The structural problem is confined primarily to the joining of panels, which makes it a general type of problem rather than one specifically relating to the use of plastics.

In exploring such secondary structural applications, the University of Michigan project staff therefore sought a panel that would be truly representative of foam-plastics technology, where the primary structural material could be kept to a minimum because of the presence of the foam and whose properties and method of production would play a major role in the resulting structural forms. This kind of product was found to be a paper-laminated polyurethane-foam board.

Although there are variations in the systems of producing such material, all use a continuous process in which the skins, introduced in the form of two paper rolls, go through the machine in a parallel fashion and at the same rate as the foam that is being produced. The liquid foam components are uniformly distributed between the two paper skins. The foam expands and adheres to the skins as the material moves through a form or nip roller. After that the foam board passes through a heat-curing operation and is cut to the desired size.

Various materials obtainable in roll form can be used for the production of the foam board. However, paper seems to be one of the best suited for this purpose, because structural and weathering properties can be added through secondary applications depending on the performance requirements of particular cases. Foam formulation, foam density, type of paper, and treatment of paper become parameters for satifying different structual criteria.

The material lends itself ideally for the development of folded-plate components because of the ease with which it can be scored and folded. A number of geometric configurations can be obtained by adding a scoring and folding operation at the end of the line that produces the board itself.

After initial investigation of various geometries a linear folded-plate component was developed (Fig. 7) to serve both as a vertical and a horizontal building element. The foam board was produced by the Union Carbide Corp. to the following specifications: board thickness, 3/8 in.; foam density, 2.6 lb/ft^3; paper skins, 69-lb kraft fiber liner board. The skins were impregnated with polyester resin, and the joints as well as the bottom chord were reinforced with glass-fiber tape. The folded-plate component was 4 ft wide and designed to span 16 ft. A two-story structure based on this structural component was erected in the courtyard of the Architectural Research Laboratory in the spring of 1964 as shown in Fig. 7.

The structural system as developed for this prototype structure consists of a series of rigid two-story bents that do not require additional bracing or shear walls to resist horizontal loading. It is important, therefore, to achieve as much joint rigidity and as much continuity of the wall components as possible.

To meet this requirement the roof component extends to the outside of the wall so that there is continuous bearing of the roof on the top of the wall. The top membrane of the roof component is extended approximately 18 in. beyond the wall space, folded down to cover the open end of the roof panel, and bonded to the exterior wall membrane with polyester resin.

For the shear joint between the floor and the wall components, one-half of a 6-in.-wide glass-fiber tape is bonded with polyester resin to the wall component in such a way that the other half remains free to be binded to the lower membrane of the floor component. The gap in the joint between the floor component (whose ends are closed by a foam-board diaphragm) and the wall component is filled with polyester resin. The result is a joint that transmits all of the load either as tension in the glass tape or as shear between the paper and the foam. The structure rests on concrete footings to which it is bonded with glass-fiber and polyester resin. The cost of the linear folded-plate bents in place was computed at $3.30 per square foot of floor area.

A variety of structural sections can be designed by using engineering formulas for elastic analysis to predict stress levels, deflections, and the like. Adequate and safe structural solutions can be obtained, provided the following restrictions are observed:

1. The stiffness of the section is computed by taking into account only the area and flexural modulus of the paper skins.
2. The paper thickness is assumed to be the average thickness prior to impregnation.
3. Deflections from permanent loads are computed by using a modulus of elasticity reduced in accordance with the expected life span of the structure.

FIG. 7. Linear folded-plate structure.

FIG. 7. (continued).

FIG. 7. (continued).

The foam board is also well suited for the realization of composite folded-shell structures. Such a structure was designed and produced by Herbert Yates, a Canadian designer. The development of this structure was encouraged by the University of Michigan project staff, who subsequently recommended its use for the housing of migrant farm workers in California.

An early sample of this structure was erected for testing purposes in the courtyard of the Architectural Research Laboratory in the summer of 1965. Figure 8 shows the process followed in erecting the structure as well as its final form.

Scored foam-board components are preassembled with tape and folded in accordion fashion for delivery to the site. There the structure is lifted, unfolded, and anchored to the floor by means of folding tabs that are part of the structure. The paper skins can be precoated, or, of preferred, coating can be added at the site. A variety of materials can be used to enclose the ends.

The structure covered 323 ft^2 of floor area, and the shell, which weighed 105 lbs could easily carry the weight of a person sitting on the ridge. Materials for the folding shell cost less than $60. Even if the structure were to be marketed at $500, it would still represent the lowest shell cost of any system with comparable performance.

C. Monocoque-Tube Structures

Such a structure was designed in 1967 by the University of Michigan project staff in collaboration with the Aerojet-General Corp. under contract from the U.S. Department of Defense for the housing of military families [3].

The structure is a slightly curved rectangular tube of sandwich construction with filament-wound glass-fiber skins and polyurethane foam as the core. Filament winding provides a method for fabricating reinforced-plastic structures. The process is similar to the spinning of a cacoon. Continuous glass filaments are coated with a binding resin and deposited on a forming surface called a mandrel. The mandrel rotates at a programmed speed, thereby permitting the impregnated glass-fiber filaments to be wound around its surface according to a preset pattern. After the desired quantity of material has been deposited on the mandrel, it is fully cured and separated from the mandrel.

The potential of using this process for building construction was first explored by the University of Michigan project staff in 1965, and a room-size unit was produced in collaboration with the Hercules Corporation.

The original concept called for a sandwich construction produced around a reusable steel-plate mandrel. However, for economic reasons a permanent mandrel was not constructed for this first experiment. Instead the mandrel was made of foam-board panel attached to a continuous

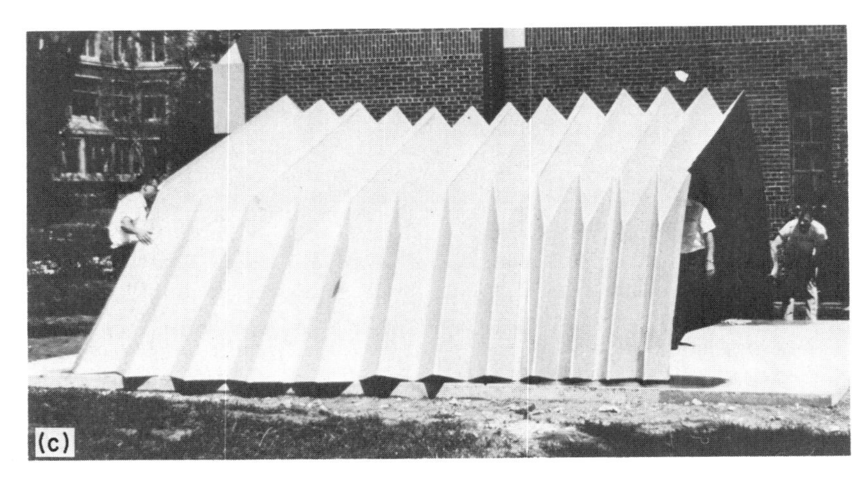

FIG. 8. Composite folded-plate structure.

FIG. 8. (continued).

FIG. 8. (continued).

steel frame and was designed to become an integral part of the structure, which was produced with only one skin on the exterior surface.

The process and product of this first experiment are shown in Fig. 9. The assembled steel frame-foam board mandrel was placed in the filament-winding machine. After completion of the winding process the structure was removed and shipped to Ann Arbor, where it was erected in the courtyard of the Architectural Research Laboratory.

In spite of some deficiencies that could be expected from an initial experiment, the structure has performed well. It clearly indicated that further development was warranted.

In designing for a second stage of application continuous production was assumed, which justifies the use of permanent mandrels. Structural, cost, and long-term performance analysis suggested the design of a monocoque rectangular tube of sandwich construction. It was determined that the tube, whose width was established at 20 ft because of design considerations, should have filament-wound skins, 0.10 to 0.15 in. thick, and a core of 3- to 4-lb/ft^3 polyurethane foam, 8 to 10 in. thick.

For good structural performance, the orientation of glass filaments in the two skins should be both longitudinal and circumferential on all surfaces (Fig. 10). The interior filament-wound surface can have sides that are either flat or curved. However, it is advisable that the exterior surface be continuously curved, at least slightly, in order to provide an adequate compaction of materials to ensure structural strength and watertightness.

The equipment proposed for the circumferential (hoop) winding consists of a horizontal turntable on which the mandrel is positioned and rotated, and a winding machine that wraps filaments around the mandrel by advancing

FIG. 9. Production of filament-wound room-size unit.

FIG. 9. (continued).

FIG. 9. (continued).

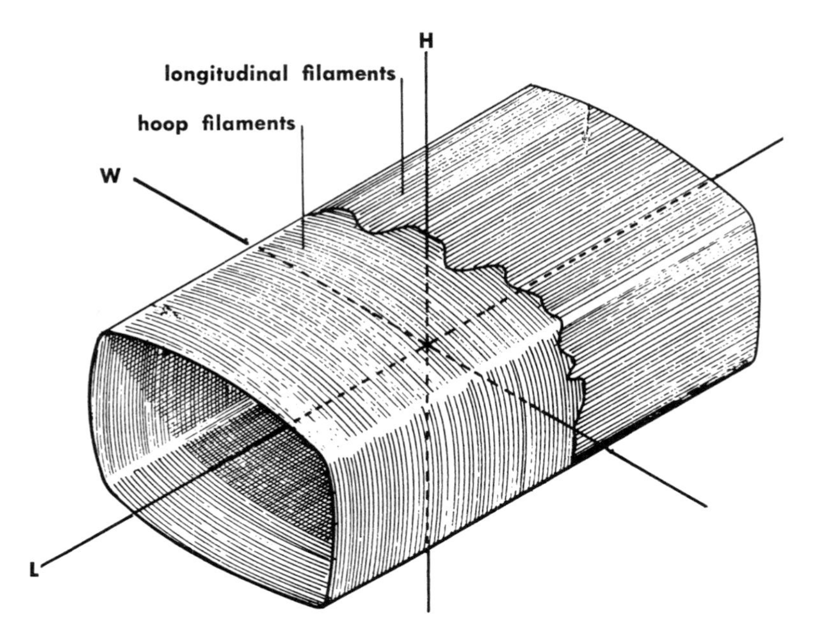

FIG. 10. Recommended orientation of glass filaments.

the filament delivery point up and down a winding tower (Fig. 11). The longitudinal filaments are applied to the circumferentially wound surface as part of the core, which is produced in the same fashion as the paper-skin foam board (Fig. 12).

The fabrication involves first the circumferential winding of the inner skin on a steel mandrel. Then core panels with the longitudinal filaments on each side are applied and adhered to this surface. The operation is completed through the winding of an outside circumferential skin. When the material is cured, the structure is removed from the mandrel.

The rectangular tubes are then fitted with end enclosures and interior components to meet dwelling design requirements (Fig. 13). A number of such designs can evolve through the juxtaposition of the tubular elements. One such possibility is shown in Fig. 14, and a grouping involving various design alternatives is illustrated in Fig. 15.

The filament-wound tubes can also be attached to vertical surfaces or frames to form high-rise complexes. One such scheme, designed by N. Gerald Rolfsen in consultation with the author, in which the tubes are cantilevered from a reinforced-concrete core element is illustrated in Fig. 16.

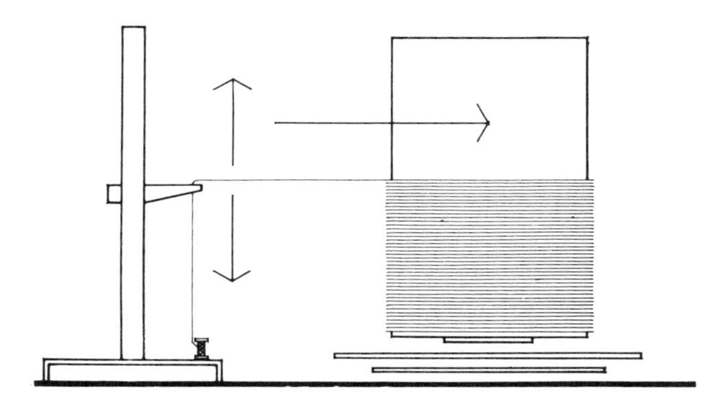

FIG. 11. Circumferential winding operation.

V. CONTRIBUTING STRUCTURAL APPLICATIONS

A. General

A number of construction uses have been made of foam plastics that do not depend on their capacity to sustain structural loads. Such uses include the forming of structures that have been designed to be constructed from other materials, such as concrete and reinforced plastics. In this case the mechanical properties of foam plastics have to be considered only to the extent that the form-giving system will have to be self-supporting until the structural materials take their final set.

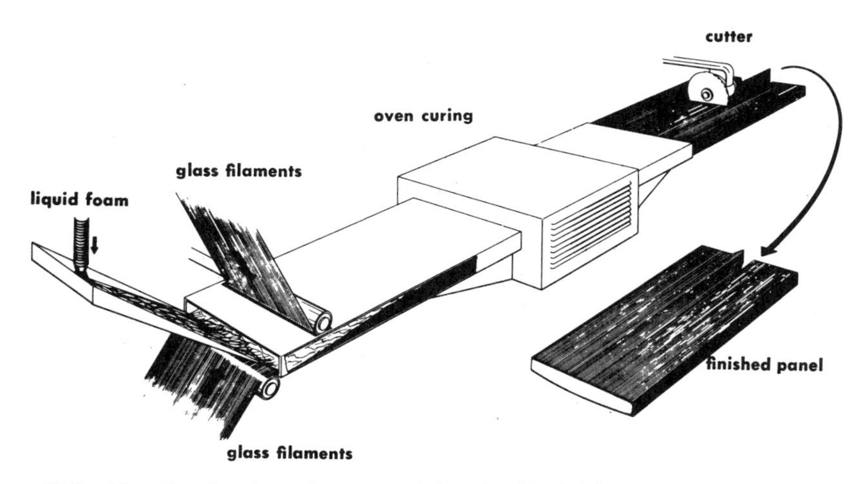

FIG. 12. Production of core with longitudinal filaments.

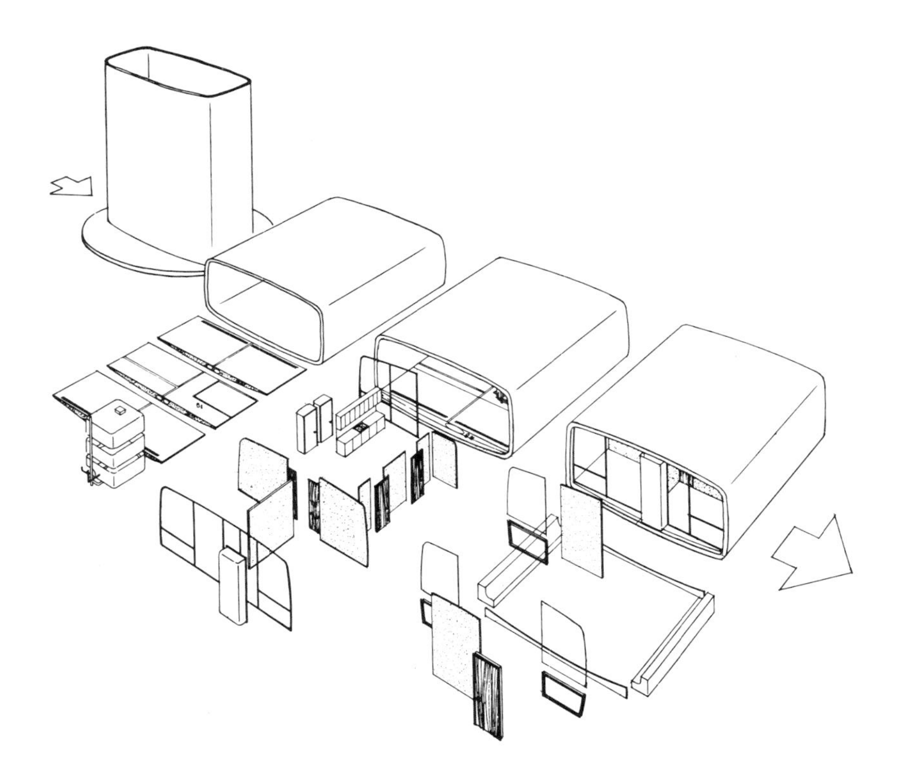

FIG. 13. Completing the tubes for occupancy. First station: fabricate filament-wound shells, cure and remove from mandrel. Second station: make cut-outs required for pipes, wires, ducts, and stairways. Install pre-assembled floor panels; make piping and wiring connections between floor panels. Install prefabricated reinforced plastic bathroom units, and connect to water and waste system in floor assembly. Third station: install interior partitions (pre-cut, pre-wired panels with base wireways to be positioned by guides molded into top surface of floor assembly). Install prehung interior doors and frames; complete wiring and installation of fixtures; install molded wall-hung storage components, including kitchen storage and counter top with sink; make piping connections for sink; mechanically fasten and bond neoprene zipper gaskets to shell for endwall enclosures; attach heating, cooling, ventilating, and electrical unit at open end of shell. Make connections to ducts and wireways in floor assembly. Fourth station: attach precast concrete supports and pre-cut closure skirt panels; install endwall enclosure panels, glass, and aluminum operating sash in zipper gaskets; install pre-hung exterior door and frame; lift unit onto trailer and haul to final location.

FIG. 14. Model of a housing unit.

Rigid-foam-plastics components supported on cables or by scaffolding have been used to form concrete shells. The foam plastics are then retained in the structure as insulation. Another possibility for the forming of reinforced plastic or concrete shells involves the use of flexible foam membranes.

B. Rigidized Flexible-Foam System

The impregnation of flexible foam-plastics membranes with a resin that causes them to become rigid after they have been stretched to the desired shape is another concept investigated by the University of Michigan project staff. With the materials at hand the idea of making a finished structural shell through the curing of resin-impregnated membranes did not materialize. What evolved was the use of this technique as a form-giving device to facilitate the construction of reinforced plastic or concrete shells.

The materials used in this experimental development of the form-giving device were (1) a reticulated flexible polurethane foam (i.e., a foam with a linear cell structure due to the removal of cell walls), (2) a moisture-curing urethane resin, and (3) a flexible polymeric adhesive that cures with the same stretch characteristics as the membrane. The latter permits the bonding of membranes to form large sheets.

After some experimentation with small components, four square 12 x 12-ft anticlastic umbrellas were produced and assembled into a pavilion at Rice University of Texas in the fall of 1964, as illustrated in Fig. 17.

Four-foot-wide strips of 3/4-in. reticulated flexible foam were bonded together to form sheets, 12 ft square. The sheets were impregnated with

FIG. 15. Design for a residential community.

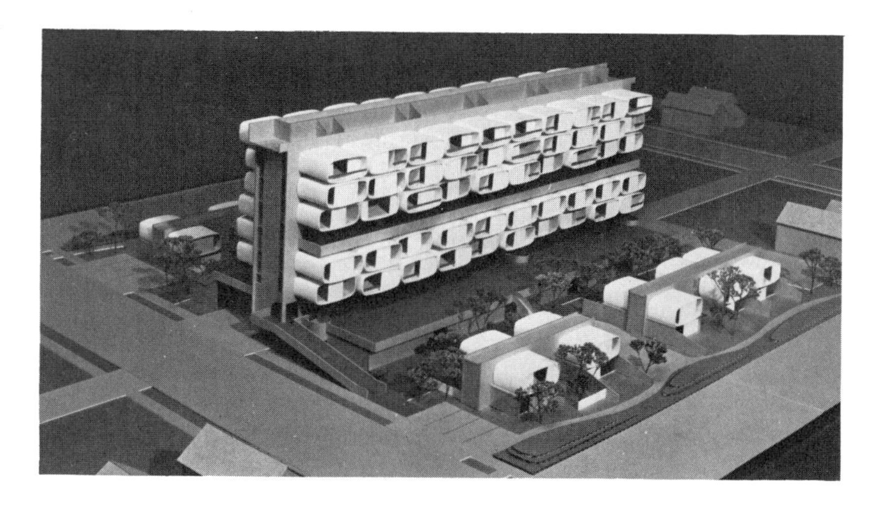

FIG. 16. Design for an apartment complex.

the moisture-curing resin and fastened continuously to a 2 x 4 in. wooden edge frame. This assembly was then fastened down to a simple template, where it was stretched to the desired shape.

After the urethane impregnant had cured, chopped glass fibers and polyester resin were sprayed on the exposed surface of the rigidized membrane, rolled down, and allowed to cure. The shell was then removed from the template and inverted for application of the interior skin.

The shells were erected atop pipe columns and joined to form an open-sided pavilion. Four other such umbrellas, but sprayed only on one side and used in tension, were assembled into a structure in the courtyard of the Architectural Research Laboratory. In this case the rigidizing of the flexible foam is not necessary.

The use of the flexible foam membrane as a form-giving device and as a core material when rigidized provides an economical technique for the production of doubly curved structures by eliminating the high cost associated with the forming of such structures. In production, membranes could be preimpregnated at a factory and protected from curing until used on the site by being sealed into vacuum bags.

VI. SUMMARY OF CONCLUSIONS

Foam plastics, when applied properly and in their own right rather than as a substitute for other materials, have a high potential for use in building. Like all building materials they exhibit some distinct limitations.

FIG. 17. Use of flexible foam for forming reinforced plastic shells.

FIG. 17. (continued).

FIG. 17. (continued).

However, through the imaginative use of these materials a number of structural concepts can be realized that will result in effective solutions from the standpoint of both performance and economy.

REFERENCES

1. A complete report on this project has been published: "Structural Potential of Foam Plastics for Housing in Underdeveloped Areas," Architectural Research Laboratory, The University of Michigan, Ann Arbor, 1966.

2. Developed by Dow Chemical Company.

3. A complete report on this project has been published: "Research on Potential of Advanced Technilogy for Housing," Architectural Research Laboratory, The University of Michigan, Ann Arbor, 1968.

Chapter 21

MILITARY AND SPACE APPLICATIONS OF CELLULAR MATERIALS

Robert J. F. Palchak

Consultant
Santa Ana, California

I. INTRODUCTION

Cellular materials have been used extensively by the military services since their introduction to the United States nearly 20 years ago. The attractiveness of these materials is due to their buoyancy characteristics, insulating properties, relatively high strength-to-weight ratio, ease of fabrication over a broad range of densities into complex shapes and patterns, and producibility "in the field." These attributes have led to the investigation of cellular materials not only as mattresses, fillers for buoys, and insulation for heating ducts but also as a means for salvaging sunken vessels, generating a "second skin" as protection against exposure to environmental thermal extremes, rapidly repairing bombed aircraft runways, and the preparation of shelters in the field.

The advent of the space age has resulted in the evaluation of cellular materials as meteoroid bumpers, as materials for the production of expandable rigidized structures, and for astronaut-recovery systems. The potential of these materials for numerous and diverse defense and support applications was early recognized, and the National Academy of Science-National Research Council Advisory Board on Military Personnel Supplies appointed a committee on foamed plastics to provide guidance in the development and applications of cellular materials and to serve as a liaison between investigators, suppliers, and users in governmental, industrial, and academic institutions. In addition to its other activities, this committee, during its tenure, cosponsored with the Army Quartermaster Command two conferences on cellular materials, and the publication of the proceedings of each [1, 2]. These conferences, probably more than any other single activity, emphasized to industry the ever increasing role of cellular materials in the defense and space program of this country. The Society of Plastics Industry subsequently devoted its twelfth annual conference to space and military applications of cellular plastics systems [3].

This chapter is by no means a complete survey of military and space applications of foams; it is rather an attempt to indicate the variety of applications that have been considered, with emphasis on the more novel usages.

II. NAVY APPLICATIONS

A. General Uses

Cellular plastics are used extensively to prevent floating objects from sinking, to salvage sunken vessels and to provide positive buoyancy for deep-submergence vessels.

Such floating objects as buoys used to support acoustical mine-sweeping gear at a prescribed depth, floats used to mark out areas that are to be swept or that have been cleared of mines, and plastic buoys used as anchor floats for seaplanes contain predominantly polystyrene foam plastic. Fuel lines frequently contain or are fabricated from either polystyrene or polyurethane foam. The voids in corroded pontoons are occasionally repaired by injecting foamed-in-place polyurethane without removing the pontoons from service [4]. Utility and personnel boats are made unsinkable by cutting and fitting cellular cellulose acetate or polystyrene foam as shown in Fig. 1 or, preferably, by foaming in place 2-lb/ft^3 polyurethane foam. The Navy also has used an unsinkable boat called the "UDT Swimmer" for use by underwater demolition teams in support of assault units. These

FIG. 1. Fitting foam in boat voids. Courtesy of U.S. Navy, Bureau of Ships.

boats can carry eight men and are constructed of 2-lb/ft^3 polystyrene foam that has been cut, shaped, and then coated with a reinforced epoxy resin. Obsolete vessels intended for use in target practice are filled with foamed plastic to make them unsinkable (Fig. 2). Since holes will not reduce buoyancy, gunners may fire on the vessels for actual hits instead of the usual near misses required for sinkable targets.

A flexible, closed-cell polyvinyl chloride foam is used as a removable liner in extremely-cold-weather jackets for both buoyancy and insulation. This is especially valuable aboard ships in arctic waters where the men might be knocked overboard by heavy seas [5].

Syntactic foams are being intensively evaluated as buoyancy void fillers in deep-diving submarines, deep-submergence search vessels, and submerged oceanographic platforms. These materials, also known as plastic floats, epoxy floats, and glass-bead floats, consist of a low-density filler material, generally hollow glass or phenolic microspheres suspended in a matrix of epoxy resin, and are more resistant than cellular polyurethanes to degradation on prolonged exposure to hydrostatic pressure and to pressure changes [5-8].

B. Nonsinkable Life Rafts

Both the Department of Defense and the National Aeronautics and Space Administration (NASA) have been engaged in the development of expandable foam-filled life rafts for emergency use at sea. Gas-inflated rafts, although successfully used for more than 20 years, are easily damaged and difficult to repair, especially when punctured below the waterline. To

FIG. 2. Section of a foam-filled obsolete destroyer. Courtesy of U.S. Navy, Bureau of Ships.

overcome this objectionable feature, a number of efforts have been undertaken, with variable degrees of success, to develop self-deployable, nonsinkable life rafts containing a foamed plastic. The following operational requirements were established by U.S. Navy personnel for a foam-generator system intended to inflate the one-man PK-2 life rafts [9].

1. Inflation must be completed within 10 sec at 25°C and within 30 sec at -30°C.
2. The system must be reliably operable over the temperature range -30 to +70°C.
3. Sufficient buoyancy must be provided to support 265 lb at sea.
4. The inflated raft must be resistant to damage.
5. The undeployed raft should occupy no more than 0.3 ft^3 and weigh no more than 7 lb.

The four alternative solutions that were evaluated by one contractor and are described here are representative of the approaches that have been considered and tested in an effort to comply with these rigid requirements. Rather extensive details are given in this section to illustrate the breadth of technical approaches available to solving problems of this sort.

1. Flexible Foams

The first approach was an evaluation of three commercially available flexible foams as a filler material for the flotation ring of the life raft (Table 1). Flexible urethane foams were selected despite their open-cell structure because they could be mechanically restrained in a compressed, folded configuration, and inflation and deployment could be achieved by natural recovery after release of the mechanical restraint. Compression-set determinations, conducted in accordance with ASTM procedures, demonstrated the superiority of the P-137 foam (Table 2). Folded-compression-set measurements were also made by folding a sample of P-137 back on itself by 180 degrees and then compressing the folded foam to 10% of its original thickness to simulate actual use conditions. Samples maintained at 70°C for up to 28 days frequently recovered without evidence of permanent set. Failures normally occurred at from 28 to 56 days. Additional compression-set tests were performed with foam samples covered with a skin of heat-sealed polyvinyl chloride to simulate the behavior in a life raft under actual operating conditions. The number and location of 0.5-in. breather holes was varied to determine the effect of various rates of air supply to the expanding foam, and the extents and rates of recovery were determined. The results of these tests also demonstrate that P-137 foam exhibited the fastest recovery, requiring 152 sec at 20°C and 720 sec at -30°C.

To overcome the more serious objection of water absorption, samples of P-137 foam were coated inside and out with a large number of hydrophobic agents, which produced a dramatic improvement in resistance to water penetration. Samples of foam, impregnated with the various water repellents, were tested by floating and partially submerging them in still water. Untreated samples sank at the rate of about 1/8 in. per day, whereas treated samples did not sink at all in 21 days. The most effective treating agents included stearic acid and zinc stearate.

Flotation tests under simulated ocean conditions, using one-third-scale model rafts with sufficient weight to represent 270 lb fullscale load, showed that flexible foam treated with zinc stearate had acceptable flotation properties, provided that the skin of the raft did not have holes both above and below the waterline. A single hole below the waterline did not seem to affect buoyancy for up to 12 days in rough water. Similar rafts that had holes both above and below the waterline, when exposed to the pumping action of vigorous waves, sank within 24 h.

TABLE 1

Properties of Commercial Flexible Foams[a]

Foam property	Foam designation		
	PA-6247-7[b]	P-12[c]	P-137[c]
Density, lb/ft^3	1.47	0.86	1.59
Tensile strength, psi	20.9	5.0	10.0
Elongation, %	335.0	175.0	175.0
Tear strength, lb/in.	3.3	1.0	1.5
Compression set (90%)[d]	8.2	--	15.0[e]

[a]Data from Ref. [9].
[b]Wyandotte Chemicals Corp.
[c]General Tire and Rubber Co.
[d]Measured according to ASTM D-1564-66T, method B.
[e]Maximum.

TABLE 2

Effect of 24-h 90% Compression of Foams[a]

Foam	Compression set at		
	70°C	25°C	-30°C
PA-6247-7[b]	83.6	9.5	0.4
P-12[c]	17.9	4.3	4.9
P-137[c]	4.7	2.0	1.6

[a]Data from Ref. [9].
[b]Wyandotte Chemicals Corp.
[c]General Tire and Rubber Co.

2. Two-Component Foamed-in-Place Systems

Two-component rigid urethane foamed-in-place systems were evaluated second. On-site production of rigid, closed-cell, low-density foams has been well documented, and an almost limitless supply of polyols and prepolymers is available for consideration. The chief drawbacks to this approach were the limited number of reactants completely miscible with a halocarbon blowing agent over the entire range of temperatures and the availability of a single formulation operable over the -30 to +70°C temperature range.

The study uncovered two systems that were completely miscible from -30 to +70°C, but neither of these was satisfactorily operable over the entire temperature range.

A system consisting of trichlorofluoromethane, a polyol (TP-440), and isocyanate (Mondur MR) produced good-quality moldable foam in the temperature range 25 to 5°C. Below 5°C the foams rose unevenly and were tacky, and above 25°C the reaction was too vigorous and the loss of blowing agent too rapid. By increasing the catalyst concentration, good-quality foams could be produced at 18°C, but further efforts to modify the formulation for use at -30°C were not successful.

In an effort to develop a formulation that would not require variable amounts of catalyst with temperature change, foam systems based on tetrakis(2-hydroxypropyl)ethylenediamine and Pluracol EDP-500 (a tetrol based on ethylenediamine and oxypropylene, of molecular weight 509) in a 70:30 solution of cyclopentane and acetone were tested. Although good-quality foams were obtained from 25 to 70°C, poor results were obtained at +18, -25, and -30°C. Added catalyst gave good foam at +18 but not at -30°C.

Several packaging designs were evaluated for use with two-component systems. Two of these are especially noteworthy because of their simplicity and ease of operation. One utilized the 3M Company Unipak concept and is shown in Fig. 3. The urethane components were sealed in polyester bags that were joined by break-away seals. Foam was produced by squeezing and kneading the bag to break the seal and mix the components. A full-scale PK-2 life raft was inflated, using the trichlorofluoromethane, Pluracol TP-440, Mondur MR formulation in eight Unipak bags, equally spaced about the perimeter of the flotation ring.

The second system consisted of a cardboard cartridge fitted with two plastic bags that contained the urethane ingredients, a piston to compress the bags, a knife edge, and a mixing nozzle. This system is shown Fig. 4. Foam is obtained by rapidly pushing the piston to the far end of the tube, thereby rupturing the bags and expelling the contents through the mixing nozzle.

3. Solid-Reactant Urethane Systems

The third approach consisted of an evaluation of solid-reactant urethane components that could be activated by heat to produce a polyurethane foam.

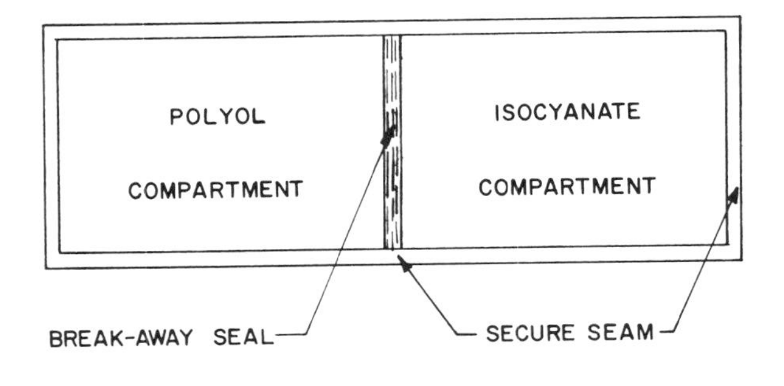

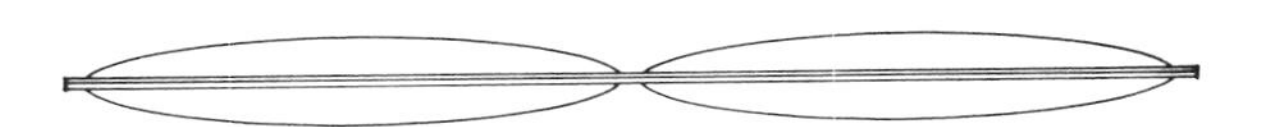

FIG. 3. Unipak concept (3M Company) for life-raft inflation. Reprinted from Ref. [9].

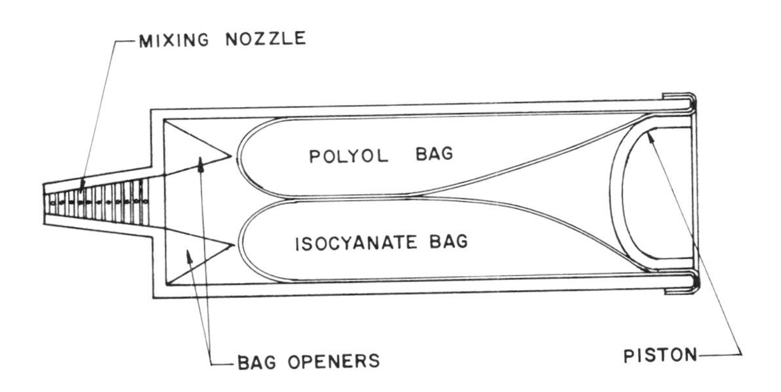

FIG. 4. Simulated mechanical foam dispenser based on a standard caulking gun. Reprinted from Ref. [9].

Such systems employ low-melting isocyanates and polyols and a heat-sensitive solid blowing agent intimately blended together. When heated, the reactants melt and polymerization proceeds. The heat simultaneously initiates the decomposition of the blowing agent to foam the polymerizing mixture. For field use heat can be provided by a pyrotechnic mixture.

This approach was attractive because it is completely self-contained, reasonably independent of environmental temperatures, and stable and inert until the pyrotechnic mixture is initiated.

Solid-reactant systems prepared under scrupulously dry conditions and then hermetically sealed were tested for thermal stability by placing them into a 71°C oven for 24 h. Mixtures containing azobisisobutyronitrile gassed noticeably under these conditions, whereas mixtures containing Celogen OT or Celogen AZ were stable. Table 3 lists typical solid-foam-producing systems, and Table 4 contains thermal-sensitivity data for solid blowing agents.

Foaming tests were conducted by spreading the reactant system onto a sheet of aluminum foil and placing this into contact with a pyrotechnic sheet prepared from iron, sulfur, and asbestos fiber, and also by placing samples into a circulating-hot-air oven.

Good foams were obtained from mixtures placed in the hot-air ovens. Mixtures placed in contact with the pyrotechnic sheet produced a char layer adjacent to the aluminum foil. A layer of foam produced immediately above the char was topped with unreacted material that was insulated from the heat by the foam. This scheme was not pursued further.

TABLE 3

Heat-Activated Solid-Foaming Systems for a Nonsinkable Life Raft[a]

Reactant	Parts by weight	
	System A	System B
Bitolylene diisocyanate	93.6	80.0
4,4'-Bis(β-hydroxyethyl)ether of bisphenol A	80.0	--
Polyol HP 410[b]	--	80.0
Dibutyltin di-2-ethylhexoate	--	0.1
Silicone DC 113[c]	4.0	4.0
Azobisisobutyronitrile	7.0	5.0 to 7.0

[a]Data from Ref. [9].
[b]α-Methyl glucoside-based polyoxypropylene polyol, molecular weight 410 (Corn Products Sales Co.).
[c]Dow Corning Co.

TABLE 4

Storage Stability of Solid Blowing Agents[a]

	Weight loss at 71°C (%)					
Time (h)	Vazo[b]	Porofor N[c]	Nitrosan[d]	Celogen OT[e]	Opex 40[f]	Celogen AZ[g]
24	--	29.03	1.88	0.85	0.18	0.169
48	--	68.30	3.17	1.26	0.34	0.169
72	99.95	--	--	--	--	--
120	--	89.1	--	1.79	--	0.169
168	--	89.4	10.2	2.05	0.84	0.169
336	--	89.6	24.04	2.39	1.66	0.105
504	--	--	36.61	3.14	2.23	0.169
840	--	--	49.81	3.67	3.07	0.169
1008	--	--	--	4.89	--	0.169

[a]Data from Ref. [9.]
[b]Azobisisobutyronitrile, E. I. duPont de Nemours & Co.
[c]Azobisisobutyronitrile, Farbenfabriken Bayer.
[d]N,N'-Dimethyl-N,N'-dinitrosoterephthalamide, E. I. duPont de Nemours & Co.
[e]Uniroyal, Inc.
[f]Dinitrosopentamethylenetetramine in inert filler, National Polychemicals, Inc.
[g]Azodicarbonamide, Uniroyal, Inc.

4. Cryogenic Solvents as Blowing Agents

The final approach to the development of a self-deployed unsinkable life raft consisted of an evaluation of foam production by venting a solution of polymer in a liquefied gas. Experiments were conducted under pressure in an ordinary Coca Cola bottle, by condensing the gas onto a weighed amount of polymer in the bottle. The bottle was capped, warmed to ambient temperature, and shaken until the polymer dissolved. Foams were obtained either by inverting the bottle and removing the cap or by piercing the cap with a nail. The best foams were produced from solutions of Lustrex 101 (polystyrene) in a mixture of trichlorofluoromethane and either dimethyl ether or methyl chloride.

Of the four approaches investigated, the elastic recovery of compressed, open-cell, flexible urethane foam appeared to most closely meet the five criteria that had been established for a life-raft-inflation system.

C. Foams in Underwater Salvage Operations

Shortly after the submarine Thresher tragedy, the U.S. Navy initiated developmental studies to modernize salvage techniques, and in 1964 the Bureau of Ships supported a program to demonstrate the feasibility of salvaging sunken vessels with foamed-in-place plastics [10]. This initial program was directed toward the development of materials and equipment to raise a vessel from a maximum depth of 375 ft. The ambient pressure and temperature at this depth are approximately 11 atm and -2°C. To overcome the extremes of pressure and temperature, highly exothermic, fairly low-density formulations were evaluated.

The first foams were prepared in the laboratory by mixing the ingredients in the chamber of a pneumatically actuated caulking gun and then ejecting them under water. Breadboard metering and mixing equipment was subsequently assembled and used to produce foams in the Navy Experimental Diving Unit at the Naval Weapons Plant in the District of Columbia. The formulation preferred because of its relatively low water absorption under salvage conditions is shown in Table 5.

TABLE 5

Typical Polyurethane Formulation for Producing Foam Under Water[a]

Reactant	Quantity
Niax La 475 polyol[b]	50 g
Pluracol TP-340 polyol[c]	50 g
Triethylenediamine	0.4 ml, 20% soln.
Dibutyltin di-2-ethylhexoate	0.1 ml
Silicone DC-201[d]	2.0 g
Water	2.8 g
Tolylene diisocyanate	88-148 g

[a] Data from Ref. [10].
[b] Union Carbide Corp.
[c] Wyandotte Chemicals Corp.
[d] Dow Corning Co.

Shortly before completion of this study the Urefroth Division of the Polytron Co. (now a division of the Olin Corp.) announced the successful salvage of a 500-ton steel barge from an 80-ft depth, using proprietary equipment and polyurethane formulations. The Navy has subsequently undertaken large-scale development programs and has raised a number of sunken craft with the aid of foamed-in-place polyurethane foams.

III. ARMY APPLICATIONS

Although the Quartermaster Command is perhaps the largest user of foamed plastics within the Army, the Signal Corps, the Army Medical Command, the Engineering R & D Command, and the Ordnance Divisions have developed applications for these versatile materials.

A. Prostheses

The Army Medical Research and Development Command has been evaluating an open-cell acrylate-acrylamide foam as a prosthetic material [11] . This foam is produced from a terpolymer composed of 90 parts of butyl acrylate, 75 parts of methyl methacrylate, 25 parts of methacrylamide, and 37 parts of polyethylmethacrylate filler. Patches comprised of woven Dacron-reinforced acrylate-acrylamide foam have been used in arterial grafts, mitral valve replacement, and esophageal replacement. The foam construction enhances biological permeability, thereby permitting rapid healing without excessive bleeding.

A foamable plastic splint has been developed for emergency field use, to immobilize injured extremities during transport to a safe permanent aid station or hospital unit [12, 13]. This splint consists of a two-component foaming system (Table 6) contained in a compartmented fabric that can be folded or rolled into a compact package weighing less than 1 lb. The ingredients produce an 18-lb/ft^3 foam within 6 min after mixing.

Cellular plastics are also used in the fabrication of porous artificial limbs for leg amputees [14]. The porous limbs permit diffusion of sweat and provide easily cleanable, lightweight prostheses. These are prepared by casting an epoxy resin, diluted with a volatile solvent, over a nylon stockinet. The solvent evaporates and leaves pores in the fabric interstices.

B. Electronic Uses

The Army Electronics Command as well as the Air Force construct antennas and signal reflectors by using foamed-in-place polyurethane and molded or machined polystyrene foam. Figure 5 is a schematic representation of a doubly convex rigid-foam reflector, and Fig. 6 is a standard concave antenna reflector constructed of foamed plastic. Foamed plastic

TABLE 6

Plastic Splint Formulations[a]

Component	Quantity
Compartment 1:	
Isofoam 15-A[b]	150 g
Compartment 2:	
Isofoam 15-W[b]	113 g
Stannous octoate	9 mg
Triethylenediamine	9 mg

[a]Data from Refs. [12] and [13].
[b]Isocyanate Products Co.

has been used to construct the supporting structure also, because of its high strength-to-weight ratio and because it can be cast into streamlined forms to reduce wind drag [15, 16].

Flat dielectric plates properly placed in front of horn antennas are known to increase the gain without increasing the physical size of the antenna. Foamed polystyrene and syntactic foams are used because of their excellent dielectric properties and relatively low densities [17].

In addition to the use of foamed plastics in insulation, mattresses, and packaging, the Army Quartermaster Command has sponsored the development of systems for the production of foamed plastics in areas where it is logistically or tactically unsatisfactory or impossible to ship the finished foam to the point of use.

One of the earliest interests in this regard was the development of energy-absorbing foam formulations to package relatively fragile items, such as electronic components, for delivery to strategic areas by air drop. The items to be dropped could be foamed in place on the aircraft, one at a time, just prior to ejection from the delivery aircraft. Two formulations with high energy-dissipation properties at low density and with very low resilience, which crushed at constant stress to a high percentage of their original height and could be produced in the field without an external heat source, were developed for this application [18] (Table 7). A simple hand-operated peristaltic pump and mixing unit, designed and constructed for use with these formulations, is illustrated in Figs. 7 and 8.

C. Shelters

The Engineering Research and Development Command utilized castable urethane compositions to construct polyurethane foamed-in-place shelters

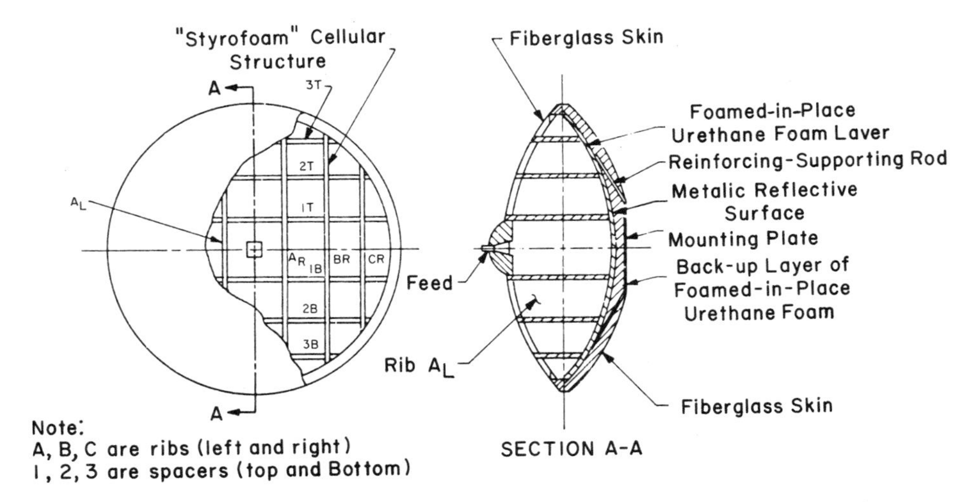

FIG. 5. Doubly convex rigid-foam reflector. Reprinted from Refs. [15] and [16].

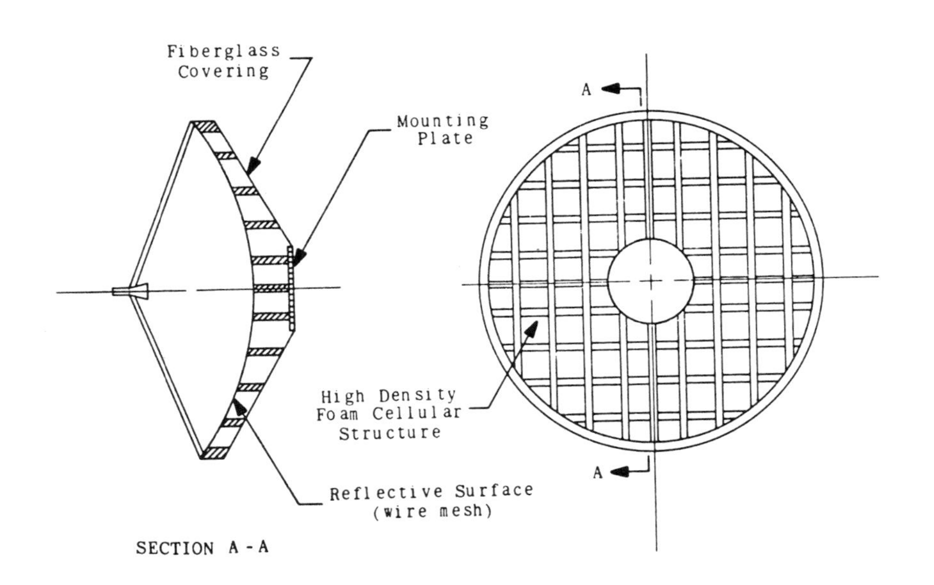

FIG. 6. Antenna reflector--standard concave cellular construction. Reprinted from Refs. [15] and [16].

TABLE 7

Energy-Dissipating Polyurethane Foams[a]

Reactant	Parts by weight	
	Formulation A	Formulation B
Tetrakis(2-hydroxypropyl)ethylenediamine	60	60
Castor oil	40	40
Diallyl phthalate	20	--
Vinyl toluene	--	20
Petromix No. 9	2	2
Witco 77/86 emulsifier	2	2
Ethyl cellulose	6	6
Water	Variable	Variable
Tolylene diisocyanate, 80:20	Variable	Variable

[a]Data from Ref. [18].

for a scientific outpost at Camp Century, Greenland [19]. The method, perfected at Fort Belvoir, Virginia, consisted of prefabricating shelter sections by casting polyurethane foaming mixtures into molds. The finished foamed modules were assembled into completed shelters. The technique was dubbed "buildings in barrels" since the foamed structures were manufactured on site from barrels of resin. This method can be used to construct complete shelter units of practically any configuration to meet any geographical need.

A second technique, developed by the Army Quartermaster Command and by the Naval Air Systems Command [20], produced foamed shelters in the field by spray application of polyurethane formulations onto large inflated bags. After the foam is cured, entrance ways are readily made by cutting through the foamed-in-place wall (see Fig. 9).

In the early 1960s the Army Quartermaster Command sought a kit that could produce a foamed plastic at extremely low ambient temperatures without an external heat source. This was intended for use by military personnel in cold-region combat zones, to provide an insulated lining for foxholes or to otherwise provide an instant, insulated, camouflaged shelter for a man or for strategic supplies. Although the kit was never developed, the feasibility of preparing foamed plastics at low ambient temperatures was demonstrated by the Lewis-acid-catalyzed polymerization of vinyl ethers in the presence of low-boiling-point solvents [21]. The best foams

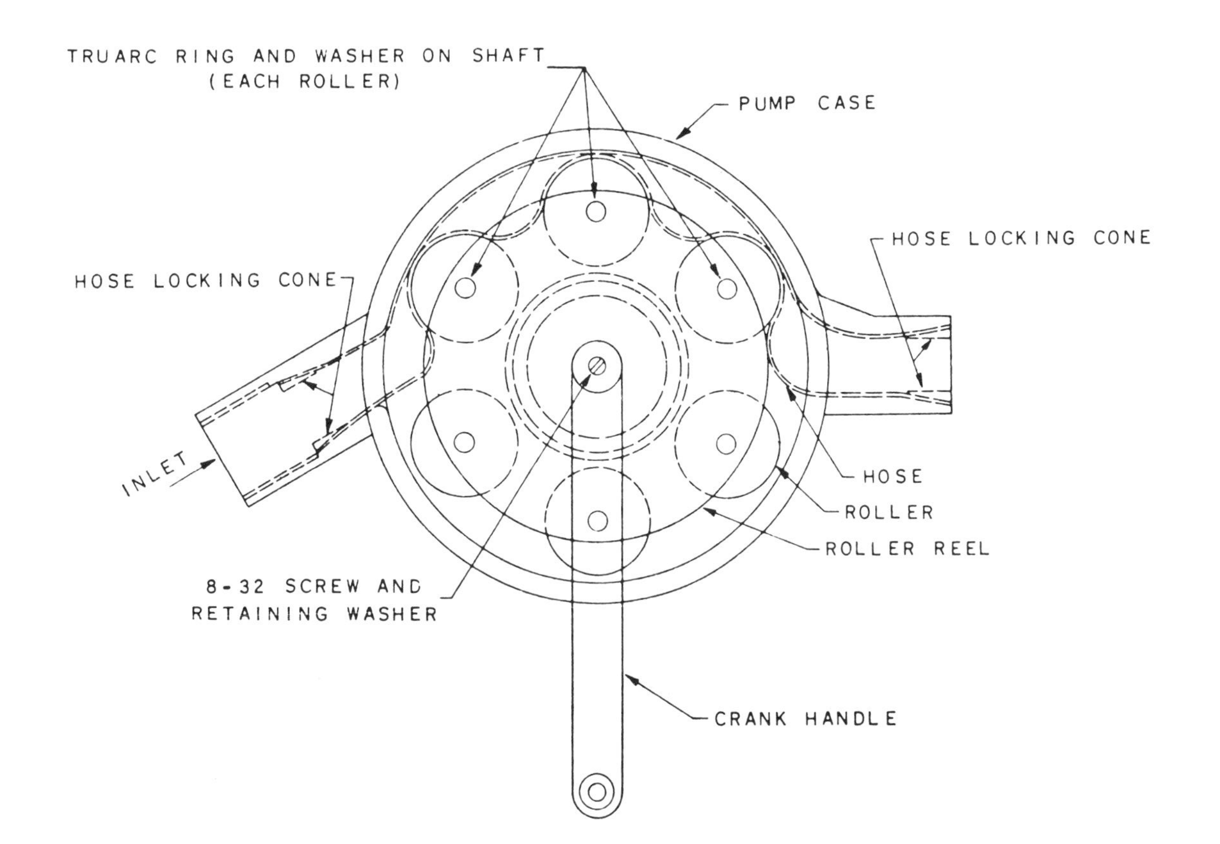

FIG. 7. Design for hand-operated foam machine. Reprinted from Ref. [18] by courtesy of Atlantic Research Corp.

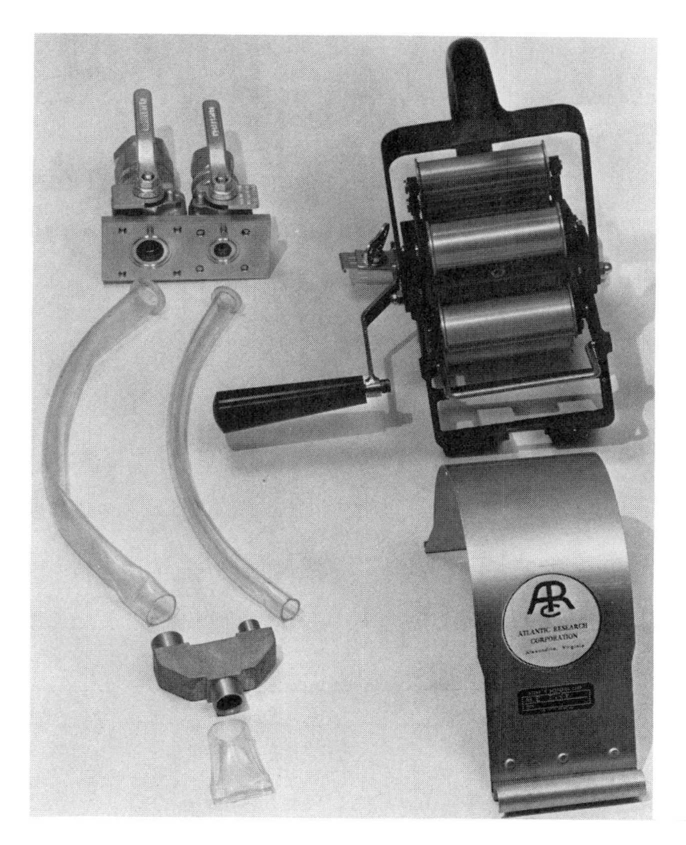

FIG. 8. Components of hand-operated foam machine. Courtesy of Atlantic Research Corp.

were obtained from the formulation shown in Table 8. After the ingredients were mixed and cooled, the catalyst, which consists of a 2.5% solution of boron trifluoride etherate in dibutyl ether, was rapidly added. Exothermic polymerization began almost immediately with the formation of a foam characterized by large, irregular cells and poor physical properties.

The feasibility of fabricating a tent-wall material that can be foamed in place under field conditions and used for the construction of shelters and storage facilities has been demonstrated under the United States-Canadian Development Sharing Program [22]. Developed was a composite kit that consisted of a layer of a heat-sensitive epoxide foamable powder sandwiched between layers of fabric that were laminated to a pyrotechnic sheet. A representative formulation is shown in Table 9. The pyrotechnic sheet contained a mixture of finely powdered iron and sulfur, bound with asbestos. Figure 10 is a photograph of the expanded structure.

FIG. 9. Air-inflated spray-foam shelter. Courtesy of Atlantic Research Corp.

TABLE 8

Ionically Catalyzed Vinyl Ether Systems for Producing Foams at Low Ambient Temperatures[a]

	Parts by volume	
Reactant	Formulation A[b]	Formulation B[b]
3,4-Dihydro-2H-pyran-2-methanol	6.0	6.0
1,4-Divinyloxybutane	6.0	6.0
Vinyl isobutyl ether	3.0	--
1,6-Divinyloxyhexane	--	6.0
Carbon tetrachloride	10.0	9.0

[a]Data from Ref. [21].

[b]Foaming was initiated by rapidly adding approximately 5 vol% of a 2.5% solution of boron trifluoride etherate in dibutyl ether to these premixed materials. No stabilizers were used.

TABLE 9

Heat-Sensitive Epoxide Foamable Powder[a]

Reactant	Parts by weight
Epon 1031[b]	100
4,4'-Diaminodiphenylsulfone	25
Azodicarbonamide	10
Diazoaminobenzene	10
Aluminum powder	40
Pluronic F-68[c]	2

[a]Data from Ref. [23].
[b]Shell Chemical Co.
[c]Wyandotte Chemicals Corp.

FIG. 10. Expanded model structure from a pyrotechnic foamable sheet with section removed showing foam wall. Courtesy of Ontario Research Foundation.

Solid-foaming systems that can be used to construct shelters in the field have also been developed under Quartermaster sponsorship. In one of these, pellets of foamable polystyrene are propelled through a blast of hot air and onto a solid surface. Expansion occurs in the hot zone, and the soft, expanded pellet adheres to the target surface (Fig. 11) [23].

A multipurpose, dormant, one-can polyurethane foaming system operable by unskilled personnel was also developed for use in the field [24]. Since the ingredients of a polyurethane foaming system normally react very rapidly when mixed together, reversible deactivation of one of the key components was essential to the development of a dormant system.

Chemically blocked isocyanates were the most promising of the various alternatives considered for this program. Certain classes of active hydrogen compounds react with isocyanates to form heat-sensitive addition products:

$$\mathrm{RNCO} + \mathrm{HR'} \rightleftharpoons \mathrm{RN(H)C(=O)R'}$$

The adducts can be reactivated usually through heat or alkaline catalysis to regenerate the original isocyanate and the blocking agent. Seventeen blocked isocyanates (Table 10) were prepared and characterized to select one or more compositions for use in a latent system. Although the bisulfite adducts decompose sharply at 45 to 55°C in aqueous solutions, they could not be decomposed by heating in a nonaqueous environment and hence were of no value to this program.

The most useful blocking agents were p-nonylphenol, ethyl acetoacetate, diethyl malonate, and tert-butanol.

The blowing agents in these formulations were hydrated salts from which the water of hydration was liberated at or near the adduct-decomposition temperature. The liberated water reacted with excess isocyanate to form the carbon dioxide blowing agent.

This study produced two useful formulations. A stable, delayed-action rigid-foam system was prepared by mixing 400 g of Voranate R-1 prepolymer (27% free –NCO), 62.4 g of Voranol RN600 polyol, and 47.6 g of tert-butanol at 20°C for 3 h. After standing overnight, the prepolymer was a low-melting-point solid with a 13% free-isocyanate content. This dry-blocked prepolymer was then used to prepare a latent polyurethane foaming system by grinding the following ingredients together in a dry nitrogen atmosphere:

Ingredient	Parts by weight
Blocked prepolymer	10.0
Boric acid	1.0
Molecular sieve CW-X105A	0.2
Silicone 6-520 (Union Carbide)	0.5

The molecular sieve was loaded with dibutyltin dilaurate. When heated at 150°C for 10 min a 2-lb/ft^3 rigid foam was produced.

FIG. 11. Hand-held heat gun for use with expandable polystyrene. Courtesy of Atlantic Research Corp.

TABLE 10

Blocked Isocyanates Considered for Use in Heat-Sensitive Foam Formulations[a]

Adduct	Melting point (°C)	Decomposition point (°C)
Di-(phenol) adduct of 3,3'-dimethyl-4,4'-biphenyl diisocyanate	205-207	165-170
Di-(p-nonylphenol) adduct of 3,3'-dimethyl-4,4'-biphenyl diisocyanate	78-95	174-180
Di-(NH_4HSO_3) adduct of TDI-80[b]	276	d
Di-(NH_4HSO_3) adduct of TDI-100[c]	276	d
Di-($KHSO_3$) adduct of TDI-80[b]	276	d
Di-($KHSO_3$) adduct of TDI-100[c]	276	d
Di-(phenol) adduct of TDI-80[b]	92-100	140-146
Di-(ϵ-caprolactam) adduct of TDI-80[b]	133-139	140-144
Di-(p-nonylphenol) adduct of TDI-80[b]	62-65	154-160
Di-(p-nonylphenol) adduct of TDI-100[c]	45-50	163-165
Di-(succinimide) adduct of TDI-100[c]	94-99	120-125
Di-(diethyl malonate) adduct of TDI-80[b]	95-97	105-110
Di-(ethyl acetoacetate) adduct of TDI-100[c]	125-130	120-125
Di-(diethyl malonate) adduct of TDI-100[c]	119-121	92-96
Di-(tert-butyl alcohol) adduct of TDI-100[c]	187-190	d
Di-($NaHSO_3$) adduct of TDI-80[b]	276	d
Di-(phenol) adduct of methylene bis(4-phenyl-isocyanate)	174-177	170-175

[a]Data from Ref. [24].
[b]Mixture of 2,4- and 2,6-tolylene diisocyanates, 80:20.
[c]2,4-Tolylene diisocyanate.
[d]Test not applicable.

A stable, one-component delayed-action foaming system that produces a fully cured flexible foam was prepared by first mixing 1000 g of Voranol CP3500 polyol and 270 g of TDI-80 at 100°C for 2 h under dry nitrogen to produce a prepolymer with a 7% free-isocyanate content. This was converted to a blocked prepolymer by stirring 800 g of 7% free-isocyanate prepolymer, 312 g of p-nonylphenol, and 8 g of triethylamine for 5 h at 150°C. This blocked prepolymer, which contained less than 0.1% free isocyanate, was incorporated as follows to produce a delayed action system:

Ingredient	Grams
p-Nonylphenol-blocked prepolymer	500
Boric acid powder	50
Linde molecular sieve CW-X105A	25
containing dibutyltin dilaurate	25
Silicone 6-520	5

When a portion of this final mixture was heated at 170°C for 10 min, a fine-pore, flexible, fully cured polyurethane foam was obtained.

D. Insulation

In the early 1950s the Quartermaster Command undertook the development of creams, ointments, and foams that, when applied to exposed areas of the skin, would offer protection against injurious effects from exposure to low temperatures. This was known as the "second-skin" concept and included an investigation of plastisol foams that could be applied to the skin in an uncured state and expanded on the hand to produce the desired protective properties [25].

The characteristics that were established for the "second skin" were as follows:

1. Thickness, approximately 1/8 in.
2. Thermal conductivity not greater than 0.35 Btu/(h)(ft^2)(°F/in.).
3. Sufficiently soft and flexible to permit good manual dexterity between temperatures of zero to -53°C.
4. Durability to permit light manual activities for a period of 8 h.
5. No water absorption.
6. Resistant to deterioration or swelling in soap, water, gasoline, and oil.
7. Simple to remove by stripping.
8. Nontoxic and nonflammable, before and after application.

Prior to application the material should be a liquid, cream, or plastic capable of being applied to body parts or clothing by spreading, dipping, spraying, painting, or by some other simple procedure requiring little or no auxiliary equipment or technical skill, at temperatures of not less than 10°C. After application the material should dry or set in 15 min or less without the use of special equipment, although the parts to which the material has been applied could be placed over a warm stove (38 to 50°C) if necessary to facilitate drying. The material must have a shelf life of 6 months or more over the temperature range -53 to +50°C.

Numerous foamed-in-place formulations were evaluated, all of which consisted of a plasticized film-forming polymer dissolved in a solvent that was also usually the blowing agent. One of the most promising formulations from this study is shown in Table 11. The coating, when spread on the back of the hand and exposed at a comfortable distance to an infrared lamp, "fluxed" in 6 min to produce a foam with a good cell structure and fair mechanical strength.

A superior foamed-in-place coating was obtained by blowing the mixture shown in Table 12 with carbon dioxide. A 500-ml Kidde pressure vessel was charged with 200 g of this formulation and pressured with a carbon dioxide cartridge. The mixture, when released through a narrow valve, formed a stable, homogeneous, liquid foam that could be applied as a coating by dipping, spraying, or spreading. A brief exposure to an infrared lamp produced a strong, small-celled foam with acceptable physical properties.

The Quartermaster Command also sponsored development of a foamable product that could be applied to the skin and clothing for protection against intense thermal exposure. The objective was a nontoxic, relatively inexpensive material that could be easily applied shortly before exposure to

TABLE 11

Second-Skin Foaming Formulation[a]

Component	Quantity
Pliovic AO[b]	7 g
Tricresyl phosphate	5 g
Freon 113	4 g
Methylene chloride	0.6 ml
Triethanolamine lauryl sulfate	0.2 ml

[a]Data from Ref. [25].
[b]Vinyl chloride copolymer dispersion, Goodyear Tire and Rubber Co.

TABLE 12

Carbon Dioxide-Blown Second-Skin Formulation[a]

Component	Parts by weight
Pliovic AO[b]	20
Sovaloid C[c]	10
Trioctyl phosphate	10
Tridecyl nitrate	2

[a]Data from Ref. [25] .
[b]Vinyl chloride copolymer dispersion, Goodyear Tire and Rubber Co.
[c]Aromatic hydrocarbon plasticizer, Socony Vacuum.

high-intensity thermal energy, such as burning gasoline. On application it should produce a permanent, multicellular, nonflammable foam with a high inherent reflectivity [26] .

Typical low-viscosity, nonflammable, easy-to-apply formulations that produce pliable, strong foams within 15 min after application are shown in Table 13. These mixtures were packaged in commercial aerosol dispensers.

Unfortunately the foamed barriers had an extremely low tensile strength, could not withstand shear, and a 1-sec exposure of a 0.25-in. foamed layer to about 29 cal/cm^2-sec caused the material to collapse with a loss of thermal protection.

E. Miscellaneous Uses

The Army Ordnance Division uses foamed polystyrene and polyurethanes as inert bomb sealants to prevent leakage of nitroglycerin and thereby reduce the sensitivity of bombs to accidental impact. Previously they used a mixture of glyceryl ester, various resins, and hydrated aluminum silicate, which became brittle at freezing temperature and flowed freely at 71°C.

Limited warfare such as we are experiencing in Southeast Asia has resulted in an increased interest in foamed plastics. In this regard the Army has expressed an interest in sticky foams to harass advancing enemy troops and vehicles, explosive foams, foams for altering the terrain to make it passable or nonpassable, and open-celled microporous foams containing chemical warfare agents. The armed forces are interested in being able to construct "instant" heliports in jungle terrain from foamed plastics and would even like to "foam the enemy in place."

TABLE 13

Heat-Foamable Thermal Barrier Cream[a]

Component	Parts by weight		
	A	B	C
Foaming stabilizers:			
Triethylamine	2.5	2.5	2.5
Palmitic acid	5	5	5
Film formers:			
5% PVP K-30[b]	7.5	40	40
5% Hydroxyethyl cellulose WP-3[c]	80	--	20
5% Elvanol[d] 52-22	--	40	20
Methocel[e]	1	9.5	9.5
Humectant: 80% sorbitol			
Propellant: Freon 12[f]	5	5	5
Blowing agent: Freon 114[g]	5	5	5

[a]Data from Ref. [26].
[b]Polyvinyl pyrrolidone, General Aniline & Film Corp.
[c]Union Carbide Corp.
[d]Polyvinyl alcohol, E. I. duPont de Nemours & Co.
[e]Methyl cellulose, Dow Chemical Co.
[f]Dichlorodifluoromethane, E. I. duPont de Nemours & Co.
[g]1,2-Dichloro-1,1,2,2-tetrafluoroethane, E. I. duPont de Nemours & Co.

Consideration has also been given to a program in which aerosol foam particles would be discharged into the atmosphere in the vicinity of a nuclear explosion in an attempt to entrap the radioactive fallout particles in the belief that the combination of light foam and heavy particles would yield a slower falling body so that radiation would decrease before reaching the ground.

IV. AIR FORCE AND SPACE APPLICATIONS

Air Force applications of foamed plastics are predominantly structural and are based on the ability to obtain moderately durable lightweight components. Foamed-in-place polyurethanes are used to fill propellers and helicopter blades, as well as in sandwich construction of various aircraft [27-32].

The excellent dielectric properties and relatively high stress capabilities of polyurethane foams make this material especially attractive for radome construction. Radomes are now constructed from polyurethane and polystyrene foams. The earliest radomes were designed and constructed to resist wind, snow, and ice loads. The large amount of conventional test data generated during the course of designing these structures to withstand the environmental loads has suggested that hardened-foam radomes capable of resisting substantial nuclear blast loads could be designed and constructed [33, 34].

Both the Air Force and NASA are investigating the uses of foamed plastics in the fabrication of expandable rigidized structures for use in outer space. These applications require the formulation of a foamed plastic in a high-vacuum environment [35, 36]. Typical of Air Force interests is the work undertaken to develop solid formulations that will foam in place by a remote command. The earliest formulations contained dianisidine diisocyanate, p,p'-bis(β-hydroxyethoxy)-2,2-diphenylpropane, trimethylolpropane, silicone DC 113, dibutyltin di-2-ethylhexoate, and boric acid, and required a relatively large heat input (temperatures of 120°C) to promote polymerization, foaming, and cure. Some of the ingredients were volatilized under the vacuum and temperature conditions of outer space [37].

Subsequent development produced vacuum-foamable, heat-initiated, one-package, predistributable formulations based on tolidine diisocyanate (TODI) and commercial liquid polyoxypropylene polyols [38]. These formulations required less heat input for foaming and curing, used reactants of lower volatility, and were made in paste or powder form (Table 14). When distributed on a woven-fabric substrate and subjected to radiant heating at pressures of 10^{-4} to 10^{-5} mm Hg, the materials reacted to yield rigid, cured foams of 2- to 5-lb/ft^3 density. Although the foaming action has been attributed to the vaporization of some of the isocyanate under the temperature and reduced-pressure conditions of the polymerization, it is more likely due to traces of moisture in the various ingredients.

Typical of the applications intended for the space foamable systems are construction of rigidized expandable lunar housing, rotating space stations, and solar energy concentrators.

To ensure personnel safety in space flight, escape capsules or space lifeboats in which the astronaut will be foamed in place are being developed [39]. In the project MOOSE concept the astronaut dons an escape unit that consists of a plastic covering and a folded heat shield. This unit is equipped with tanks containing foaming plastic, a mixer oxygen supply, recovery aids, a retro rocket, an orbital measurement package, and survival gear. After leaving the spacecraft, the astronaut inflates the heat shield and plastic covering and then foams himself in place in 1-lb/ft^3 polyurethane foam. He then visually orients himself to the earth and measures his

TABLE 14

Heat-Activated Paste System for Foaming in the Vacuum of Outer Space[a]

Reactant	Parts by weight
Tolidine diisocyanate	80
Polyol HP-410[b]	80
Dibutyltin di-2-ethylhexoate	0.1
Silica-nucleating agent	20
Silicone DC 113[c]	4

[a]Data from Ref. [38].
[b]Corn Products Co.
[c]Dow Corning Co.

TABLE 15

Astronaut Foam-in-Place Salvage System[a]

Reactant	Parts by weight
Pleogen 4050 resin[b]	100
Pleogen 4012 propolymer[c]	112
Prepolymer XD 1078[d]	58
Benzyldimethylamine	0.5
Water	2
Fluorocarbon 113	3
Silicone SF 1034[e]	2

[a]Data from Ref. [39].
[b]Polyester resin, Molrez Division of American Petrochemical Corp.
[c]Polyester prepolymer, Molrez.
[d]Castor-oil prepolymer, Spencer Kellog Co.
[e]Cell-size control agent, General Electric, Silicon Products Division.

altitude and direction of flight by using an optical sight mounted on a retro rocket. With this information, he is able to aim and fire the rocket motor to achieve reentry.

Prior to reentry a high-intensity flare is fired, and a radio beacon is activated. During reentry the elastomeric heat shield ablates, protecting the man from the thermal environment while the very-low-density plastic cushions him against the deceleration and landing shock. At about 30,000 ft a parachute is automatically opened by a pressure-activated switch. Impact velocity at sea level is estimated at less than 30 ft/sec. The total time from retro-rocket initiation to impact is estimated at 45 min, and the distance traveled is 10,000 nautical miles to earth. It is anticipated that after impact the man will release himself from his foam package, extract his survival kit, and await rescue. Semirigid 2.5-lb/ft^3 polyurethane foams were developed for this application (Table 15), and live humans were successfully encapsulated in the laboratory to demonstrate wearability and ease of removal of the foam sheath, but recovery tests from outer space have not as yet been reported.

REFERENCES

1. Foamed Plastics, Proceedings of a Conference (April 1963), PB 181576, U.S. Department of Commerce, Office of Technical Services, Washington, D.C.
2. Cellular Plastics, Proceedings of a Conference (U.S. Department of Commerce, Washington, D.C., April 1966), National Academy of Sciences-National Research Council Publ. No. 1462, 1967.
3. "Space and Military Applications of Cellular Plastics Systems," 12th Annual Conference, Cellular Plastics Division, Society of Plastic Industry, New York, 1967.
4. W. Bourne, in Ref. 1 , pp. 215-222.
5. Selection of Buoyant Materials for Deep Submergence Search Vehicle, (Proteus, Inc., first report, October 1965), AD487-688, U.S. Department of Commerce, Office of Technical Services, Washington, D.C.
6. U.S. Naval Applied Science Laboratory, Brooklyn, N.Y., Report on the Development of Buoyant Materials for Deep Submergence (October 1963), AD420382, U.S. Department of Commerce, Office of Technical Services, Washington, D.C.
7. General Dynamics Electric Boat Division, Buoyancy Materials for Salvage Operations (first quarterly Rept., 1966), AD801667, U.S. Department of Commerce, Office of Technical Services, Washington, D.C.
8. E. C. Hobaica and S. D. Cook, in Ref. [3], Section 4-E.

9. I. O. Salyer, J. Schwendemen, and C. E. McClung, in Ref. [3], Section 4-A; available in AD647373, U.S. Department of Commerce, Office of Technical Services, Washington, D.C., 1967.
10. Atlantic Research Corp., Feasibility Study of Flotation of Sunken Vessels by Injection of Foamed Flotation Material (final report), AD455535, U.S. Department of Commerce, Office of Technical Services, Washington, D.C., 1965.
11. U.S. Army Medical Biomechanical Research Laboratory, Walter Reed Army Medical Center, Development of Foamed Acrylate-Amide Terpolymer (summary Rept. TR6712), AD664131, U.S. Department of Commerce, Office of Technical Services, Washington, D.C., 1967.
12. U.S. Army Medical Biochemical Research Laboratory, Walter Reed Army Medical Center, Evaluation of Isofoam PE 15 as a Splint Material, AD428666, U.S. Department of Commerce, Office of Technical Services, Washington, D.C., 1963.
13. J. T. Hill, D. R. Ingenito, F. Leonard, and P. M. Margetis, U.S. Pat. 3,373,741, March, 1968.
14. J. T. Hill and H. G. Mouhot, in Ref. [3], Section 2-A.
15. Plastics Research Laboratory, Massachusetts Institute of Technology, A Study of Rigid Polyurethane Foam as a Structural Material for Radomes for Lincoln Laboratory, AD271567, U.S. Department of Commerce, Office of Technical Services, Washington, D.C., 1961.
16. Lincoln Laboratories, Massachusetts Institute of Technology, Preliminary Investigations on the Mechanical Design of Hardened Foam Radomes (AFESD TDR62-90), AD275713, U.S. Department of Commerce, Office of Technical Services, Washington, D.C., 1962.
17. Army Signal Research and Development Laboratories, Fort Monmouth, N.J., The Dielectric Plate Antenna (ASRD2 TR2240), AD276888, U.S. Department of Commerce, Office of Technical Services, Washington, D.C., 1962.
18. R. J. Palchak, in Ref. [1], pp. 189-196.
19. S. B. Swenson, in Ref. [3], Section 66, also in Ref. [1], p. 251.
20. W. L. Mackie, Pacific Missile Range, Point Mugu, Calif., A Method of Fabricating a Foamed Plastic Shelter, AD667721, U.S. Department of Commerce, Office of Technical Services, Washington, D.C., 1968.
21. R. J. Palchak, Atlantic Research Corp., Development of Low Density, High Strength Foamed Plastic Via Ionic Polymerization (final report), AD600247, U.S. Department of Commerce, Office of Technical Services, Washington, D.C., 1964.
22. T. R. Francis, M. A. Jones, and M. P. Thorne, in Ref. [1], pp. 138-151.

23. Atlantic Research Corp., Development of a Heat Foamable Resin System and an Apparatus for Producing the Foam, AD452046, U.S. Department of Commerce, Office of Technical Services, Washington, D.C., October, 1963.
24. A. R. Reilly and M. Zwoninski, General Mills Corp., Investigation of the Use of Isocyanate Adducts in Urethane Foam, AD277420, U.S. Department of Commerce, Office of Technical Services, Washington, D.C., 1962.
25. R. Brantley and H. Meyer, Occidental College Chem. Dept., Investigation of an Insulating "Second Skin" (final report), AD28997, U.S. Department of Commerce, Office of Technical Services, Washington, D.C., 1955.
26. J. J. McBride, J. D. Medler, A. Petrunti, and F. Testa, Evans Research and Development Co., Development of a Foam Product for Protection Against Thermal Effects, final report under contract DA19-129-AM-382 to Quartermaster Research & Development Command, 1956.
27. B. Neal and W. J. Edwards, National Aeronautical Establishment (Canada), An Improved Method for Producing Foam-Filled Fiberglass-Reinforced Plastic Propeller Blades, (Aeronautical Rept. No. 6R-346), AD283-935, U.S. Department of Commerce, Office of Technical Services, Washington, D.C., 1962.
28. North American Aviation, Inc., Sandwich Rocket Motor Case Program, AD274-069 and AD282-105, U.S. Department of Commerce, Office of Technical Services, Washington, D.C., 1962.
29. Hexcel Products Co., Application of Composite Constructions with Honeycomb and Foam Cores, AD268-594, U.S. Department of Commerce, Office of Technical Services, Washington, D.C., 1960.
30. Aerospace Corp., Trends and Future Developments in Aerospace Materials, AD273-476, U.S. Department of Commerce, Office of Technical Services, Washington, D.C., 1961.
31. S. Litvak, U.S. Air Force Non-Metallic Materials Lab., Wright-Patterson Air Force Base, Conference on Structural Plastics, Adhesives and Filament Wound Composites, AD408-781, U.S. Department of Commerce, Office of Technical Services, Washington, D.C., 1962.
32. D. Weyer, J. R. Russell, and K. R. Hoffman, Dow Corning Corp., Development of a Heat Resistant Foamed-in-Place Low-Density Silicone Resin Core Material, AD63-264, U.S. Department of Commerce, Office of Technical Services, Washington, D.C., 1955.
33. Aeronca Manufacturing Corp., Investigation of Foam Plastic Techniques for Hardening Ground Radomes, AD608-473, U.S. Department of Commerce, Office of Technical Services, Washington, D.C., 1964.

34. P. J. Campagna, H. M. Preston, and N. E. Wahl, Cornell Aeronautical Labs., Inc., Development of Heat-Resistant Foamed-in-Place Dielectric Core Materials for Sandwich Radomes, AD48-277, U.S. Department of Commerce, Office of Technical Services, Washington, D.C., 1954.
35. Air Force Aeropropulsion Laboratories, Conference Transactions--Aerospace Expandable Structures, 1963.
36. Air Force Aeropropulsion Laboratories, Conference Transactions--Aerospace Expandable Structures, AD631-406, U.S. Department of Commerce, Office of Technical Services, Washington, D.C., 1965.
37. S. Schwartz, in Ref. [1], pp. 126-137.
38. J. M. Butler and L. E. Erbaugh, Monsanto Research Corp., Solid Reactant Polyurethane Foam Compound, AD609-215, U.S. Department of Commerce, Office of Technical Services, Washington, D.C., 1964.
39. J. H. Quillinan, in Ref. [1], p. 197.

AUTHOR INDEX

Numbers in brackets are reference numbers and indicate that an author's work is referred to although his name is not cited in the text. Underlined numbers give the page on which the complete reference is listed.

A

B

C

G

H

S

Y

Z

SUBJECT INDEX

V